Springer Proceedings in Complexity

Springer Proceedings in Complexity publishes proceedings from scholarly meetings on all topics relating to the interdisciplinary studies of complex systems science. Springer welcomes book ideas from authors. The series is indexed in Scopus.

Proposals must include the following:

- name, place and date of the scientific meeting
- a link to the committees (local organization, international advisors etc.)
- scientific description of the meeting
- list of invited/plenary speakers
- an estimate of the planned proceedings book parameters (number of pages/articles, requested number of bulk copies, submission deadline)

Submit your proposals to: Hisako.Niko@springer.com

Clemens Mensink · Oriol Jorba
Editors

Air Pollution Modeling and its Application XXVIII

Springer

Editors
Clemens Mensink
VITO NV
Mol, Belgium

Oriol Jorba
Barcelona Supercomputing Center
Barcelona, Spain

ISSN 2213-8684 ISSN 2213-8692 (electronic)
Springer Proceedings in Complexity
ISBN 978-3-031-12785-4 ISBN 978-3-031-12786-1 (eBook)
https://doi.org/10.1007/978-3-031-12786-1

This Springer imprint is published by the registered company Springer Nature Switzerland AG
The registered company address is: Gewerbestrasse 11, 6330 Cham, Switzerland

Organizing Committee

Members of the Scientific Committee for the 38th International Technical Meeting on Air Pollution Modeling and its Application

Clemens Mensink, Belgium (chair)
Ekaterina Batchvarova, Bulgaria
Wanmin Gong, Canada
Ari Karppinen, Finland
Laurent Deguillaume, France
Volker Matthias, Germany
Maria Kanakidou, Greece
Silvia Trini Castelli, Italy
Hilde Fagerli, Norway
Ana Isabel Miranda, Portugal
Oriol Jorba, Spain
Renske Timmermans, The Netherlands
Ulas Im, Turkey
Rohit Mathur, USA

Honorary Life Members

Chris De Wispelaere, Belgium
Douw Steyn, Canada
Sven-Erik Gryning, Denmark
Werner Klug, Germany
Carlos Borrego, Portugal
Han van Dop, The Netherlands

Preface

The 38th edition of the International Technical Meeting (ITM) on Air Pollution Modeling and its Application took place in Barcelona, Spain, from 18 to 22 October and was organized by the Barcelona Supercomputing Center—Centro Nacional de Supercomputación (BSC) and the Universitat Politècnica de Catalunya (UPC) as local hosts and VITO in Belgium (pilot country).

Shortly after the celebration of ITM's 50th anniversary in September 2019 in Hamburg, the prospects of having a next ITM conference did not look too good, given the restrictions imposed on us by the COVID-19 pandemic. It was very challenging to bring participants together in Barcelona and at the same time allow scientists facing travel restrictions to participate as well.

Therefore, a unique hybrid meeting format was set up with on one hand scientists participating onsite in the UPC auditorium in Barcelona and on the other hand the online contributions from participants from all over the world, including countries such as Brazil, Canada, Iran, and USA. In this way the international and transatlantic co-operation—that is so typical for ITM—could be secured.

The ITM series of conferences were initiated in 1969 under the auspices of the North Atlantic Treaty Organization's Committee on Challenges of Modern Society (NATO-CCMS). The ITM series became independent of NATO since 2013. Over its long history, the ITM conference series provided a broadly independent international meeting of the highest scientific stature. ITM has become one of the most prominent forums for discussing the latest scientific developments and applications related to air pollution modelling at scales ranging from local to global.

An attractive feature of the conference remains the Early Career Award, which is the contest for the best paper or poster award for researchers under 35. Almost 40 young scientists subscribed for the 11th edition of the Early Career Award.

This volume contains the papers and posters presented at the 38th ITM. It was attended by 115 participants from 24 countries. Keynotes were given by Jose M. Baldasano (Universitat Politècnica de Catalunya, UPC), Wanmin Gong (Environment and Climate Change Canada), Renske Timmermans (Climate, Air and Sustainability, TNO), and Carlos Pérez García-Pando (ICREA, BSC).

On behalf of the Steering Committee we like to thank all who contributed to a successful meeting. We like to acknowledge the key note speakers for their outstanding contributions, the session chairpersons and rapporteurs for their efforts, María Gonçalves-Ageitos (UPC) as co-organizer and main contact point with UPC, and the institutions that organized and sponsored this conference: the teams from BSC and UPC for their organizational support of ITM, the Technical University of Catalonia for financially supporting the event, and VITO in Belgium for financially supporting the Early Career Awards and some of the graduate students to participate in this conference.

The 39th ITM conference will take place in Chapel Hill, North Carolina, USA, 22–26 May 2023.

Mol, Belgium

Clemens Mensink
Chair of the Steering Committee

Barcelona, Spain

Oriol Jorba
Conference Organizer

History of the International Technical Meeting (ITM) on Air Pollution Modeling and its Application

Pilot Studies

1969–1974, Air Pollution Pilot Study *(Pilot Country: United States of America)*
1975–1979, Air Pollution Assessment Methodology and Modeling (*Pilot Country: Germany)*
1980–1984, Air Pollution Control strategies and Impact Modeling (*Pilot Country: Germany)*

Pilot Follow-Up Meetings

Pilot Country—United States of America (R. A. McCormick, L. E. Niemeyer)
February 1971, Eindhoven, The Netherlands, First Conference on Low Pollution Power Systems Development
July 1971, Paris, France, Second Meeting of the Expert Panel on Air Pollution Modeling

NATO/CCMS International Technical Meetings (ITM) on Air Pollution Modeling and its Application

Subsequent meetings were supported by the NATO Committee for Challenges to Modern Society and were designated NATO/CCMS International Technical Meetings (ITM) on Air Pollution Modeling and its Application.

October 1972, Paris, France, 3rd ITM
May 1973, Oberursel, Federal Republic of Germany, 4th ITM
June 1974, Roskilde, Denmark, 5th ITM

Pilot Country—Germany (Erich Weber)
September 1975, Frankfurt, Federal Republic of Germany, 6th ITM
September 1976, Airlie House, USA, 7th ITM
September 1977, Louvain-la-Neuve, Belgium, 8th ITM
August 1978, Toronto, Canada, 9th ITM
October 1979, Rome, Italy, 10th ITM

Pilot Country—Belgium (Chris De Wispelaere)
November 1980, Amsterdam, The Netherlands, 11th ITM
September 1981, Menlo Park, California, USA, 12th ITM
September 1982, Ile des Embiez, France, 13th, ITM
September 1983, Copenhagen, Denmark, 14th ITM
April 1985, St. Louis, Missouri, USA, 15th ITM

Pilot Country—The Netherlands (Han van Dop)
April 1987, Lindau, Federal Republic of Germany, 16th ITM
September 1988, Cambridge, United Kingdom, 17th ITM
May 1990, Vancouver, BC, Canada, 18th ITM
September 1991, Ierapetra, Greece, 19th, ITM

Pilot Country—Denmark (Sven-Erik Gryning)
November 1993, Valencia, Spain, 20th ITM
November 1995, Baltimore, Maryland, USA, 21st ITM
May 1997, Clermont-Ferrand, France, 22nd ITM
September, 1998, Varna, Bulgaria, 23rd ITM
May 2000, Boulder, Colorado, USA, 24th ITM

Pilot Country—Portugal (Carlos Borrego)
September 2001, Louvain-la-Neuve, Belgium, 25th ITM
May 2003, Istanbul, Turkey, 26th ITM
October 2004, Banff, Canada, 27th ITM
May 2006, Leipzig, Germany, 28th ITM

NATO/SPS International Technical Meetings (ITM) on Air Pollution Modeling and its Application

In 2007 NATO's Committee for Challenges to Modern Society was disbanded and replaced by NATO's Committee on Science for Peace and Security (NATO/SPS), which continued its support for the ITM.

September 2007, Aveiro, Portugal, 29th ITM

Pilot Country—Canada (Douw Steyn)
May 2009, San Francisco, California, USA, 30th ITM
September 2010, Torino, Italy, 31st ITM
May 2012, Utrecht, The Netherlands, 32nd ITM

International Technical Meetings (ITM) on Air Pollution Modeling and its Application

In 2012 the NATO Committee on Science for Peace and Security refocused its mandate, and the ITM became independent of NATO/SPS support.

September 2013, Miami, USA, 33rd ITM
May 2015, Montpellier, France, 34th ITM

Pilot Country—Belgium (Clemens Mensink)
October 2016, Chania (Crete), Greece, 35th ITM
May 2018, Ottawa, Canada, 36th ITM
September 2019, Hamburg, Germany, 37th ITM
October 2021, Barcelona, Spain, 38th ITM

Contents

Contributors

Adani M. Laboratory of Atmospheric Pollution, Italian National Agency for New Technologies, Energy and Sustainable Economic Development – ENEA, Bologna, Italy

Aghito M. Norwegian Meteorological Institute, Oslo, Norway

Agresti Valentina Sustainable Development and Energy Sources Department, RSE Spa, Milano, Italy

Aittamaa R. University of Oulu, Oulu, Finland

Akyuz Ezgi Eurasia Institute of Earth Sciences, Istanbul Technical University, Istanbul, Turkey

Alarcón Marta Department of Physics, EEBE, Barcelona, Spain

Albuquerque T. T. A. Department of Sanitary and Environmental Engineering, Federal University of Minas Gerais, Av. Pres. Antônio Carlos, Belo Horizonte, Brazil

Alegre Ll. Autoritat del Transport Metropolità, Barcelona, Spain

Alivernini A. Council for Agricultural Research and Economics (CREA), Research Centre for Forestry and Wood, Rome, Italy

Alonso C. Generalitat de Catalunya, Barcelona, Spain

Amorati Roberta ARPAE, viale Silvani 6, 40122, Bologna, Italy

André M. COSYS Univ Gustave Eiffel, IFSTTAR, Lyon, France

Appel K. Wyat Center for Environmental Measurement and Modeling, US Environmental Protection Agency, Durham, NC, USA

Arampatzi P. DANAOS, Limassol, Cyprus

Armand Patrick DAM, CEA, Arpajon, DIF, France

Arndt Jan Institute of Coastal Environmental Chemistry, Helmholtz-Zentrum Hereon, Geesthacht, Germany;
Department of Chemical Transport Modeling, Helmholtz-Zentrum hereon GmbH, Institute of Coastal Research, Geesthacht, Germany

Arunachalam Saravanan Institute for the Environment, University of North Carolina at Chapel Hill, Chapel Hill, NC, USA

Badeke Ronny Department of Chemical Transport Modeling, Helmholtz-Zentrum hereon GmbH, Institute of Coastal Research, Geesthacht, Germany;
Institute of Coastal Environmental Chemistry, Helmholtz-Zentrum Hereon, Geesthacht, Germany

Badia Alba Institute of Environmental Science and Technology (ICTA), Universitat Autònoma de Barcelona (UAB), Barcelona, Spain

Bagkis Evangelos Environmental Informatics Research Group, School of Mechanical Engineering, Aristotle University of Thessaloniki, Thessaloniki, Greece

Baldasano José M. Emeritus Full Professor of Environmental Engineering, Technical University of Catalonia (UPC), Barcelona, Spain;
Barcelona Supercomputing Centre (BSC), Barcelona, Spain

Bande Stefano ARPA Piemonte, Torino, Italy

Baraldi R. Istituto Per La Bioeconomia (CNR), Bologna, Italy

Barmpas F. Aristotle University of Thessaloniki, Thessaloniki, Greece

Barnaba Francesca Consiglio Nazionale delle Ricerche-Istituto di Scienze dell'Atmosfera e del Clima (CNR-ISAC), Bologna, Italy

Barra Marc Institute of Atmospheric Physics, University Mainz, Mainz, Germany

Basart Sara Barcelona Supercomputing Center (BSC), Barcelona, Spain

Bash Jesse Center for Environmental Measurement and Modeling, US Environmental Protection Agency, Durham, NC, USA

Beagley Stephen Air Quality Research Division, Science and Technology Branch, Environment and Climate Change Canada, Toronto, ON, Canada

Belmonte Jordina Department of Animal Biology, Plant Biology and Ecology, Faculty of Bioscience, Universitat Autònoma de Bellaterra, Bellaterra, Spain;
Institute of Environmental Sciences and Technology (ICTA-UAB), Universitat Autònoma de Bellaterra, Bellaterra, Spain

Benavides J. Barcelona Supercomputing Center, BSC, Barcelona, Spain

Benincasa Francesco Barcelona Supercomputing Center (BSC), Barcelona, Spain

Bergas-Massó Elisa Barcelona Supercomputing Center (BSC), Barcelona, Spain; Universitat Politècnica de Catalunya (UPC), Barcelona, Spain

Bloss William James School of Geography, Earth & Environmental Sciences, University of Birmingham, Birmingham, UK

Bonafè Giovanni ARPA-FVG, Palmanova, Italy

Bonasoni Paolo CNR Institute of Atmospheric Sciences and Climate, Bologna, Italy

Borge Rafael Department of Chemical and Environmental Engineering, ETSII – Universidad Politécnica de Madrid (UPM), Madrid, España;
Laboratory of Environmental Modelling, Department of Chemical and Environmental Engineering, Escuela Técnica Superior de Ingenieros Industriales, Universidad Politécnica de Madrid (UPM), Madrid, Spain

Borken-Kleefeld J. International Institute of Applied Systems Analysis, Laxenburg, Austria

Bowdalo Dene Barcelona Supercomputing Center (BSC), Barcelona, Spain

Breivik O. Norwegian Meteorological Institute, Oslo, Norway

Bretonnière Pierre-Antoine Barcelona Supercomputing Center (BSC), Barcelona, Spain

Briganti G. Laboratory of Atmospheric Pollution, Italian National Agency for New Technologies, Energy and Sustainable Economic Development – ENEA, Bologna, Italy

Broström G. University of Gothenburg, Gothenburg, Sweden

Bruffaerts Nicolas Belgian Scientific Institute of Public Health (SCIENSANO), Brussels, Belgium

Buñuel Arnau Barcelona Supercomputing Center (BSC), Barcelona, Spain

Cai Xiaoming School of Geography, Earth & Environmental Sciences, University of Birmingham, Birmingham, UK

Calgaro L. University of Venice, Venice, Italy

Cappelletti A. Laboratory of Atmospheric Pollution, Italian National Agency for New Technologies, Energy and Sustainable Economic Development – ENEA, Bologna, Italy

Carlino G. SIMULARIA Srl, Turin, Italy

Carvalho Ana Cristina Swedish Meteorological and Hydrological Institute, SMHI FoUmmk, Norrköping, Sweden

Carvalho David CESAM and Department of Physics, University of Aveiro, Aveiro, Portugal

Casas-Castillo M. Carmen Department of Physics, ESEIAAT, Universitat Politècnica de Catalunya·BarcelonaTech, Terrassa, Spain

Castrillo Miguel Barcelona Supercomputing Center (BSC), Barcelona, Spain

Chaillou C. Aramco Fuel Research Center, Aramco Overseas Company, Rueil-Malmaison, France

Chen Zhao-Yue ISGLOBAL, Barcelona, Spain;
State Key Laboratory of Organ Failure Research, Department of Biostatistics, Guangdong Provincial Key Laboratory of Tropical Disease Research, School of Public Health, Southern Medical University, Guangzhou, China;
Universitat Pompeu Fabra (UPF), Barcelona, Spain

Ciccioli P. Istituto Per I Sistemi Agricoli E Forestali del Mediterraneo (CNR), Perugia, Italy;
Istituto Per I Sistemi Biologici (CNR), ADR CNR, 1, Montelibretti, Monterorotondo Scalo, Italy

Cingiroglu Fulya Eurasia Institute of Earth Sciences, Istanbul Technical University, Istanbul, Turkey

Coelho Sílvia CESAM and Department of Environment and Planning, University of Aveiro, Aveiro, Portugal

Colette Augustin INERIS, Verneuil-en-Halatte, France

Cremona G. Laboratory of Atmospheric Pollution, Italian National Agency for New Technologies, Energy and Sustainable Economic Development – ENEA, Bologna, Italy

Crespin Pierre ISSeP, Liège, Belgium

Cuevas Emilio Izaña Atmospheric Research Center-State Meteorological Agency, Santa Cruz de Tenerife, Spain;
Izaña Atmospheric Research Center, AEMET, Santa Cruz de Tenerife, Spain

Dammers Enrico Department on Climate, Air and Sustainability, TNO, Utrecht, The Netherlands

de Andrés J. M. Department of Chemical and Environmental Engineering, ETSII – Universidad Politécnica de Madrid (UPM), Madrid, España

de Cámara Estibaliz Sáez Faculty of Engineering Bilbao, University of the Basque Country (UPV/EHU), Bilbao, Spain

De Causmaecker Karen Royal Meteorological Institute of Belgium (KMI), Ukkel, Belgium

de la Paz David Department of Chemical and Environmental Engineering, ETSII – Universidad Politécnica de Madrid (UPM), Madrid, España;
Laboratory of Environmental Modelling, Department of Chemical and Environmental Engineering, Escuela Técnica Superior de Ingenieros Industriales, Universidad Politécnica de Madrid (UPM), Madrid, Spain

de la Viesca P. Generalitat de Catalunya, Barcelona, Spain

De Linares Concepción Department of Botany, Faculty of Sciences, Universidad de Granada, Bellaterra, Spain

Delcloo Andy W. Royal Meteorological Institute of Belgium, Ukkel, Brussels, Belgium;
Department of Physics and Astronomy, Ghent University, Ghent, Belgium;
Royal Meteorological Institute of Belgium (KMI), Ukkel, Belgium and Department of Physics and Astronomy, Ghent University, Ghent, Belgium

Denby Bruce Rolstad The Norwegian Meteorological Institute, Oslo, Norway

Descombes Gaël INERIS, Verneuil-en-Halatte, France

Di Tomaso Enza Barcelona Supercomputing Center (BSC), Barcelona, Spain

Duchenne Christophe DAM, CEA, Arpajon, DIF, France

Dury Marie ISSeP, Liège, Belgium

D'Anna B. Aix-Marseille Univ, CNRS, LCE, Marseille, France

D'Elia I. Laboratory of Atmospheric Pollution, Italian National Agency for New Technologies, Energy and Sustainable Economic Development – ENEA, Bologna, Italy

D'Isidoro M. Laboratory of Atmospheric Pollution, Italian National Agency for New Technologies, Energy and Sustainable Economic Development – ENEA, Bologna, Italy

Ellermann Thomas Department of Environmental Science, Aarhus University, Roskilde, Denmark

Emrich A. L. Graduate Program in Sanitation, Environment and Water Resources, Federal University of Minas Gerais, Av. Pres. Antônio Carlos, Belo Horizonte, Brazil

Enciso S. Barcelona Supercomputing Center, BSC, Barcelona, Spain

Escribano Jerónimo Barcelona Supercomputing Center (BSC), Barcelona, Spain

Fagerli Hilde MetNorway, Oslo, Norway

Fallmann Joachim City of Heidelberg, Office of Environmental Protection, Trade Supervision and Energy, Heidelberg, Germany;
Institute of Meteorology and Climate Research, Karlsruhe Institute of Technology, Karlsruhe, Germany

Fares S. Council for Agricultural Research and Economics (CREA), Research Centre for Forestry and Wood, Rome, Italy

Feldner Josefine Department of Chemical Transport Modeling, Helmholtz-Zentrum hereon GmbH, Institute of Coastal Research, Geesthacht, Germany;
Institute of Coastal Environmental Chemistry, Helmholtz-Zentrum Hereon, Geesthacht, Germany

Ferrari Fabrizio TerrAria srl, Milano, Italy

Ferreira Joana CESAM and Department of Environment and Planning, University of Aveiro, Aveiro, Portugal

Finardi S. ARIANET Srl, Milan, Italy;
ARIANET, Palermo, Italy

Fink Lea Department of Chemical Transport Modeling, Helmholtz-Zentrum hereon GmbH, Institute of Coastal Research, Geesthacht, Germany;
Institute of Coastal Environmental Chemistry, Helmholtz-Zentrum Hereon, Geesthacht, Germany

Foley Kristen Center for Environmental Measurement and Modeling, US Environmental Protection Agency, Durham, NC, USA

Formenti Paola LISA, UMR CNRS 7583, Université Paris-Est-Créteil, Université de Paris, Institut Pierre-Simon Laplace (IPSL), Créteil, France

Fridell E. Swedish Environmental Research Institute, Stockholm, Sweden

Galvão E. S. Department of Environmental Engineering, Federal University of Espírito Santo, Vitória, Espírito Santo, Brazil

Gama Carla CESAM & Department of Environment and Planning, University of Aveiro, Aveiro, Portugal

Gangoiti Gotzon Faculty of Engineering Bilbao, University of the Basque Country (UPV/EHU), Bilbao, Spain

García-Pando Carlos Pérez Barcelona Supercomputing Center (BSC), Barcelona, Spain;
Catalan Institution for Research and Advanced Studies, ICREA, Barcelona, Spain

García Jose Antonio Faculty of Engineering Bilbao, University of the Basque Country (UPV/EHU), Bilbao, Spain

Ghahreman Roya Air Quality Research Division, Science and Technology Branch, Environment and Climate Change Canada, Toronto, ON, Canada

Gharibzadeh Maryam Institute of Geophysics, University of Tehran, Tehran, Iran

Gilliam Robert Center for Environmental Measurement and Modeling, US Environmental Protection Agency, Durham, NC, USA

Ginoux Paul Geophysical Fluid Dynamics Laboratory (GFDL), Princeton, NJ, USA

Giubilato E. University of Venice, Venice, Italy

Gondikas A. Creative Nano, Athens, Greece

Gong Wanmin Air Quality Research Division, Science and Technology Branch, Environment and Climate Change Canada, Toronto, ON, Canada

Gonçalves-Ageitos María Barcelona Supercomputing Center (BSC), Barcelona, Spain;
Universitat Politècnica de Catalunya (UPC), Barcelona, Spain;
Department of Project and Construction Engineering, Universitat Politècnica de Catalunya – BarcelonaTech (UPC), Barcelona, Spain;
Universitat Politècnica de Catalunya - BarcelonaTech (UPC), Department of Project and Construction Engineering, Terrassa, Spain

Granberg M. Swedish Environmental Research Institute, Stockholm, Sweden

Grigoriadis A. Aristotle University of Thessaloniki, Thessaloniki, Greece

Grönholm T. Finnish Meteorological Institute (FMI), Helsinki, Finland

Guevara M. Barcelona Supercomputing Center, BSC, Barcelona, Spain

Guinart X. Generalitat de Catalunya, Barcelona, Spain

Guzikowski Jakub Faculty of Earth Sciences and Environmental Management, University of Wroclaw, Wroclaw, Poland

Handley Jonathan Cambridge Environmental Research Consultants, Cambridge, UK

Hassellöv I.-M. Chalmers University of Technology, Gothenburg, Sweden

Hassellöv M. University of Gothenburg, Gothenburg, Sweden

Henderson Barron Office of Air Quality Planning and Standards, U.S. Environmental Protection Agency, Research Triangle Park, NC, USA

Hernandez I. Generalitat de Catalunya, Barcelona, Spain

Hertel Ole Department of Ecoscience, Aarhus University, Roskilde, Denmark

Hoebelke Lucie Belgian Scientific Institute of Public Health (SCIENSANO), Brussels, Belgium

Hogrefe Christian Office of Research and Development, U.S. Environmental Protection Agency, Research Triangle Park, NC, USA;
Center for Environmental Measurement and Modeling, US Environmental Protection Agency, Durham, NC, USA

Hole L. R. Norwegian Meteorological Institute, Oslo, Norway

Hood Christina Cambridge Environmental Research Consultants, Cambridge, UK

Hudson M. University of Southampton, Southampton, UK

Hõrrak Urmas Institute of Physics, University of Tartu, Tartu, Estonia

Itahashi Syuichi Central Research Institute of Electric Power Industry, Tokyo, Japan

Jaakkola J. J. K. University of Oulu, Oulu, Finland;
Finnish Meteorological Institute (FMI), Helsinki, Finland

Jalkanen J.-P. Finnish Meteorological Institute (FMI), Helsinki, Finland

Jin Jie-Qi State Key Laboratory of Organ Failure Research, Department of Biostatistics, Guangdong Provincial Key Laboratory of Tropical Disease Research, School of Public Health, Southern Medical University, Guangzhou, China;
Universitat Pompeu Fabra (UPF), Barcelona, Spain

Joassin Pascal ISSeP, Liège, Belgium

Johansson Lasse Finnish Meteorological Institute, Helsinki, Finland

Johnson Kate Cambridge Environmental Research Consultants, Cambridge, UK

Jorba Oriol Barcelona Supercomputing Center (BSC), Barcelona, Spain

Jung Daeun Laboratory of Environmental Modelling, Department of Chemical and Environmental Engineering, Escuela Técnica Superior de Ingenieros Industriales, Universidad Politécnica de Madrid (UPM), Madrid, Spain

Junninen Heikki Institute of Physics, University of Tartu, Tartu, Estonia

Kaasik Marko Institute of Physics, University of Tartu, Tartu, Estonia;
Estonian Environmental Research Centre, Tallinn, Estonia

Kaitaranta J. HELCOM, Helsinki, Finland

Kang Daiwen Office of Research and Development, U.S. Environmental Protection Agency, Research Triangle Park, NC, USA

Karatzas Kostas Environmental Informatics Research Group, School of Mechanical Engineering, Aristotle University of Thessaloniki, Thessaloniki, Greece

Karppinen Ari Finnish Meteorological Institute, Helsinki, Finland

Kassandros Theodosios Environmental Informatics Research Group, School of Mechanical Engineering, Aristotle University of Thessaloniki, Thessaloniki, Greece

Kaynak Burcak Department of Environmental Engineering, School of Civil Engineering, Istanbul Technical University, Istanbul, Turkey

Khan Jibran Department of Environmental Science, and Danish Big Data Centre for Environment and Health (BERTHA), Aarhus University, Roskilde, Denmark

Kiihamäki S.-P. University of Oulu, Oulu, Finland

Kim Y. CEREA, École des Ponts, EDF R &D, Île-de-France, France

Kirchner Ricard Department of Physics, ESEIAAT, Universitat Politècnica de Catalunya·BarcelonaTech, Terrassa, Spain

Kiss Raido Institute of Physics, University of Tartu, Tartu, Estonia; Estonian Environmental Research Centre, Tallinn, Estonia

Klose Martina Barcelona Supercomputing Center (BSC), Barcelona, Spain; Karlsruhe Institute of Technology (KIT), Institute of Meteorology and Climate Research (IMK-TRO), Department Troposphere Research, Karlsruhe, Germany

Kouznetsov Rostislav Finnish Meteorological Institute (FMI), Helsinki, Finland

Krasakopoulou E. University of the Aegean, Mytilene, Greece

Kryza Maciej Faculty of Earth Sciences and Environmental Management, University of Wroclaw, Wroclaw, Poland

Ktoris A. Maritime Institute of Eastern Mediterranean, Larnaca, Cyprus

Kukkonen J. Finnish Meteorological Institute (FMI), Helsinki, Finland; University of Hertfordshire, Hatfield, UK

Kumar Vinod Max-Planck Institute for Chemistry, Mainz, Germany

Larrieu C. Aramco Fuel Research Center, Aramco Overseas Company, Rueil-Malmaison, France

Lenartz Fabian ISSeP, Liège, Belgium

Link Eliza-Maria Department of Chemical Transport Modeling, Helmholtz-Zentrum hereon GmbH, Institute of Coastal Research, Geesthacht, Germany

Lonati Giovanni Department of Civil and Environmental Engineering, Politecnico di Milano, Milano, Italy

Lopes Myriam CESAM and Department of Environment and Planning, University of Aveiro, Aveiro, Portugal

Lugon L. CEREA, École des Ponts, EDF R &D, Île-de-France, France

López Franco Barcelona Supercomputing Center (BSC), Barcelona, Spain; Department of Project and Construction Engineering, Universitat Politècnica de Catalunya –BarcelonaTech (UPC), Barcelona, Spain

Macchia Francesca Barcelona Supercomputing Center (BSC), Barcelona, Spain

Maffeis Giuseppe TerrAria srl, Milano, Italy

Magnusson K. Swedish Environmental Research Institute, Stockholm, Sweden

Maia B. L. V. Graduate Program in Sanitation, Environment and Water Resources, Federal University of Minas Gerais, Av. Pres. Antônio Carlos, Belo Horizonte, Brazil

Majamäki E. Finnish Meteorological Institute (FMI), Helsinki, Finland

Malguzzi Piero CNR Institute of Atmospheric Sciences and Climate, Bologna, Italy

Mangold Alexander Royal Meteorological Institute of Belgium (KMI), Ukkel, Belgium

Maragkidou A. Finnish Meteorological Institute (FMI), Helsinki, Finland

Marcomini A. University of Venice, Venice, Italy

Marques B. Aix-Marseille Univ, CNRS, LCE, Marseille, France;
French Agency for Ecological Transition, Angers, France

Marques D. D. Graduate Program in Sanitation, Environment and Water Resources, Federal University of Minas Gerais, Av. Pres. Antônio Carlos, Belo Horizonte, Brazil

Mateu J. Barcelona Supercomputing Center, BSC, Barcelona, Spain

Mathur Rohit Office of Research and Development, U.S. Environmental Protection Agency, Research Triangle Park, NC, USA;
Center for Environmental Measurement and Modeling, US Environmental Protection Agency, Durham, NC, USA

Matthias Volker Department of Chemical Transport Modeling, Helmholtz-Zentrum hereon GmbH, Institute of Coastal Research, Geesthacht, Germany;
Institute of Coastal Environmental Chemistry, Helmholtz-Zentrum Hereon, Geesthacht, Germany

Mazzeo Andrea School of Geography, Earth & Environmental Sciences, University of Birmingham, Birmingham, UK

Miller Ron L. NASA Goddard Institute for Space Studies, New York, NY, USA

Miranda Ana Isabel CESAM and Department of Environment and Planning, University of Aveiro, Aveiro, Portugal

Mircea M. Laboratory of Atmospheric Pollution, Italian National Agency for New Technologies, Energy and Sustainable Economic Development—ENEA, Bologna, Italy

Moldanova J. Swedish Environmental Research Institute, Stockholm, Sweden

Mona Lucia Consiglio Nazionale delle Ricerche-Istituto di Metodologie per l'Analisi Ambientale (CNR-IMAA), Rome, Italy

Moniruzzaman Chowdhury Institute for the Environment, University of North Carolina at Chapel Hill, Chapel Hill, NC, USA

Montané Gilbert Barcelona Supercomputing Center (BSC), Barcelona, Spain

Monteiro A. University of Aveiro, Aveiro, Portugal

Mu Qing The Norwegian Meteorological Institute, Oslo, Norway

Myriokefalitakis Stelios Institute for Environmental Research and Sustainable Development (IERSD), National Observatory of Athens, Penteli, Greece

Mytilinaios Michail Consiglio Nazionale delle Ricerche-Istituto di Metodologie per l'Analisi Ambientale (CNR-IMAA), Rome, Italy

Napelenok Sergey Office of Research and Development, U.S. Environmental Protection Agency, Research Triangle Park, NC, USA

Narros A. Department of Chemical and Environmental Engineering, ETSII – Universidad Politécnica de Madrid (UPM), Madrid, España

Navarro-Barboza Hector Barcelona Supercomputing Center (BSC), Barcelona, Spain

Neophytou M. Maritime Institute of Eastern Mediterranean, Larnaca, Cyprus

Nicolas F. HELCOM, Helsinki, Finland

Niemi Jarkko Helsinki Region Environmental Authority HSY, Helsinki, Finland

Noe Steffen Institute of Forestry and Engineering, Estonian University of Life Sciences, Tartu, Estonia

Notario Alberto Physical Chemistry Department, Faculty of Chemical Science and Technologies, Universidad de Castilla-La Mancha, Ciudad Real, Spain

Ntziachristos L. Aristotle University of Thessaloniki, Thessaloniki, Greece

Obiso Vincenzo Barcelona Supercomputing Center (BSC), Barcelona, Spain

Oikonomou F. DANAOS, Limassol, Cyprus

Olid Miriam Barcelona Supercomputing Center (BSC), Barcelona, Spain

Ou Chun-Quan State Key Laboratory of Organ Failure Research, Department of Biostatistics, Guangdong Provincial Key Laboratory of Tropical Disease Research, School of Public Health, Southern Medical University, Guangzhou, China; Universitat Pompeu Fabra (UPF), Barcelona, Spain

Ozdemir U. A. University of Hertfordshire, Hatfield, UK

Pandey Gavendra Institute for the Environment, University of North Carolina at Chapel Hill, Chapel Hill, NC, USA

Pandolfi Marco Institute of Environmental Assessment and Water Research, IDAEA-CSIC), Barcelona, Spain

Pay M. T. Universitat de Barcelona, Barcelona, Spain

Pepe N. ARIANET Srl, Milan, Italy;
ARIANET, Palermo, Italy

Periago Cristina Department of Physics, EEBE, Barcelona, Spain

Petetin H. Barcelona Supercomputing Center, BSC, Barcelona, Spain

Petrik Ronny Department of Chemical Transport Modeling, Helmholtz-Zentrum hereon GmbH, Institute of Coastal Research, Geesthacht, Germany;
Institute of Coastal Environmental Chemistry, Helmholtz-Zentrum Hereon, Geesthacht, Germany

Petrovic M. Catalan Institute for Water Research, Girona, Spain

Piccoli Andrea Department of Civil and Environmental Engineering, Politecnico di Milano, Milano, Italy;
Sustainable Development and Energy Sources Department, RSE Spa, Milano, Italy

Piersanti A. Laboratory of Atmospheric Pollution, Italian National Agency for New Technologies, Energy and Sustainable Economic Development – ENEA, Bologna, Italy

Pirovano Guido Sustainable Development and Energy Sources Department, RSE Spa, Milano, Italy

Pleim Jonathan Center for Environmental Measurement and Modeling, US Environmental Protection Agency, Durham, NC, USA

Poom Age Helsinki Institute of Urban and Regional Studies (Urbaria), Helsinki, Finland

Prandi R. SIMULARIA Srl, Turin, Italy

Pérez E. M. Generalitat de Catalunya, Barcelona, Spain

Quante Markus Department of Chemical Transport Modeling, Helmholtz-Zentrum hereon GmbH, Institute of Coastal Research, Geesthacht, Germany;
Institute of Coastal Environmental Chemistry, Helmholtz-Zentrum Hereon, Geesthacht, Germany

Ramacher Martin Otto Paul Department of Chemical Transport Modeling, Helmholtz-Zentrum hereon GmbH, Institute of Coastal Research, Geesthacht, Germany

Ramos M. C. A. F. Graduate Program in Sanitation, Environment and Water Resources, Federal University of Minas Gerais, Av. Pres. Antônio Carlos, Belo Horizonte, Brazil

Ranjbar Saadatabadi Abbas Atmospheric Science and Meteorological Research Center (ASMERC), Tehran, Iran

Remut Arkadiusz Faculty of Earth Sciences and Environmental Management, University of Wroclaw, Wroclaw, Poland

Ribeiro A. K. C. Graduate Program in Sanitation, Environment and Water Resources, Federal University of Minas Gerais, Av. Pres. Antônio Carlos, Belo Horizonte, Brazil

Robertson Lennart Swedish Meteorological and Hydrological Institute, SMHI FoUmmk, Norrköping, Sweden

Rodriguez-Mozaz S. Catalan Institute for Water Research, Girona, Spain

Rodriguez D. Barcelona Supercomputing Center, BSC, Barcelona, Spain

Rodríguez-Solà Raül Department of Physics, ETSEIB, Universitat Politècnica de Catalunya·BarcelonaTech, Barcelona, Spain

Roustan Y. CEREA, École des Ponts, EDF R &D, Île-de-France, France

Russo F. Laboratory of Atmospheric Pollution, Italian National Agency for New Technologies, Energy and Sustainable Economic Development – ENEA, Bologna, Italy

Russo M. A. University of Aveiro, Aveiro, Portugal

Sanyer X. Autoritat del Transport Metropolità, Barcelona, Spain

Sarica T. CEREA, École des Ponts, EDF R &D, Île-de-France, France

Sartelet K. CEREA, École des Ponts, EDF R &D, Île-de-France, France

Sarwar Golam Office of Research and Development, U.S. Environmental Protection Agency, Research Triangle Park, NC, USA

Sawiński Tymoteusz Faculty of Earth Sciences and Environmental Management, University of Wroclaw, Wroclaw, Poland

Sbert Macià Mut Centre de Recerca Matemàtica, Faculty of Science, Universitat Autònoma de Barcelona (UAB), Barcelona, Spain

Schaap Martijn Department on Climate, Air and Sustainability, TNO, Utrecht, The Netherlands

Schutgens Nick Department of Earth Science, Vrije Universiteit Amsterdam, Amsterdam, Netherlands

Schwarzkopf Daniel Department of Chemical Transport Modeling, Helmholtz-Zentrum hereon GmbH, Institute of Coastal Research, Geesthacht, Germany;
Institute of Coastal Environmental Chemistry, Helmholtz-Zentrum Hereon, Geesthacht, Germany

Schwede Donna Center for Environmental Measurement and Modeling, US Environmental Protection Agency, Durham, NC, USA

Segers Arjo Department on Climate, Air and Sustainability, TNO, Utrecht, The Netherlands;
MetNorway, Oslo, Norway

Serradell K. Barcelona Supercomputing Center, BSC, Barcelona, Spain

Silibello C. ARIANET Srl, Milan, Italy;
ARIANET, Palermo, Italy

Sofiev Mikhail Finnish Meteorological Institute (FMI), Helsinki, Finland

Sokhi R. S. University of Hertfordshire, Hatfield, UK

Soret A. Barcelona Supercomputing Center, BSC, Barcelona, Spain

Sorrentino B. Laboratory of Atmospheric Pollution, Italian National Agency for New Technologies, Energy and Sustainable Economic Development – ENEA, Bologna, Italy

Sousse Rubén Barcelona Supercomputing Center (BSC), Barcelona, Spain

Stocker Jenny Cambridge Environmental Research Consultants, Cambridge, UK

Stortini Michele ARPAE, viale Silvani 6, 40122, Bologna, Italy

Tammekivi Terje Institute of Physics, University of Tartu, Tartu, Estonia

Tamminen J. Finnish Meteorological Institute (FMI), Helsinki, Finland

Tena C. Barcelona Supercomputing Center, BSC, Barcelona, Spain

Terrado M. Barcelona Supercomputing Center, BSC, Barcelona, Spain

Thomas Manu Anna Swedish Meteorological and Hydrological Institute, SMHI FoUmmk, Norrköping, Sweden

Timmermans Renske Department on Climate, Air and Sustainability, TNO, Utrecht, The Netherlands

Timonen Hilkka Finnish Meteorological Institute, Helsinki, Finland

Tinarelli Gianni ARIANET Srl, Milan, Italy

Tiwari P. R. University of Hertfordshire, Hatfield, UK

Torre-Pascual Eduardo Faculty of Engineering Bilbao, University of the Basque Country (UPV/EHU), Bilbao, Spain

Torres V. A. Graduate Program in Sanitation, Environment and Water Resources, Federal University of Minas Gerais, Av. Pres. Antônio Carlos, Belo Horizonte, Brazil

Tost Holger Institute of Atmospheric Physics, University Mainz, Mainz, Germany

Trini Castelli Silvia CNR Institute of Atmospheric Sciences and Climate, Turin, Italy

Tsegas G. Aristotle University of Thessaloniki, Thessaloniki, Greece

Uboldi Francesco ARIANET Srl, Milan, Italy

Unal Alper Eurasia Institute of Earth Sciences, Istanbul Technical University, Istanbul, Turkey

Uppstu Andreas Finnish Meteorological Institute (FMI), Helsinki, Finland

Valdebenito Alvaro MetNorway, Oslo, Norway

Venkatram Akula University of California at Riverside, Riverside, CA, USA

Verstraeten Willem W. Royal Meteorological Institute of Belgium, Ukkel, Brussels, Belgium

Vidal Veronica Institute of Environmental Science and Technology (ICTA), Universitat Autònoma de Barcelona (UAB), Barcelona, Spain;
Departament d'Arquitectura de Computadors i Sistemes Operatius (CAOS), Escola d'Enginyeria, Universitat Autònoma de Barcelona (UAB), Barcelona, Spain

Villalba Gara Institute of Environmental Science and Technology (ICTA), Universitat Autònoma de Barcelona (UAB), Barcelona, Spain;
Department of Chemical, Biological and Environmental Engineering, Universitat Autònoma de Barcelona (UAB), Barcelona, Spain

Votsis Athanasios Department of Governance and Technology for Sustainability (BMS-CSTM), University of Twente, Enschede, Netherlands;
Finnish Meteorological Institute (FMI), Weather and Climate Change Impact Research, Helsinki, Finland

Wedemann Ralf Department of Chemical Transport Modeling, Helmholtz-Zentrum hereon GmbH, Institute of Coastal Research, Geesthacht, Germany;
Institute of Coastal Environmental Chemistry, Helmholtz-Zentrum Hereon, Geesthacht, Germany

Werner Ernest State Meteorological Agency of Spain, Barcelona, Spain;
State Meteorological Agency (AEMET), Madrid, Spain

Werner Malgorzata Faculty of Earth Sciences and Environmental Management, University of Wroclaw, Wroclaw, Poland

Williams I. University of Southampton, Southampton, UK

Wolstencroft Mark Birmingham City Council, Birmingham, UK

Xing Jia Tsinghua University, Beijing, China

Ytreberg E. Chalmers University of Technology, Gothenburg, Sweden

Zapata-Restrepo L. University of Southampton, Southampton, UK

Zappitelli I. Council for Agricultural Research and Economics (CREA), Research Centre for Forestry and Wood, Rome, Italy

Zervakis V. University of the Aegean, Mytilene, Greece

Zhong Jian School of Geography, Earth & Environmental Sciences, University of Birmingham, Birmingham, UK

Zuazo Iñaki Faculty of Engineering Bilbao, University of the Basque Country (UPV/EHU), Bilbao, Spain

Local and Urban Scale Modelling

Air Quality Forecast Systems from the Regional Scale to the Urban Street

José M. Baldasano

Abstract The development of air quality forecasting systems (AQFS), the elements that compose them, their objective, the basic functions they perform and the requirements they currently have, as well as their conditions of use, are briefly presented. Today, AQFS advanced systems combine a suite of state-of-the-art models (NWP and CTM), high-resolution emissions inventories, satellite observations, and surface measurements of weather variables and the most relevant pollutants to provide forecasts and analysis of air pollution, from the global to the local scale, but especially at the regional scale, with a horizontal grid resolution reaching 1 km^2, moving towards street level, for countries, regions and particularly urban areas in most areas of the world. This development has been possible thanks to the exponential growth of computer systems.

Keywords Air quality forecasting systems · Air quality · Air management

1 Introduction

The history of air quality forecast systems (AQFS) can be said to go back to the 1960s, the first forecasts of air pollution potential using numerical weather prediction (NWP) models to forecast weather conditions that lead to poor air quality (Niemeyer, 1960).

Air pollution models have played and continue to play a critical role in scientific understanding and management support for air quality improvement. The application of air quality models—its scientific basis, already had years of development- expands from the second half of the sixties of the twentieth century, with the boom in the development of Thermal Power Plants—especially in the United States and Europe—to support the growing demand for electricity (Slade, 1968).

J. M. Baldasano (✉)
Emeritus Full Professor of Environmental Engineering, Technical University of Catalonia (UPC), Barcelona, Spain
e-mail: jose.baldasano@upc.edu

Barcelona Supercomputing Centre (BSC), Barcelona, Spain

C. Mensink and O. Jorba (eds.), *Air Pollution Modeling and its Application XXVIII*, Springer Proceedings in Complexity, https://doi.org/10.1007/978-3-031-12786-1_1

In all these years, it is absolutely necessary to mention and remember the role played by the series of conferences: "*International Technical Meeting on Air Pollution Modelling and its Applications (ITM)*" since 1969. Throughout its long history, the ITM conferences have established themselves as one of the most prominent international forums to discuss developments, improvements and applications related to air quality modelling at scales ranging from local to global [https://itm2021.vito.be/en/about-itm].

Specifically dedicated to AQFS, it is also necessary to mention the series of "International Workshops on Air Quality Forecasting Research (IWAQFR)" whose objective is to provide a forum to discuss scientific issues and advances related to air quality forecasting [https://community.wmo.int/meetings/10th-international-workshop-air-quality-forecasting-research-iwaqfr].

Starting in the 1970s, various AQFS techniques and tools—also called RT-AQF: Real Time Air Quality Forecast—were developed to forecast air pollution, especially focused on urban areas. Mainly based on empirical approaches and statistical models trained or adapted to historical meteorological and air quality data (Zhang et al., 2012).

For the development and application of the current AQFS, three fundamental facts occurred in the last 30 years of the twentieth century:

1. The origin of air pollution went from being mostly caused by emissions from chimneys to being caused mainly by traffic in cities, causing a paradigm shift.
2. An accelerated process of development and improvement of meteorological and air dispersion models. In particular, chemical transport models (CTM) coupled to high-resolution mesoscale numerical weather prediction (NWP). Developments in local scale and street level air quality models continue.
3. An exponential increase in computing capacity and data management, going from being managed manually to being done automatically.

With the beginning of the twenty-first century, the basic characteristics of the current AQFS are already mature, the works of Brandt et al. (2000) and from Tilmes et al. (2002) in Europe and NOAA (2001) in the United States already indicate a comprehensive review of the science related to air quality forecasting.

Over the last two decades, AQFS systems have been developed and put into operation to understand and forecast the spatial and temporal pattern of air pollutants from local to global scales. AQFS systems provide a more holistic view of air pollution for strategies and plans to improve air quality and its daily management. Chemical Transport Models (CTM) coupled to Weather Forecast Models (NWP) are commonly used today for 3D air quality forecasting. AQFS systems based on NWP and CTM have developed rapidly in recent years, strongly aided by progress in computational power. It is very common to use the NWP WRF model together with different CTM (CMAQ, CHIMERE, etc.) running in nested mode up to an internal domain with a grid size of 1 km; with initializations derived from NCEP and ECMWF. With forecast times that are between 48 and 72 h.

The development and improvement of NWP and CTM, along with computing technologies, have enabled the implementation of more powerful and comprehensive

AQFS systems, such as coupled offline and online meteorological and chemical models. In the last ten years, several articles have been published that carry out an in-depth review of the situation and evolution of the AQFS: Menut and Bassagnet (2010), Balk et al. (2011), Zhang et al. (2012), Roueil and Bassagnet (2013), Baklanov et al. (2014); and Baklanov and Zhang (2020) that thoroughly examine the principles and development of AQFS systems and identify areas for improvement.

It can be pointed out that the AQFS have entered a stage of maturity, where they have not only evolved to cover from the regional to the local and street scale, but have also expanded the territorial domain towards continental and global scales. In the European context, the CAMS system: Copernicus Atmosphere Monitoring System [https://atmosphere.copernicus.eu/]; in United States, the US National Air Quality Forecast Capability (NAQFC), developed by the National Oceanic and Atmospheric Administration (NOAA) in partnership with the Environmental Protection Agency (EPA) [https://www.weather.gov/sti/stimodeling_airquality], and within the framework of the World Meteorological Organization (WMO): Global Atmosphere Watch Program (GAW) the GAFIS: Global to Local Air Quality Forecast System Inventory [https://community.wmo.int/activity-areas/gaw/science-for-services/gafis].

2 AQFS: Objective, Basic Functions and Requirements

The main objective and raison d'être of an AQFS is to ensure the best air quality forecast. However, it is necessary to point out two important clarifications:

1. An important point is that an AQFS is not just a set of models. An AQFS is essentially a COMPUTER SYSTEM that manages models—basic and essential-, input/output data, graphics, databases, web pages, computer communications, etc., all in real time and automatically. It is obviously based on a set of models, but it is much more than a set of models, since it must ensure each day—with the greatest possible precision—the air quality forecast for the next few days, to ensure with the best information, decision-making in air quality management, especially in situations of pollution episodes.
2. An AQFS is a powerful tool in understanding and managing air quality, and it complements in-depth air quality measurement networks, on the temporal scale but especially on the spatial scale.

The most common forecast products found in an AQFS are listed below:

- Dynamic maps by domain: 2D and 3D
  - Emissions
  - Meteorology
  - Air quality: O_3, NO_2, CO, SO_2, PM_{10}, $PM_{2.5}$, Benzene, …
- 2D temporal graphs and tables of forecast air quality values at 24 and 48 h (especially at air quality stations)

- Air quality indicators
- Air quality concentration
- Air quality index

- Forecast evaluation: meteorological and air quality (quasi in real time), for different domains:
  - Biweekly and monthly series
  - Annual evaluation (statistics and graphs)
- Satellite images:
  - Air quality
  - Meteorological

In order to provide all this broad set of products, it is necessary to define a whole set of different functions and requirements, this list allows to view the current complexity of an AQFS:

- Domain's and Topography
- Pollutants considered
- Chemical speciation
- Grid spatial resolution
- Land cover database and digital terrain model
- Number and distribution of vertical layers and coupling
- Temporal projection
- Weather model (NWP)
- Chemical Transport Model (CTM)
- Chemical mechanism
- Emissions model
- Initial and boundary conditions (Meteorological and Chemical)
- Air quality network connection
- Post-processing data
- Evaluation criteria
- Prognostic evaluation in near real time (NRT) versus surface observations, vertical profile, satellites
- Display products
- Database management
- Computational resources
- System operational management
- Communication: web, …

But above all and fundamentally, it is absolutely necessary to integrate a human team with different professional profiles.

This leads to another of the products generated by the continuous operation of an AQFS, since keeping an AQFS in operation implies carrying out daily simulations that accumulate, resulting in the generation of historical databases, and the need for computer servers to store and provide this service. This also makes it an important

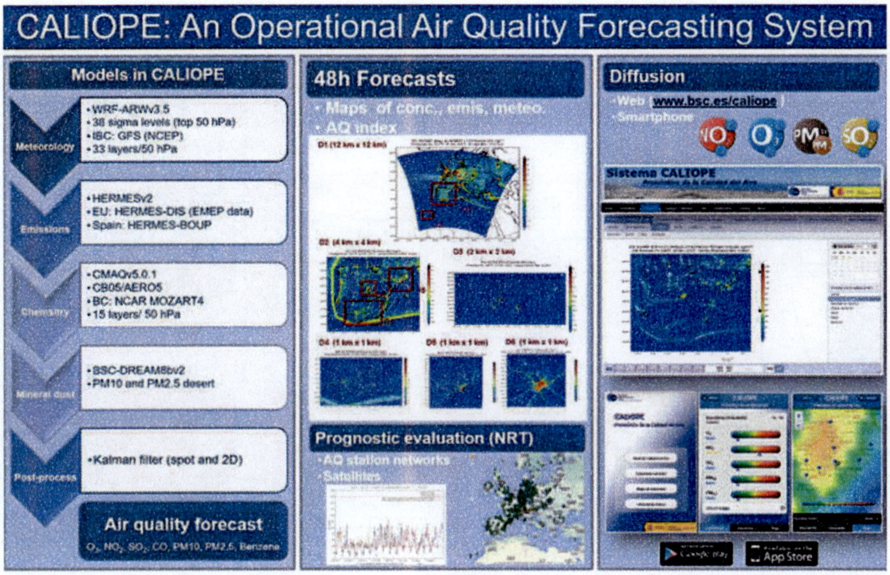

Fig. 1 Main elements of an AQFS, example of the CALIOPE system

information source tool for diagnostic studies of weather and pollution events and patterns over time.

Figure 1 shows the main elements that currently make up an AQFS, case of the CALIOPE system (Baldasano et al., 2011; http://www.bsc.es/caliope/es).

Another significant aspect to point out refers to the different perspectives that exist between the user and the developer and administrator of an AQFS, both perspectives are necessary to evaluate its functionality. From the user's perspective, there is the friendliness of the web environment, the functionality of the system, the information it offers, the potential handling of information, the description of the system, etc.; that is, the visible and public part of the system. While for the system developer and administrator, responsible for the hidden part of the system, it is essential to ensure its temporary operation 365/7/24 with the best forecast and quality possible over time, in its ability to constantly improve the different elements and system requirements (see the Fig. 2).

3 The Need for AQFS

An AQFS cannot, by itself, solve air pollution problems, but air quality forecasts, if reliable and accurate enough, play an important and complementary role as part of a quality management system of the air that works in conjunction with more traditional

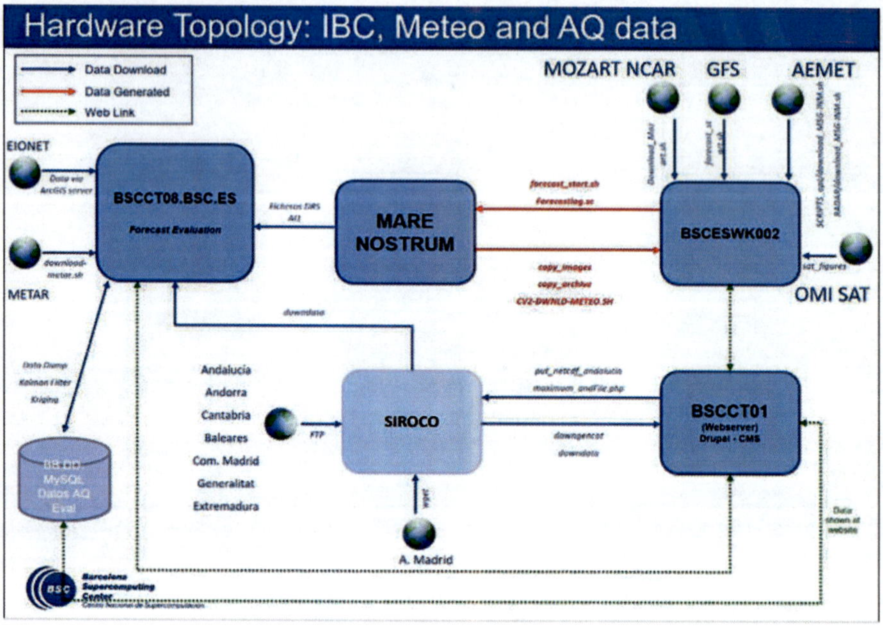

Fig. 2 Hardware topology of an AQFS, example of the CALIOPE system (Baldasano et al., 2013)

approaches. Dabberdt et al. (2004) already defined the needs of users regarding air quality forecasts (see the Fig. 3).

4 Future Directions

All the elements of the system mentioned are important and fundamental to obtain the best quality of the forecast, but they can be pointed out as more significant:

- The resolution of the horizontal grid and the vertical layer's distribution.
- The coupling between the meteorological model (NWP) and the chemical transport model (CTM).
- The initial conditions derived from the global models.
- Chemical mechanisms: chemical species and particulate matter.
- Emission sources versus Pollutants and Speciation.
- High resolution emission system.

The use of post-process techniques-Karman filter, for example-clearly improves the forecast (Sicardi et al., 2012). An alternative considered widely used in improving the results of the NWP model is data assimilation (DA). But it is necessary to take into account that the computation times increase exponentially with respect to the relative gain, compared to more affordable techniques.

TABLE 1. User needs for air quality forecasts.

Sector	Group or organization	Forecast products: Regional daily maximum	Local or city predictions	Air quality index (AQI)	Forecast uncertainty	Visibility predictions	Human exposure	Forecast period: Short range (0–48 h)	Mid-range (48–168 h)	Seasonal predictions
Public	Sensitive groups	✓	✓	✓			✓	✓	✓	
	General public	✓	✓	✓			✓	✓	✓	
	Outdoor workers/recreation	✓	✓	✓			✓	✓	✓	
Decision makers	Air quality agencies									
	- Episodic or special sampling		✓		✓	✓		✓	✓	
	- Emission-reduction programs	✓	✓	✓	✓			✓	✓	
	- Health/air awareness	✓	✓	✓			✓	✓	✓	
	Emergency response	✓	✓		✓		✓	✓	✓	
	Industry (e.g., power)	✓	✓		✓	✓		✓		✓
	Aviation	✓				✓		✓		
	Transportation		✓			✓		✓		
	Health care	✓	✓	✓			✓	✓	✓	
	Environment	✓	✓	✓	✓	✓	✓	✓	✓	✓
	Forecast and park service	✓	✓		✓	✓		✓	✓	
Media	Television weathercasters	✓	✓	✓		✓	✓	✓	✓	
	Internet content providers	✓	✓	✓		✓	✓	✓	✓	
	Newspapers	✓	✓	✓		✓	✓	✓	✓	
Researchers	Air quality scientist	✓	✓	✓		✓	✓	✓	✓	✓
	Air quality regulators	✓	✓	✓	✓	✓	✓	✓	✓	✓
	Field measurement studies	✓			✓	✓		✓	✓	✓

Fig. 3 The need for AQFS (Dabberdt et al., 2004)

The key trends for improving AQFS are improving current methods and systems and developing new methods, including data assimilation and fusion algorithms, machine learning methods, and bias correction techniques. The following list, based on and modified from the work of Baklanov and Zhang (2020), indicates several possible directions:

- Online coupling of atmospheric and chemical dynamics models.
- Better representation of the processes and reactions of aerosols, their interactions with clouds and radiation.
- PBL and mixed layer characterization
- Better assimilation and fusion of data, machine learning and artificial intelligence methods.
- Towards a better modelling of the Earth system.
- Multi-scale prediction approach.
- Sub-seasonal to seasonal forecast.
- Forecast based on impacts and adaptation systems to the needs to improve air quality
- Data integration from observations, models and satellites.
- Need to reach citizenship.

5 Conclusions

After a development and application of more than sixty years. Today, the AQFS advanced systems combine a set of state-of-the-art models (NWP and CTM), high-resolution emissions inventories, satellite observations and surface measurements of meteorological variables and the most relevant pollutants to provide forecasts and analysis of air pollution, from the global to the local scale, but especially at the regional scale, with horizontal grid resolution reaching 1 km^2, advancing towards the street level, for countries, regions and particularly urban areas in most areas of the world.

This development has been possible thanks to the exponential growth of computer systems, which have been progressively providing computing capacity, storage, and means of data visualization and management.

The AQFS essentially represent a compendium of science, computing, and technology. They currently generate information on the air quality that we will have to breathe in the coming days for decision-making from a supercomputer to a smartphone.

References

Baklanov, A., Schlünzen, K., Suppan, P., Baldasano, J. M., Brunner, D., Aksoyoglu, S., Carmichael, G., Douros, J., Flemming, J., Forkel, R., Galmarini, S., Gauss, M., Grell, G., Hirtl, M., Joffre, S., Jorba, O., Kaas, E., Kaasik, M., Kallos, G., et al. (2014). Online coupled regional meteorology chemistry models in Europe: Current status and prospects. *Atmospheric Chemistry and Physics, 14*, 317–398. https://doi.org/10.5194/acp-14-317-2014

Baklanov, A., & Zhang, Y. (2020). Advances in air quality modeling and forecasting. *Global Transitions, 2*, 261–270. https://doi.org/10.1016/j.glt.2020.11.001

Baldasano, J. M., Pay, M. T., Arévalo, G., & Gassó, S. (2013, October 8–9). Evaluation of chemical initial and boundary on ozone concentrations in the Spanish CALIOPE AQFS. In *5th International Workshop on Air Quality Forecasting Research (IWAQFR)*, Santiago de Chile (Chile).

Baldasano, J. M., Pay, M. T., Jorba, O., Gassó, S., & Jiménez-Guerrero, P. (2011). An annual assessment of air quality with the CALIOPE modeling system over Spain. *Science of the Total Environment, 409*, 2163–2178. https://doi.org/10.1016/j.scitotenv.2011.01.041

Balk, T., Kukkonen, J., Karatzas, K., Bassoukos, T., & Epitropou, V. (2011). A European open access chemical weather forecasting portal. *Atmospheric Environment, 45*, 6917–6922. https://doi.org/10.1016/j.atmosenv.2010.09.058

Brandt, J., Christensen, J. H., Frohn, L. M., Palmgren, F., Berkowicz, R., & Zlatev, Z. (2000). Operational air pollution forecasts from European to local scale. *Atmospheric Environment, 35-S1*, S91–S98.

Dabberdt, et al. (2004, April). Meteorological research needs for improved air quality forecasting. *Bulletin of the American Meteorological Society (BAMS)*, 563–586.

Menut, L., & Bassagnet, B. (2010). Atmospheric composition forecasting in Europe. *Annals of Geophysics, 28*, 61–74. www.ann-geophys.net/28/61/2010/

Niemeyer, L. E. (1960). Forecasting air pollution potential. *Monthly Weather Review, 88*(3), 88–96.

NOAA. (2001). *Air quality forecasting. A review of federal programs and research needs* (26 pp.). NOAA Aeronomy Laboratory.

Roueil, L., & Bassagnet, B. (2013). *ETC/ACM Technical Paper 2013/11 FAIRMODE* (103 pp.).

Sicardi, V., Ortiz, J., Rincón, A., Jorba, O., Pay, M. T., Gassó, S., & Baldasano, J. M. (2012). Assessment of Kalman filter bias-adjustment technique to improve the simulation of ground-level ozone over Spain. *Science of the Total Environment, 416*, 329–342. https://doi.org/10.1016/j.scitotenv.2011.11.050

Slade, D. H. (Ed.). (1968). *Meteorology and atomic energy 1968* (459 pp.). U.S. Atomic Energy Commission.

Tilmes, S., Brandt, J., Flatøy, F., Bergström, R., Flemming, J., Langner, J., Christensen, J. H., Frohn, L., Hov, Ø., Jacobsen, I., Reimer, E., Stern, R., & Zimmermann, J. (2002). Comparison of five Eulerian air pollution forecasting systems for the summer of 1999 using the German ozone monitoring data. *Journal of Atmospheric Chemistry, 42*, 91–121.

Zhang, Y., Bocquet, M., Mallet, V., Seigneur, C., & Baklanov, A. (2012). Real-time air quality forecasting, Part I: History, techniques, and current status. *Atmospheric Environment, 60*, 632–655. https://doi.org/10.1016/j.atmosenv.2012.06.031

Questions and Answers

Clemens Mensink: Since AI, Machine Learning, Neural networks, etc. do offer powerful tools in air quality forecasting at low cost, how do you see the future for deterministic air quality forecasting systems?

Answer: In recent years, there is a clear emergence of new techniques based on AI and machine learning, which have opened up new forms of research in this field with new perspectives. It should be noted that although AI-based techniques have made clearly progress in air quality forecasting, they have not yet been implemented on a large scale due to concerns about stability and interpretability. I believe that both approaches will continue to develop and coexist in the coming years.

Giovanni Bonafé: Considering the trend of the AQ forecasting system to go to finer and finer resolution, could we expect the next frontier to be the indoor AQ forecast?

Answer: At the moment the efforts and developments are focused on the street level scale.

Urban Vegetation Effects on Meteorology and Air Quality: A Comparison of Three European Cities

M. Mircea, Rafael Borge, S. Finardi, S. Fares, G. Briganti, M. D'Isidoro, G. Cremona, F. Russo, A. Cappelletti, I. D'Elia, M. Adani, A. Piersanti, B. Sorrentino, David de la Paz, J. M. de Andrés, A. Narros, C. Silibello, N. Pepe, I. Zappitelli, A. Alivernini, R. Prandi, and G. Carlino

Abstract Nature based solutions (NBS) have been put forward as an effective approach for improving human well-being in the cities during the last decade and many of them make use of vegetation (trees, grass, etc.). The role of vegetation in regulating air temperature and moisture availability is recognized together with its capacity to remove pollutants. Yet, cities lack of quantitative evaluation of these

---

M. Mircea (✉) · G. Briganti · M. D'Isidoro · G. Cremona · F. Russo · A. Cappelletti · I. D'Elia · M. Adani · A. Piersanti · B. Sorrentino
Laboratory of Atmospheric Pollution, Italian National Agency for New Technologies, Energy and Sustainable Economic Development – ENEA, 40129 Bologna, Italy
e-mail: mihaela.mircea@enea.it

G. Briganti
e-mail: gino.briganti@enea.it

M. D'Isidoro
e-mail: massimo.disidoro@enea.it

G. Cremona
e-mail: giuseppe.cremona@enea.it

F. Russo
e-mail: felicita.russo@enea.it

A. Cappelletti
e-mail: andrea.cappelletti@enea.it

I. D'Elia
e-mail: ilaria.delia@enea.it

M. Adani
e-mail: mario.adani@enea.it

A. Piersanti
e-mail: antonio.piersanti@enea.it

B. Sorrentino
e-mail: beatrice.sorrentino@enea.it

R. Borge · D. de la Paz · J. M. de Andrés · A. Narros
Department of Chemical and Environmental Engineering, ETSII – Universidad Politécnica de Madrid (UPM), Madrid 28040, España
e-mail: rafael.borge@upm.es

C. Mensink and O. Jorba (eds.), *Air Pollution Modeling and its Application XXVIII*, Springer Proceedings in Complexity, https://doi.org/10.1007/978-3-031-12786-1_2

effects that may set the basis for the assessment of the impacts of new NBS. This study conducted within the framework of Life VEG-GAP project shows the impact of vegetation on temperature and, further, on air concentrations and deposition of ozone (O_3), particulate matter (PM10) and nitrogen dioxide (NO_2) in three European cities: Bologna, Madrid and Milan. The simulations were carried out with two state-of-the-art air quality modelling systems using the same meteorological model WRF with building effect parametrization (BEP) to account for the differences in urban morphology, but different chemical transport models (CTM): CMAQ for Madrid and FARM for Bologna and Milan. The emissions of biogenic volatile organic compounds (BVOC) were produced with the same species-specific model, called PSEM. The simulations were carried out over several nested domains from European to city level where the spatial resolution was 1 km^2. The results show that vegetation effects on air temperature, pollutant concentrations and depositions depend on the characteristics of the city (morphology, geographic location and size), of the vegetation (extension and species) and on local chemical conditions (the cocktail of anthropogenic emissions, dispersion conditions, etc.). By applying this one-atmosphere approach for vegetation-meteorology-chemistry, the combined effect of pollutants removal,

D. de la Paz
e-mail: david.delapaz@upm.es

J. M. de Andrés
e-mail: juanmanuel.deandres@upm.es

A. Narros
e-mail: adolfo.narros@upm.es

S. Finardi · C. Silibello · N. Pepe
ARIANET Srl, 20128 Milan, Italy
e-mail: s.finardi@aria-net.it

C. Silibello
e-mail: c.silibello@aria-net.it

N. Pepe
e-mail: n.pepe@aria-net.it

S. Fares · I. Zappitelli · A. Alivernini
Council for Agricultural Research and Economics (CREA), Research Centre for Forestry and Wood, 00166 Rome, Italy
e-mail: silvano.fares@cnr.it

I. Zappitelli
e-mail: ilaria.zappitelli@crea.gov.it

A. Alivernini
e-mail: alessandro.alivernini@crea.gov.it

R. Prandi · G. Carlino
SIMULARIA Srl, Via Sant'Antonio da Padova 12, 10121 Turin, Italy
e-mail: r.prandi@simularia.it

G. Carlino
e-mail: g.carlino@simularia.it

BVOC emissions distribution as well as changes of wind patterns, temperature, etc., leads to contrasting time-dependent variability patterns of vegetation effects across the cities.

Keywords Air quality · Vegetation · Air temperature · Ozone · Particulate matter

1 Introduction

Urban vegetation provides many ecosystem services which influence the atmospheric processes responsible for air quality and climate variations. In addition to its direct benefit for the human well-being by offering opportunities for outdoor recreation, education, socialization, etc., green spaces indirectly contributes to human health by changing the air temperature and chemical composition. This study shows for the first time how the current vegetation affects the urban environment, quantifying its contribution to thermal comfort and air quality, in three European cities, Bologna, Madrid and Milan, for a summer day during anticyclonic weather conditions.

2 Materials and Methods

The effect of vegetation on temperature, air concentration and deposition of ozone (O_3), particulate matter (PM10) and nitrogen dioxide (NO_2) was evaluated with state-of-the-art air quality modelling systems (AQMS) since they are the only tools able to assess the effects of vegetation on the atmosphere in a comprehensive and realistic manner and are largely used to support Air Quality Plans at regional and local levels as well as at national level in support to NEC Directive.

The simulations were carried out over the cities at 1 km^2 spatial resolution and were nested in simulations over European and national/regional domains. Two AQMS were used, the Atmospheric Modelling System of MINNI project (AMS-MINNI; Mircea et al., 2014, 2016) for Italy and WRF-SMOKE-CMAQ (Borge et al., 2008) for Spain. AMS-MINNI is composed of the meteorological model WRF v3.9.1.1 (Skamarock & Klemp, 2008), the emission processor EMMA (Emission MAnager, ARIA/ARIANET, 2008) and the chemical transport model (CTM) FARM (Flexible Air Quality Regional Model, Silibello et al., 2008). The Spanish AQMS uses WRF v4.1.2, the Community Multiscale Air Quality (CMAQ) CTM (Byun & Schere, 2006) and the emissions processor SMOKE (Sparse Matrix Operator Kernel Emissions).

The simulations at European and national/regional scales were made by ENEA and UPM using same anthropogenic emission inventory, CAMS-REGAP_v2.2.1, provided by TNO for 2015. Boundary conditions for meteorological and chemical conditions at European scale were recovered from European Centre for Medium-Range Weather Forecasts (ECMWF) and, respectively, from CAMS Copernicus platform. The validation of simulations used observations from European Environment

Agency and from national/regional/local environmental agencies. Emissions from Spanish sources in the nested domains are a combination of different official emission inventories: the National Emission Inventory (MITECO, 2019) and the local emission inventory of the Madrid City (AM, 2019). The national emission inventory distributed by ISPRA (Italian Institute for Environmental Protection and Research) with provincial level (NUTS3) detail (where NUTS stands for Nomenclature of territorial units for statistics, the hierarchical system for dividing up the territory of the European Union) was used for the Italian domain, while local emission inventories provided by Regional Environmental Agencies (ARPA) of Lombardy and Emilia Romagna Regions were used for Milan and Bologna. All the simulations used biogenic volatile organic compounds (BVOC) emissions produced with PSEM model (Silibello et al., 2017). BVOC emission fluxes were computed using vegetation maps built from the tree inventories provided by the partner municipalities and from ancillary information on the regional forest cover, complemented by European reference CORINE land cover.

The effect of current vegetation on atmospheric characteristics was estimated as the difference between a simulation with vegetation (SVR—simulation vegetation real) and a simulation without urban vegetation (SVN—simulations vegetation null). However, SVN did not exclude the peri-urban forests located in the Northern part of Madrid (Cuenca Alta del Manzanares Regional Park) and in the South-Eastern part of Bologna (Colli Bolognesi) respectively, as well as the agricultural area located south and west of Milan (Parco agricolo sud). The CTMs and meteorological model used the same approach for both scenarios: the vegetation was replaced by bare soil only in the urbanized area of the city. The CTM simulations included all the detailed information on vegetation cover extension and characteristics, the effect of vegetation on meteorology and detailed species-dependent BVOC emissions.

3 Results

Figures 1 and 2 show the differences between SVR and SVN simulations for the domains including the cities with a spatial resolution of 1 km^2. The cities are located approximately in the middle of the domain to avoid significant impact of the boundary conditions. The negative values indicate a decrease of temperature, concentrations and dry depositions due to vegetation and the positive values indicate the opposite. Differences in temperature and concentrations are shown as daily averages while in depositions and BVOC emissions are shown as daily totals. In the three cities, the actual vegetation has a cooling effect but its magnitude and extension varies inside/around the city core and among the cities (Fig. 1). The temperature is reduced more than 0.5 °C in some areas of Madrid and Milan but below 0.4 °C in Bologna. As expected, there is no cooling in the areas of peri-urban forests of Bologna and Madrid since their effect was not investigated while a low decrease of temperature is observed south-west of Milan. Increases of temperature observable mainly outside the conurbation boundaries can be attributed to the variation of local atmospheric

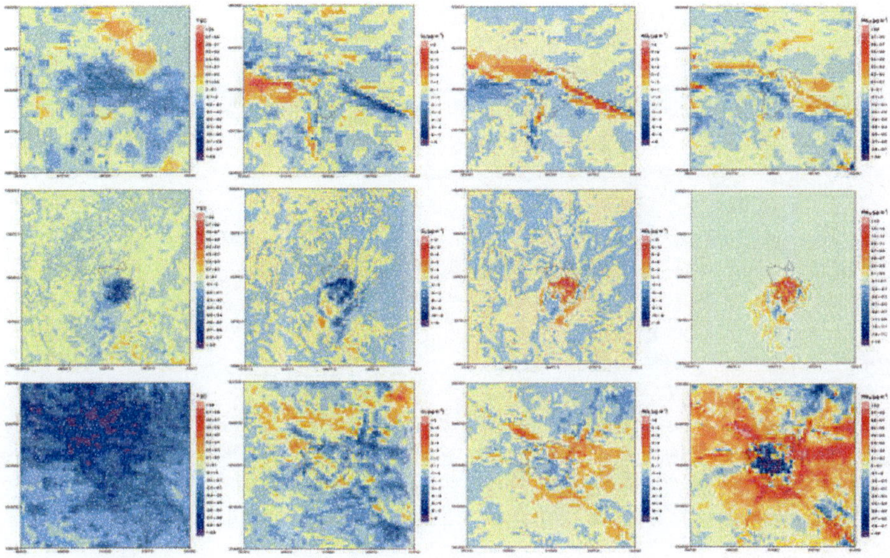

Fig. 1 Differences of daily average (SVR-SVN) temperature (°C) (1st column) and air concentrations ($\mu g/m^3$) of O_3 (2nd column), NO_2 (3rd column) and PM10 (4th column) for 13 July 2015 and for Bologna (1st row), Madrid (2nd row) and Milan (3rd row)

circulation induced by the urban vegetation and generally show transient features with different space location in different days.

The decrease of air temperature due to vegetation leads to both increases and decreases of ozone concentrations according to the local chemical composition: the Madrid and Milan areas where O_3 decreases correspond to an increase of NO_2 and PM10, possibly due to the reduction of wind speed induced by vegetation. The opposite effect also occurs. In Bologna, a substantial decrease of O_3 is observed mainly along A14 highway while in the other two cities the decrease is not so clearly related to traffic emissions. For all three pollutants, the differences in concentrations vary among the cities, with the highest changes observed in Madrid. This effect and the small variations observed outside of the cities are due to meteorology changes associated to vegetation removal. The highest differences in the BVOC emissions and dry deposition patterns due to the overall vegetation cover is observed within the city limits except for the O_3 in Bologna (Fig. 2). For all three cities, the amount of pollutants removed by dry deposition is higher for O_3 than for NO_2. The highest amount of PM10 is removed in Milan while comparable values are observed in the areas of Bologna and Madrid. The magnitude of differences for all pollutants is characterised by positive values substantially higher than the negative ones pointing out that the efficiency of vegetation in filtering the air locally is more important than the negative effects due to changes in meteorology and BVOC emission. However, in all the cities, the combination of meteorological changes with perturbations of local chemistry as well as with urban morphology leads also to areas where SVR dry

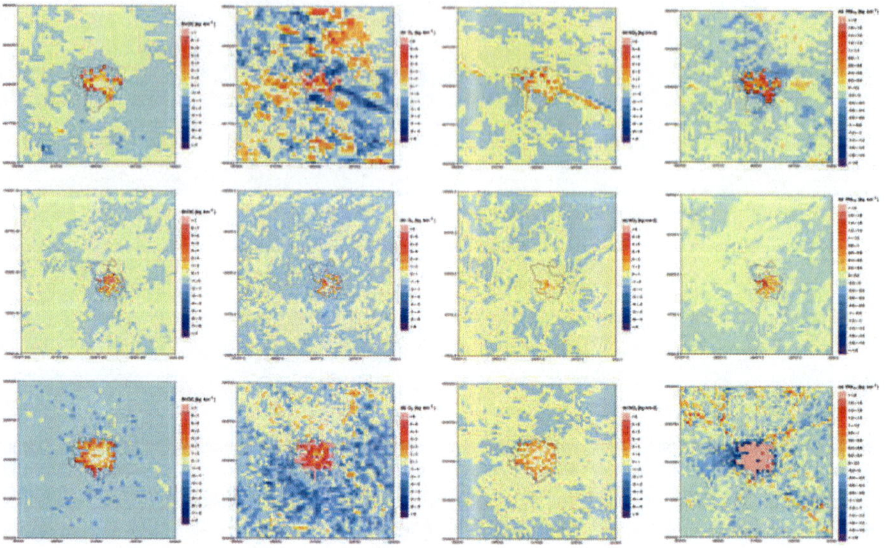

Fig. 2 Differences (SVR-SVN) of daily sum of BVOC emissions (kg/km^2) (1st column) and of dry depositions (kg/km^2) of O_3 (2nd column), NO_2 (3rd column) and PM10 (4th column) for 13 July 2015 and for Bologna (1st row), Madrid (2nd row) and Milan (3rd row)

deposition is lower than that for SVN. Interesting to note similarities in emissions and deposition pattern driven by vegetation cover for a given city and differences among cities.

Moreover, deposition patterns vary with pollutant type since the deposition models for gases and particulate matter use different approaches to evaluate interactions with vegetation.

4 Conclusions

The coherence among AQMSs' results in different cities reinforces their capacity to support air quality and urban planning decision processes that include both vegetation increase and anthropogenic emissions reductions. It also points out the necessity to couple the assessments of air quality with the changes in the meteorological conditions induced by new interventions in urban structure. AQMSs may also support a deeper understanding of the differences in temperature, air concentrations and deposition of pollutants such as ozone (O_3), particulate matter (PM10) and nitrogen dioxide (NO_2) inside and around the cities due to urban green.

Acknowledgements This study was developed under the project Life VEG:GAP (https://www.lifeveggap.eu/) which was c-funded by European Union Life Program in 2018, Grant Number LIFE18 PRE IT 003.

References

ARIA/ARIANET. (2008). EMMA (EMGR/make) User Manual, R2008.99. Arianet, Milano, Italy.

Borge, R., Lumbreras, J., & Rodríguez, E. (2008). Development of a high-resolution emission inventory for Spain using the SMOKE modelling system: A case study for the years 2000 and 2010. *Environmental Modelling & Software, 23*, 1026–1044.

Byun, D., & Schere, K. L. (2006). Review of the governing equations, computational algorithms, and other components of the models-3 community multiscale air quality (CMAQ) modeling system. *Applied Mechanics Reviews, 59*, 51–77.

Mircea, M., Grigoras, G., D'Isidoro, M., Righini, G., Adani, M., Briganti, G., Cremona, G., Ciancarella, L., Cappelletti, A., Calori, G., Cionni, I., Finardi, S., Larsen, B., Pace, G., Perrino, C., Piersanti, A., Silibello, C., Vitali, L., & Zanini, G. (2016). Impact of grid resolution on aerosol predictions: a case study over Italy. *Aerosol and Air Quality Research, 16*, 1253–1267.

Mircea, M., Ciancarella, L., Briganti, G., Calori, G., Cappelletti, A., Cionni, I., Costa, M., Cremona, G., D'Isidoro, M., Finardi, S., Pace, G., Piersanti, A., Righini, G., Silibello, C., Vitali, L., & Zanini, G. (2014). Assessment of the AMS-MINNI system capabilities to predict air quality over Italy for the calendar year 2005. *Atmospheric Environment, 84*, 178–188.

Silibello, C., Calori, G., Brusasca, G., Giudici, A., Angelino, E., Fossati, G., et al. (2008). Modelling of PM10 concentrations over Milano urban area using two aerosol modules. *Environmental Modelling and Software, 23*, 333–343.

Silibello, C., Baraldi, R., Rapparini, F., Facini, O., Neri, L., Brilli, F., Fares, S., Finardi, S., Magliulo, E., Ciccioli, P., & Ciccioli, P. (2017). Modelling of biogenic volatile organic compounds emissions over Italy. In *HARMO 2017—18th International Conference on Harmonisation within Atmospheric Dispersion Modelling for Regulatory Purposes, Proceedings*, 2017-October (pp. 14–18).

Skamarock, W. C., & Klemp, J. B. (2008). A time-split nonhydrostatic atmospheric model for weather research and forecasting applications. *Journal of Computational Physics, 227*, 3465–3485.

Questions and Answers

Question (online): Will we be able in the next future to give indications for the choice of trees to plant in urban areas, in an air pollution mitigation perspective?

Answer: Atmospheric Modelling Systems may already provide more reliable information for planting trees than "static" approaches used until now. However, more applications in various cities and new developments of current parametrizations used for both for meteorological and air quality predictions in relation to vegetated surfaces are needed to better estimate vegetation effects on atmosphere and climate.

The Impact of Imperviousness on Boundary Layer Mixing, Air Pollution and Cloud Formation over Urban Areas

Joachim Fallmann, Vinod Kumar, Marc Barra, and Holger Tost

Abstract Urbanization requires the replacement of natural land into impervious surfaces, which in turn leads to surface sealing and a loss of natural soil functions. Next to a modification of the surface hydrological properties, warming of concrete materials arouses a respective warming of the air aloft. This study analyses the impact of increased sensible heat flux from heated building materials onto the thermodynamic and chemical features of the urban boundary layer and hence the urban heat island intensity. The state-of-the-art coupled chemistry-climate model MECO(n) combined with the COSMO urban canopy model TERRA_URB is facilitated on a km-scale to analyse the interplay between chemical and thermodynamic processes in the urban canopy layer, linked to modification of surface properties. Two case studies for projected scenarios in future urban planning directions—urban sprawl and re-densification—are compared for a 10-days heat wave period during July 2018 with respect to the greater Rhine-Main metropolitan area. Summarising the key outcomes of the study, firstly a 10% increase in surface imperviousness resulted in a respective warming of surface temperature by 0.7 K and 2 m air temperature by 0.2 K within the core urban area. 50% total increase of surface sealing was responsible for a 3 K and 1 K spatially averaged warming respectively. For the same scenario, increased boundary layer mixing in warmed street canyons lead to a reduction of surface NO_2 concentration by 10–20% during the day and an increase of about 10% over night, when denser canyons reinforced blocking of air flows. Lastly, modifications of surface properties have also been recorded at about 1500–2000 m hight within the boundary layer mirrored by a change in cloud properties. Here, increased

J. Fallmann (✉)
City of Heidelberg, Office of Environmental Protection, Trade Supervision and Energy, Heidelberg, Germany
e-mail: joachim.fallmann@heidelberg.de

Institute of Meteorology and Climate Research, Karlsruhe Institute of Technology, Karlsruhe, Germany

V. Kumar
Max-Planck Institute for Chemistry, Mainz, Germany

M. Barra · H. Tost
Institute of Atmospheric Physics, University Mainz, Mainz, Germany

C. Mensink and O. Jorba (eds.), *Air Pollution Modeling and its Application XXVIII*, Springer Proceedings in Complexity, https://doi.org/10.1007/978-3-031-12786-1_3

turbulent vertical fluxes transporting warmer air higher up into the urban boundary layer, trigger the formation of low-level urban induced clouds.

1 Introduction

Urbanization is one of the key drivers for surface sealing and hence the transformation of natural into impervious land. Although only covering a small percentage of total land cover, urban entities are responsible for the vast amount of emissions and atmospheric pollution. Urban air pollution is considered as one of the most relevant causes of premature deaths in Europe and climate change aggravates health problems of the urban dwellers. Climate extremes such as hot days or tropical nights are most likely to increase at all spatial scales from global to regional to local levels. The urban morphology further amplifies these problems by reduced evaporation, heat accumulation by building structures and heat trapping in the urban canyon (Schau-Noppel et al., 2020). Vast amounts of studies confirm that green and blue infrastructures are effective measures to mitigate to rising temperature levels. However, air pollution levels are controlled by thermally driven flows arising from changes in the surface properties, e.g. when heat mitigation strategies alter surface sensible heat fluxes. The interplay between urban boundary layer dynamics and air pollution is in the focus of that study. For this reason, two scenarios of expected urban developments are executed, using a state-of-the-art coupled chemistry-climate modelling system at high resolution. Using that model configuration for an extreme heat event in July 2018 for the Rhine-Main urban agglomeration analyses the regional effect of surface sealing on both thermal and chemical properties of the urban boundary layer.

2 Data and Methods

In this study we use the regional climate-chemistry model MECO(n) (MESSy-fied ECHAM and COSMO models nested n times) (Kerkweg & Jöckel, 2012) which consists of a one-way online coupling between the global earth system model EMAC) and the regional climate model COSMO-CLM/MESSy. The urban canopy parametrization within MECO is achieved via the urban-canopy land-surface scheme TERRA-URB (Wouters et al., 2015 and others).

In MECO(n) we consider one global (~200 km) and two regional nests n with 0.44° and 0.03° horizontal resolution and 50 pressure hybrid levels. The latter covers Central Europe and the Rhine-Main region in particular (Fig. 1a). For the MECO_2 (~3 km) simulation domain TNO-MACCIII emissions for the year 2011 are used. A Schematic of the model system is depicted in Fig. 1b. More details are found in Fallmann et al. (2021).

For model evaluation, we use selected urban background stations located within the model domain and passive radiometer measurements within the urban centre

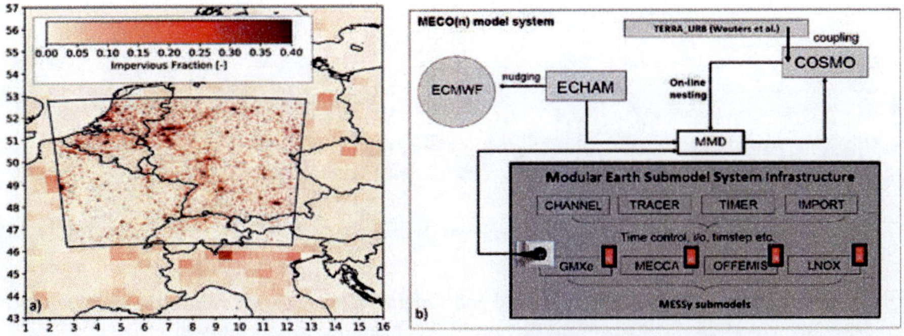

Fig. 1 MECO_2 domain (3 km) embedded in extracted are of the MECO_1 domain (0.44°) showing ISA (impervious surface area) (**a**) and schematic of model system MECO(n) (**b**)

of Mainz. The latter is important for assessing the thermal structure of the urban boundary layer. The model is nudged towards ERA-Interim reanalysis and runs for the period 1–10 July 2018.

3 Results and Discussion

For the evaluation of the urban canopy model, two model simulations are compared whereas Terra_3km considers TERRA_URB coupling and a reference (NoTerra_3km) which does not. Figure 2a shows hourly mean differences of measured and simulated NO_2 concentrations for 8 urban background stations, with red being the urban and blue the reference simulation. The shading indicates the standard deviation between the locations and the bold lines depict the station mean. One observation point is compared to the closest 3×3 km grid cell according to the nearest neighbour method. With regard to 2-m air temperature (not shown) Terra_3km reduces the underestimation of negative bias from −1.6 K (NoTerra_3km) to −0.8 K, averaged over 8 urban background stations. Bias reductions are also found for 10-m wind speed and 10-m relative humidity (more details see Fallmann et al., 2021). For NO_2, the mean bias averaged over the model period is reduced from 6.61 to 4.48 ppbv (Fig. 2a), which mostly is true for the evening hours.

For the single station Mainz-Mombach, the relative bias in peak concentration decreases from 25% (NoTerra_3km) to 18% (Terra_3km). Moreover, data from a passive microwave radiometer, located in the centre of the urban area of Mainz is used to address the potential improvement of the boundary layer representation by the coupled compared to the uncoupled model. Figure 2b highlights the temperature profile on 4 July 2100 h at the grid cell closest to the radiometer location. With Terra_3km (red line) being closer to the observations (grey lines—in minute interval), the improvement in surface air chemistry can be explained by improved boundary layer physics described by TERRA_URB.

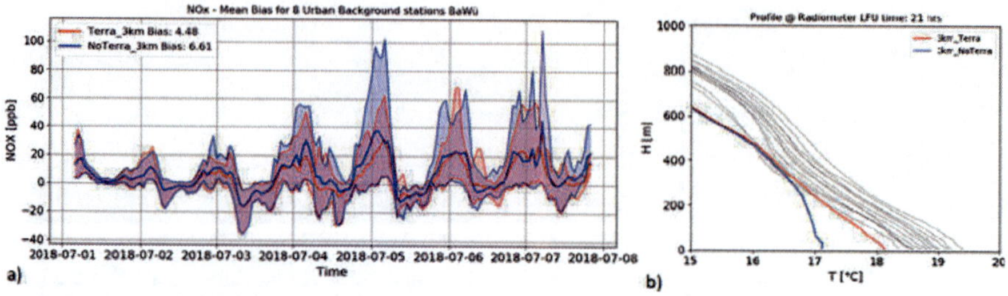

Fig. 2 Hourly mean differences of measured and simulated NO_2 concentrations for 8 urban background stations (**a**) and vertical temperature profiles at central urban location (**b**)

The model configuration further allows for the analysis of different urban developments, by modifying the parameter 'impervious surface fraction (ISA)' in the geographical input data. We focus on two scenarios, a re-densification of the urban centre (ISA_plus) or a push towards the close vicinity, symbolizing an urban sprawl (ISA_minus). For urban grid cells with ISA 40%, ISA_plus increases the impervious surface area fraction per grid cell by 50%, equivalent to a 50% decrease of natural land cover in the same cell. ISA_minus considers an increase (plus 80%) in the closest rural surrounding and a respective decrease (−30%) in the urban centre.

The level of imperviousness in one grid cell (percentage impervious as fraction 0–1) controls the energy balance and the overall fluxes from the surface into the overlying atmosphere. Initially, higher surface sealing results in reduction of latent heat and respective heating of the ground. In the case of the core urban area of the City of Mainz, that increase amounts to up to 4 K at noon, averaged over all core urban grid cells (ISA > 40%). This results in the simultaneous warming of the air aloft which is most pronounced in the evening, when heat is trapped within the urban canopy or released from warmed surfaced. Figure 3 shows the difference between Terra3km to ISA_plus (red) and ISA_minus (green) for both 2-m temperature (a) and NO_2 (b).

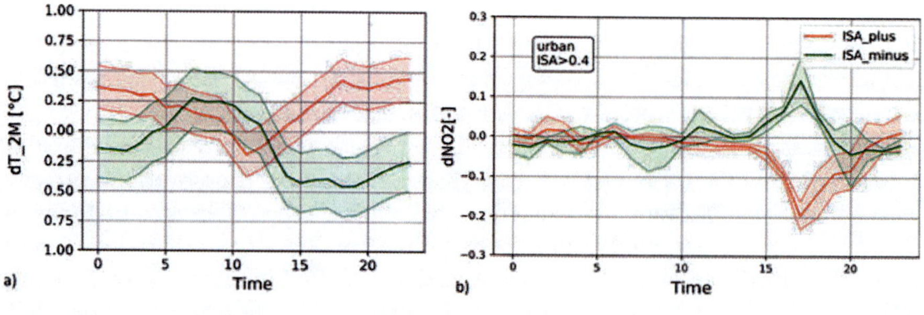

Fig. 3 Average diurnal cycle for the period 1–10 July 2018 as mean of urban grid cell with the criteria ISA > 0.4 as difference [ISA_plus − Terra_3km] (red) and [ISA_minus − Terra_3km] for 2-m temperature (**a**) and NO_2 (**b**). Shading indicates standard deviation of hour per day

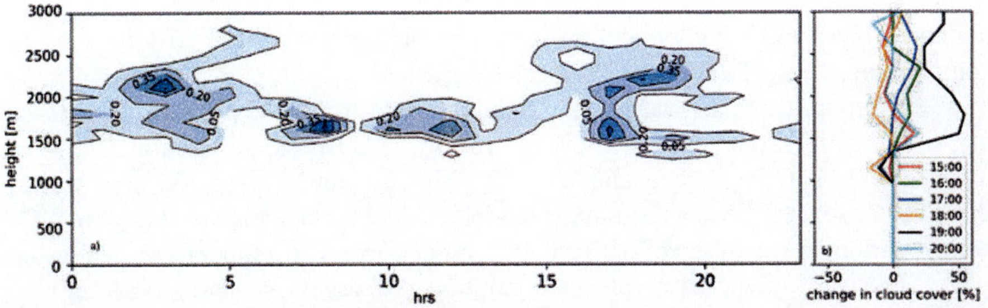

Fig. 4 Hovmoller plot of cloud cover [ISA_plus − Terra_3km] (**a**) and profiles in cloud cover change [ISA_plus − Terra_3km] (**b**) for different times of the day. Both 2 July 2018 at central urban grid cell

In denser urban canopies (ISA_plus), we find a general warming trend (red line in Fig. 3a), as building structure and urban canyons accumulate and trap and release heat into the urban atmosphere via sensible heat flux. In the evening and nighttime, the lack of evaporative cooling through less vegetation per grid cell explains the release of sensible heat. In the morning between 10 and 11 h however, when the sun's incidence angle is flat, larger parts of the surface witness a shadowing by the neighboring buildings, resulting in a relative cooling trend. That effect also explains the positive trend when street canyons get broader in ISA_minus (green line in Fig. 3a). Hence, the air warms from 5 to 7 h relatively to the reference Terra3km and cools from 10 to 5 h. Modification in thermal structures of the lower atmosphere result in changes in turbulent mixing within the urban boundary layer. Hence, the impact on near surface air pollutants (e.g. NO_2—Fig. 3b) mirrors the changes in 2 m-temperature (Fig. 3a). That means, an increase of temperature in the evening results in an increase of vertical mixing which in turn dilutes near surface NO_2 (Fig. 3b). This is also true for the respective increase when near surface mixing diminishes caused by local temperature reduction. Both relative NO_2 increase and reduction is in the order of $\pm 20\%$—with respect to average diurnal cycle (from 1 to 10 June 2018) and grid cells with ISA > 40%.

On the other hand, increasing turbulent vertical fluxes transport warmer air higher up into the urban boundary layer, which in turn leads to the formation of low-level clouds over the warmed urban area (Fig. 4a). During calm and clear days, these clouds persist throughout the day (Fig. 4a), spreading from 1500 m up to 2500 m. The highest relative change in cloud cover (Fig. 4b) coincides with the peak in temperature increase (Fig. 3a) at about 1900 h.

4 Conclusion

This study indicates that a sophisticated representation of the urban canopy improves the simulation of surface temperature and thus the thermal structure of the urban boundary layer and the chemical properties within. With about 8–10% increase in computational cost, TERRA_URB is relatively cheap.

We find that urban air warms at a maximum rate of 0.2 °C per 0.8 °C increase in surface temperature. Per 10% change in imperviousness, simulations show an increase of 0.17 °C for surface- and 0.05 °C for air temperature respectively. However, these values seem to be low absolutely, model results expect a relative change of NO_2 concentration in the order of 20%, with a reduction for a temperature increasing (ISA_plus) and an increase for a temperature decreasing scenario (ISA_minus) respectively. A second indicator of surface-boundary layer interactions is pictured by the increase of low level clouds over a relatively warmer surface in ISA_plus (higher imperviousness) (see also Theeuwes et al., 2019).

It has to be noted however, that only using NO_2 concentration as reference might not be enough for revealing the full width of chemistry-dynamics interactions in urban areas. Focusing on a clear summer period with weak winds and weak synoptic forcing, we potentially also select the simplest synoptic setup. We however could show small-scale features caused by changing the surface properties with regard to possible directions in future urban planning.

We conclude from our findings that urban planning strategies reducing inner city air temperature are most effective when combined to measures reducing traffic emissions, particularly relevant for evening and night-time emission peaks.

References

Fallmann, J., Barra, M., Kumar, V., & Tost, H. (2021). Impact of urban imperviousness on boundary layer meteorology and air chemistry on a regional scale. *Meteorologische Zeitschrift,* 349–367.

Kerkweg, A., & Jöckel, P. (2012). The 1-way on-line coupled atmospheric chemistry model system MECO (n)-Part 1: Description of the limited-area atmospheric chemistry model COSMO/MESSy. *Geoscientific Model Development, 5,* 87–110.

Schau-Noppel, H., Kossmann, M., & Buchholz, S. (2020). Meteorological information for climate-proof urban planning—The example of KLIMPRAX. *Urban Climate, 32,* 100614.

Theeuwes, N. E., Barlow, J., Teuling, A., Grimmond, S., & Kotthaus, S. (2019). Persistent cloud cover over mega-cities linked to surface heat release. *npj Climate and Atmospheric Science, 2*(1), 1.

Wouters, H., Demuzere, M., de Ridder, K., & van Lipzig, N. P. (2015). The impact of impervious water-storage parametrization on urban climate modelling. *Urban Climate, 11,* 24–50.

Question

Clemens Mensink: You mentioned urban fluxes. Can you comment on what fluxes you considered or calculated?

Answer: I have looked more detailed to sensible, latent heat and turbulence flux only-coming from the model output directly.

Comparison of Different Modeling Strategies for Estimating Long-Term $PM_{2.5}$ Exposure Using MAIAC (Multiangle Implementation of Atmospheric Correction) AOD in China

Zhao-Yue Chen, Jie-Qi Jin, and Chun-Quan Ou

Abstract The evidence on long-term effects on health is vital for establishing a causal association and assessing the disease burden related to particulate air pollution. However, the remaining problem of aerosol retrievals' severe absences strikes a serious blow at long-term $PM_{2.5}$ exposure estimation. The skyrocketed sample size with finer spatial resolution of aerosol products further increases the difficulty to resolve this problem. For finding out a less time-consuming, higher coverage and more precise AOD interpolation, three strategies were proposed and compared in mainland China in 2014–2018, including DI (Directly Interpolate long-term averaged AOD then predict long-term $PM_{2.5}$), AODS2L (integrate Short-term interpolated AOD into Long-term AOD then predict Long-term $PM_{2.5}$) and PMS2L (integrate Short-term $PM_{2.5}$ predicted by Short-term AOD into Long-term $PM_{2.5}$). AOD coverage varies widely in spatial (0–84.56%) and temporal (10.17–22.57%) distribution, while the places with monsoon climate (Southeast China) and summer (typhoon season) had the lowest coverage. For the monthly or annual $PM_{2.5}$ estimation, three strategies can fix almost all of missing gaps with the coverage rising from 15.65% to 98.55%–98.79% and provide accurate long-term $PM_{2.5}$ predictions (monthly CV R^2: 0.91, 0.92 and 0.93; annual CV R^2: 0.83, 0.85 and 0.92 for DI, AODS2L and PMS2L, respectively). The well-behaved strategies of AODS2L and PMS2L spent much more running time. How to balance the model performance and consuming time is a key point in long-term air pollution research. For the large-scale population-based study, using DI strategy can save time and guarantee certain accuracy. For the individual-based study, PMS2L is the more accurate and flexible for individual exposure but the most time-consuming choice.

Z.-Y. Chen (✉)
ISGLOBAL, Barcelona, Spain
e-mail: Zhaoyue.chen@isglobal.org

Z.-Y. Chen · J.-Q. Jin · C.-Q. Ou
State Key Laboratory of Organ Failure Research, Department of Biostatistics, Guangdong Provincial Key Laboratory of Tropical Disease Research, School of Public Health, Southern Medical University, Guangzhou 510515, China

Universitat Pompeu Fabra (UPF), Barcelona, Spain

C. Mensink and O. Jorba (eds.), *Air Pollution Modeling and its Application XXVIII*, Springer Proceedings in Complexity, https://doi.org/10.1007/978-3-031-12786-1_4

Keywords Machine learning · Aerosol optical depth · Missing replacement · Long-term · $PM_{2.5}$

1 Introduction

Recently, the satellite aerosol optical depth (AOD) provide the chance of estimating individual $PM_{2.5}$ exposures within his residential region, and the spatial resolution becomes higher and higher (from 0.5° to 1 km) (Goldberg et al., 2019; He & Huang, 2018; Just et al., 2018). However, the existing satellite technique inevitably leads to the high-missing rate of satellite-retrieved AOD, due to orbit patterns, cloudiness, snowy or surface reflectivity (Chen et al., 2019; Kahn et al., 2009). Therefore, it makes less sense to compare the long-term (i.e., annual) averaged AOD in different grids, because each grid has different missing days (Lv et al., 2017). Even in the same grid, different time-series integrity in AOD and ground-level $PM_{2.5}$ greatly weakened their relationship in a long period (Kumar, 2010; Paciorek & Liu, 2009).

To our best knowledge, few studies had conducted AOD miss-filling process in long-term $PM_{2.5}$ estimation. The high-resolution AOD dataset raises questions to find out practical solutions for less running time and better performance: Should we directly interpolate long-term averaged AOD then build the model? Or interpolate daily AOD then get long-term averaged AOD? Or directly aggregate short-term $PM_{2.5}$ estimation into long-term averaged $PM_{2.5}$. All these questions are hunger for further discussion, which is of critical importance to long-term health assessment of air pollution in cohort studies. And this study was to compare different strategies (Fig. 1) for long-term $PM_{2.5}$ estimation with 6-year 1 km-resolution MAIAC AOD, for discovering an appropriate way to get long-term $PM_{2.5}$ exposure estimation, and further discuss how different missing-filled methods will affect long-term $PM_{2.5}$ exposure estimation.

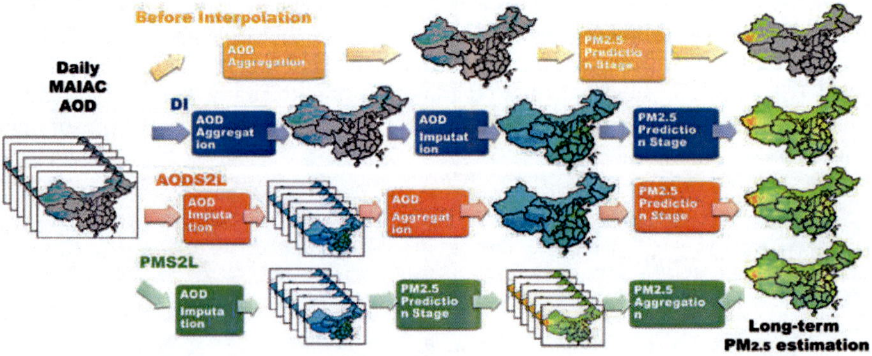

Fig. 1 Framework of the three strategies to estimate long-term $PM_{2.5}$ exposure with MAIAC AOD

2 Results

2.1 Descriptive Analysis

From 2013 to 2018, the monthly coverage of MAIAC AOD (median = 15.65%) varies greatly by region (from 0 to 84.56%), and the density-populated Southeast China had relatively lower MAIAC AOD coverage rather than the Northwest China in any season. For temporal, the monthly coverage of MAIAC AOD in China varied from 10.17% (June, 2015) to 22.57% (October, 2015), and summer usually had lowest coverage. Over four months were with average coverages less than 14% (June (12.15%), February (12.93%), August (13.37%), and July (13.42%)), while October-December were three highest-coverage months (20.89%, 18.90% and 18.04%, respectively).

From 2014 to 2018, the monthly median $PM_{2.5}$ concentrations measured at monitoring sites gradually decreased from 45.60 $\mu g/m^3$ (IQR = 24.66) to 32.14 $\mu g/m^3$ (IQR = 24.07). The $PM_{2.5}$ concentrations were always higher in the cold season (October–April). For these sites, the median and interquartile range (IQR) of MAIAC AOD was around 0.33 (0.36).

2.2 Performance in Monthly Model

For monthly AOD missed-filling, the DI strategy took less training time but had poorer interpolated performance than AODS2L and PMS2L, because missed-filling technique conducted at daily level generally takes longer times. And all the strategies can eliminate almost all missing gaps to obtain full coverage. In AODS2L and PMS2L, it has the highest correlation (r = 0.34) between XG-AOD and monthly $PM_{2.5}$.

For estimating monthly $PM_{2.5}$ exposure (Fig. 2), the worst performance was to use the PMS2L strategy without AOD interpolation (CV $R^2 = 0.54$), followed by AODS2L strategy without AOD interpolation (CV $R^2 = 0.84$), and the performance were remarkably improved after RF or XG interpolation. The CV R^2 were similar for three strategies (0.91–0.93), while the PMS2L had 20% smaller CV RMSE than DI and AODS2L.

2.3 Performance in Annual Model

For annual AOD missed-filling, the DI strategy still took less training time but poorer interpolated performance in missed-filled models, compared with AODS2L and PMS2L. For estimating annual $PM_{2.5}$ exposure (Fig. 3), the PMS2L had the best

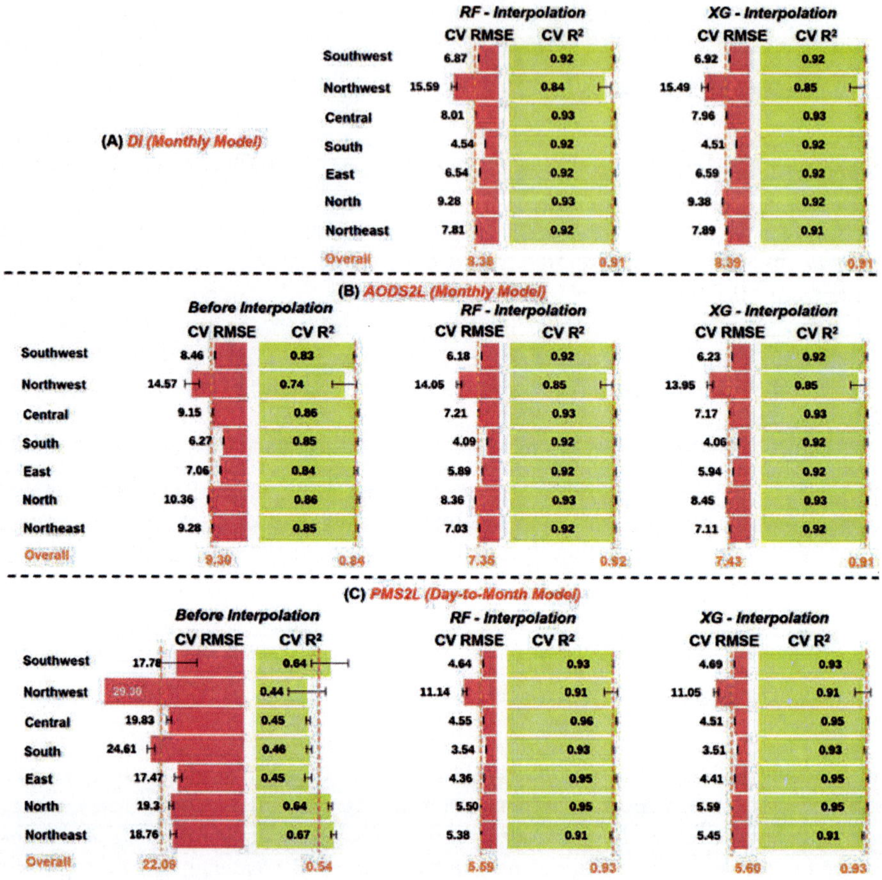

Fig. 2 The cross-validation performance of three strategies in estimating monthly $PM_{2.5}$ exposure using MAIAC AOD (Bar: 95% CI)

performance (CV $R^2 = 0.91$–0.92) after RF or XG interpolation but took the longest running time (over 41.9 times as DI, 1.3 times as AODS2L) among three strategies.

2.4 Comparison in the Lowest Coverage Month

In the lowest coverage month (June, 2015), only 9.95% of grids in China had more than 50% completeness before AOD interpolation, but the monthly DI, AODS2L and PMS2L-filled AOD (with RF-interpolation) could get almost full coverage (about 97.63%, 96.53% and 96.53%). And the median (IQR) levels of original, DI-filled, AODS2L-filled and PMS2L-filled AOD were 0.17 (0.17), 0.28 (0.20), 0.26 (0.17) and 0.26 (0.17), respectively.

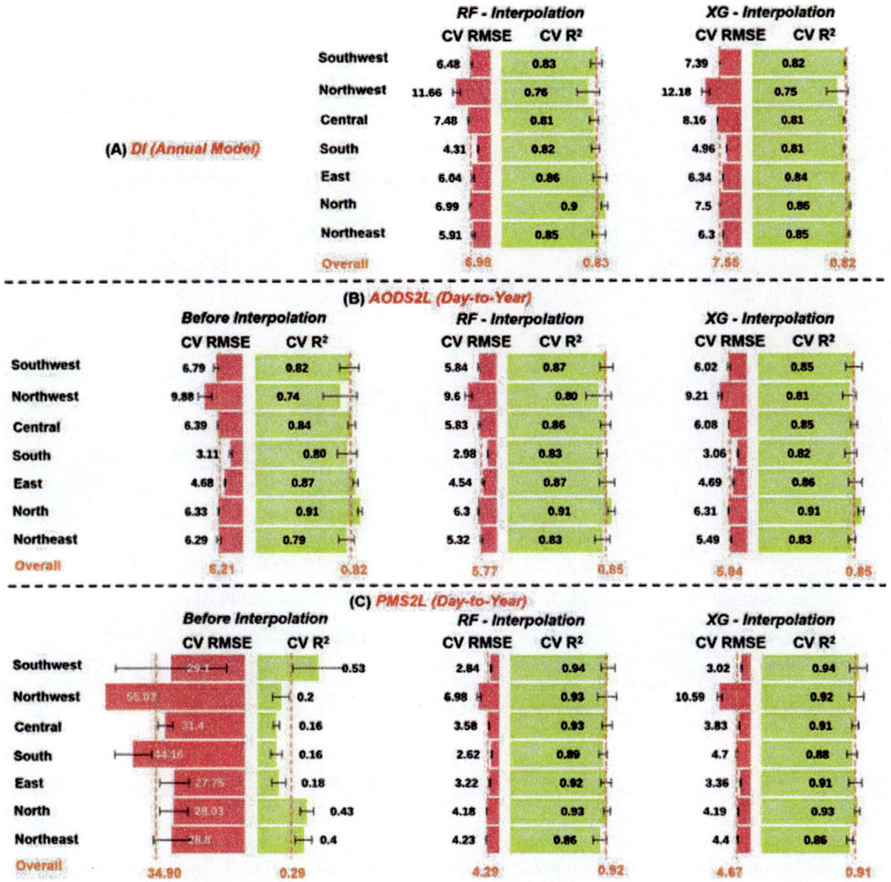

Fig. 3 The cross-validation performance of three strategies in estimating annual $PM_{2.5}$ exposure with MAIAC AOD: DI, AODS2L and PMS2L (Bar: 95% CI)

In the meanwhile, the coverage of $PM_{2.5}$ predicted by original and three strategies were 9.84%, 96.56%, 95.62% and 95.78% (Fig. 4), and the median (IQR) levels (μg/m^3) were 36.06 (26.21), 31.70 (17.20), 29.56 (16.28) and 28.82 (15.78), respectively.

3 Conclusions

The AOD interpolation is the last thing to ignore in estimating long-term $PM_{2.5}$ exposure based on satellite data. We proposed and thoroughly compared three different strategies (DI, AODS2L and PMS2L). All three strategies can provide almost a full coverage of $PM_{2.5}$ exposure at 1 km solution. Better methods tend to take longer

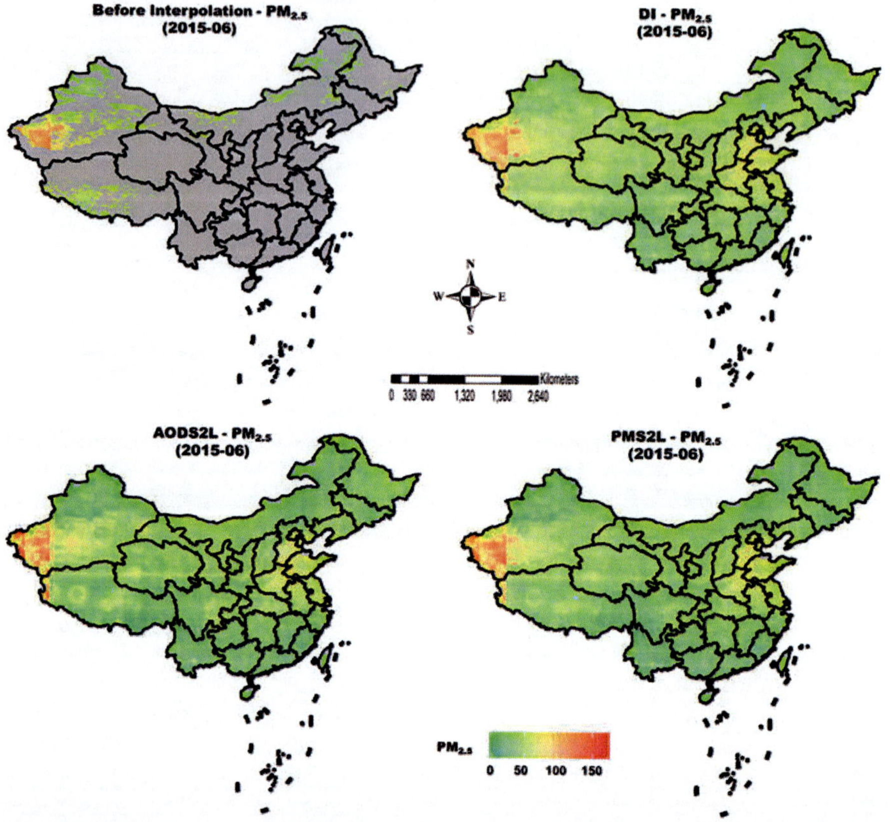

Fig. 4 The Grid surfaces of the original and three strategies (DI, AODS2L and PMS2L) predicted $PM_{2.5}$ in the lowest coverage month (2015–06)

among three proposed strategies, so how to balance the model performance and consuming time generally becomes a key point in further studies related to long-term exposure to air pollution. For a large-scale population-based study, using DI strategy can save time and guarantee accuracy to some extent. For an individual-based study, PMS2L is the more accurate and flexible for individual exposure but the most time-consuming choice.

Acknowledgements This study was supported by National Nature Science Foundation of China [81573249], Nature Science Foundation of Guangdong Province [2016A030313530].

References

Chen, Z.-Y., Zhang, T.-H., Zhang, R., Zhu, Z.-M., Yang, J., Chen, P.-Y., Ou, C.-Q., & Guo, Y. (2019). Extreme gradient boosting model to estimate $PM_{2.5}$ concentrations with missing-filled satellite data in China. *Atmospheric Environment*. https://doi.org/10.1016/j.atmosenv.2019.01.027

Goldberg, D. L., Gupta, P., Wang, K., Jena, C., Zhang, Y., Lu, Z., & Streets, D. G. (2019). Using gap-filled MAIAC AOD and WRF-Chem to estimate daily $PM_{2.5}$ concentrations at 1 km resolution in the Eastern United States. *Atmospheric Environment, 199*, 443–452.

He, Q., & Huang, B. (2018). Satellite-based mapping of daily high-resolution ground $PM_{2.5}$ in China via space-time regression modeling. *Remote Sensing of Environment, 206*(December 2017), 72–83. https://doi.org/10.1016/j.rse.2017.12.018

Just, A. C., De Carli, M. M., Shtein, A., Dorman, M., Lyapustin, A., & Kloog, I. (2018). Correcting measurement error in satellite aerosol optical depth with machine learning for modeling $PM_{2.5}$ in the Northeastern USA. *Remote Sensing, 10*(5). https://doi.org/10.3390/rs10050803

Kahn, R. A., Nelson, D. L., Garay, M. J., Levy, R. C., Bull, M. A., Diner, D. J., Martonchik, J. V., Paradise, S. R., Hansen, E. G., & Remer, L. A. (2009). MISR aerosol product attributes and statistical comparisons with MODIS. *IEEE Transactions on Geoscience and Remote Sensing, 47*(12), 4095–4114. https://doi.org/10.1109/TGRS.2009.2023115

Kumar, N. (2010). What can affect AOD-PM(2.5) association? *Environmental Health Perspectives, 118*(3), D22206. https://doi.org/10.1289/ehp.0901732

Lv, B., Hu, Y., Chang, H. H., Russell, A. G., Cai, J., Xu, B., & Bai, Y. (2017). Daily estimation of ground-level $PM_{2.5}$ concentrations at 4 km resolution over Beijing-Tianjin-Hebei by fusing MODIS AOD and ground observations. *Science of the Total Environment, 580*, 235–244.

Paciorek, C. J., & Liu, Y. (2009). Limitations of remotely sensed aerosol as a spatial proxy for fine particulate matter. *Environmental Health Perspectives, 117*(6), 904–909. https://doi.org/10.1289/ehp.0800360

Question and Answer

QUESTIONER: Clemens Mensink

QUESTION: Can your methodology still work when there are big changes in the air quality situation? E.g., due to covid-19 in China?

ANSWER: Thanks for your comment. Our methodology are mainly depended on the satellite AOD data and meteorological data. If the relationship between satellite data (or meteorological data) and ground-level air pollutant data remain stable when some big changes (social factors) happened, our models can work well. In the future, we can test them in a longer study period, because the $PM_{2.5}$ stations in China were built in recent 6–7 years, which is not long enough to test them. Furthermore, our model did not include any human-activity related factors, due to the inaccessible data in China. Thus, I thought our model are better to evaluate or update, before some big changes have happened.

Developments of SPRAY Lagrangian Particle Dispersion Model for Tracing the Origin of Odour Nuisance

Silvia Trini Castelli, Gianni Tinarelli, Francesco Uboldi, Piero Malguzzi, and Paolo Bonasoni

Abstract SMART meteo-dispersive modelling suite is interfaced with the NOSE WEB-APP alert-system, in order to track the origin of odour nuisance. New developments in the retroSPRAY model to use the alert-system output as input, and then simulate the backward puffs from potential odour sources, are described and tested to be implemented in SMART. A case study is presented and discussed.

1 Introduction

An alert-system, NOSE—Network for Odours Sensitivity (https://nose-cnr.arpa.sicilia.it/), has been realized by CNR-ISAC and ARPA Sicilia with the aim of tracking episodes of odour nuisance, through a citizen-science approach. To identify the area of possible sources, deterministic back-trajectories are presently traced using MOLOCH meteorological model (Malguzzi et al., 2006). Improvements of the modelling module in NOSE are in process with the integration of the meteo-dispersive modelling suite SMART (Spray-Moloch Atmospheric Regional Tool, Bisignano et al., 2020; Trini Castelli et al., 2020), recently developed. In this framework, new and original developments are in progress for SPRAY Lagrangian particle

S. Trini Castelli (✉)
CNR Institute of Atmospheric Sciences and Climate, Turin, Italy
e-mail: s.trinicastelli@isac.cnr.it

G. Tinarelli · F. Uboldi
ARIANET Srl, Milan, Italy
e-mail: g.tinarelli@aria-net.it

F. Uboldi
e-mail: f.uboldi@aria-net.it

P. Malguzzi · P. Bonasoni
CNR Institute of Atmospheric Sciences and Climate, Bologna, Italy
e-mail: p.malguzzi@isac.cnr.it

P. Bonasoni
e-mail: p.bonasoni@isac.cnr.it

C. Mensink and O. Jorba (eds.), *Air Pollution Modeling and its Application XXVIII*, Springer Proceedings in Complexity, https://doi.org/10.1007/978-3-031-12786-1_5

dispersion model, starting from a version of the model that includes the option for back-trajectories, retroSPRAY (Armand et al., 2013; Tinarelli et al., 2018).

In retroSPRAY the location of an unknown source is estimated from concentration observations with two approaches: in the first case, plausible source locations are estimated by identifying areas with maximum spatial and temporal consistency among backward trajectories from each sensor; in the second, a variational method is used to minimize the objective function at each grid-box reached by backward trajectories, providing information on the source location and related uncertainty.

The new challenge is using the signals from citizens in place of observed concentrations as input receptors for retroSpray. The warnings received through the NOSE WEB-APP are clearly sparse in space and time, yet can be considered as a moving in space/time receptor grid. A clever clustering of the warnings is the first step to generate proper 'receptors' for the back-trajectories. The novel methodologies in SMART and NOSE and the preliminary results obtained in a case study are presented and discussed. In parallel, new modules in SPRAY for treating the specific characteristics of the odour dispersion in forward-mode have been designed. The odour nuisance is characterized by concentration fluctuations and peaks. A peak-to-mean approach is now available in SPRAY to realistically reconstruct the concentration peaks of the emitted substances producing the odour nuisance.

2 The NOSE Case Study and the Simulations

The NOSE system has been implemented for the region of the Siracusa Province in Sicily (Italy), including a large industrial area located along the coast and the Augusta port (Fig. 1). The case study considers the odour nuisance event recorded on the 13th of April 2020, when almost 770 signals were registered in NOSE WEB-APP in four hours, between 8 and 12 AM. During this event the main reports concerned the smell of hydrocarbons, burnt and solvents, with declared difficulty in breathing, burning and irritation in the throat and headaches felt by citizens.

In Table 1 an example of the file with the signals' details as output by NOSE WEB-APP and to be used in input to retroSPRAY model is reported. In Fig. 2 an example of the clustering of the signals as elaborated by NOSE WEB-APP for the full 8–12 AM episode is shown; in Fig. 3 an example of a half-hour distribution of the signals counting with their location and intensity, as post-processed for retroSPRAY input, is plotted. For the elaboration of this type of information to prepare the input to the dispersion model, a 500-m-spacing grid is defined on the domain. Then, thresholds for a minimum number of signals with minimum odour intensity is determined for each grid cell and each time, in order to identify the cells that can be considered as sensible 'receptor' areas for the release of backward stochastic trajectories.

Based on the threshold criteria, in this case study three receptor cells have been determined, as in Fig. 4. For the SMART simulations, the 3D meteorological fields from MOLOCH atmospheric model have been processed by ARAMIS turbulence and

Fig. 1 Study domain for the Siracusa region area (Sicily, Italy) (from GoogleMaps)

Table 1 Example of the format for the citizen signals output from NOSE WEB-APP

Date and hour	13/04/2020 13:08	13/04/2020 12:56	13/04/2020 ….
Lat	37.272	37.268	….
Lon	15.178	15.175	….
Location name	Augusta	Augusta	….
Odour intensity	3	5	….
Perceived smell id	2	3	….
Perceived smell name	Hydrocarbons	Sulfate	….
Perceived smell comment	Bla bla	Bla bla	….
Type of upset	Headache	Burning sensation	….
Symptom a	1	0	….
Symptom b	0	0	….
Symptom c	0	1	….
Symptom d	1	1	….
Symptom comments	Bla bla	Bla bla	….

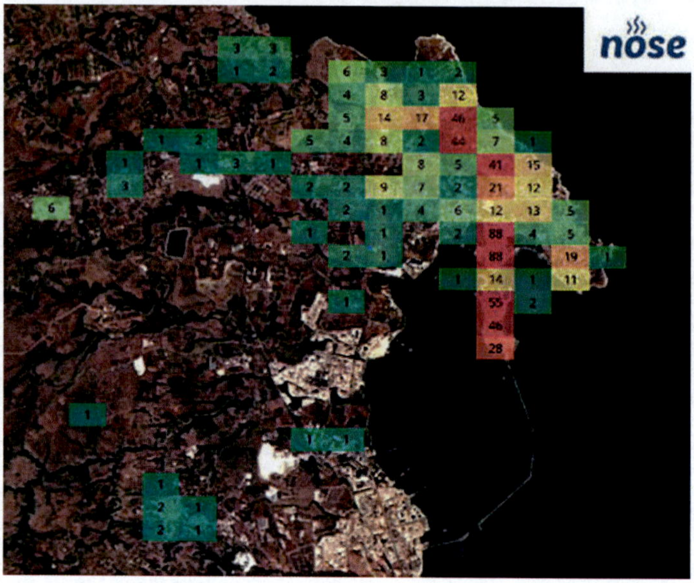

Fig. 2 Map of the number of the signals as clustered in NOSE WEB-APP, 8–12 AM on 13/04/2020

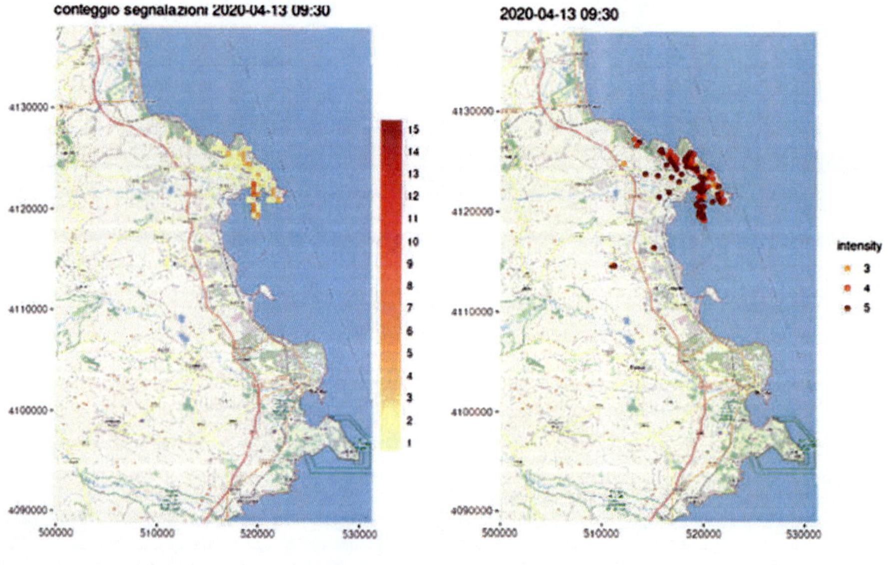

Fig. 3 Maps of the number of NOSE WEB-APP signals in the area (left) and of their related intensity (right) on the 13th of April 2020 at 9:30 local time

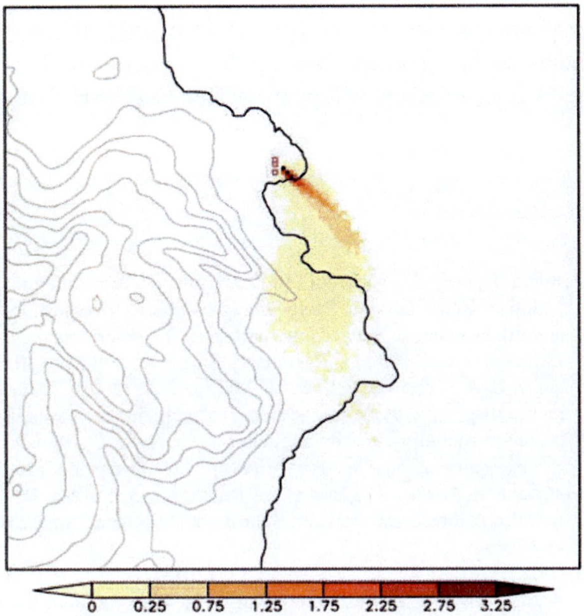

Fig. 4 Map identifying the area where potential sources may be located for the event of the 13th of April 2020

boundary-layer parameterization code to elaborate the input for retroSPRAY. Simulations with retroSPRAY are then performed by releasing, from the identified receptor cells, a series of retro-puffs at each time interval in which a significant number of signals are collected, typically at a 30' time frequency. The retro-concentration fields generated by the retro-puffs are then combined both at emission and receptor times, through a process that calculates their products (representing a logical AND operation) and their sums (representing a logical OR operation) in order to build final maps describing the region where possible sources can be potentially located and their related probability. In Fig. 4 the final map of the area where potential sources may be located is reported, over a simplified topography output by the model run. In the colour scale, lighter colours indicate locations where the emission should be very high to reach the receptors, thus darker colours indicate areas where emitting sources may be more likely expected.

3 Conclusions

The newly developed approach implemented to use, after appropriate processing, citizens' warnings from NOSE WEB-APP as input to retroSPRAY model, demonstrated to be promising and applicable. This has been confirmed by performing test simulations in forward mode, where potential releasing sources have been placed in different locations in the areas identified as more or less possible origin of odour nuisance, and also in the region outside them. It was seen that the sources placed in the 'most probable' areas were in fact affecting the receptor locations during the main hours of the recorded warnings. The new modules are being integrated in SMART modelling suite and will be interfaced to NOSE alert system in near future.

References

Armand, P., Olry, C., Albergel, A., Duchenne, C., & Moussafir, J. (2013). Development and application of Retro-Spray, a backward atmospheric transport and dispersion model at the regional and urban scale. In *Proceedings of the 15th Conference on Harmonisation Within Atmospheric Dispersion Modelling for Regulatory Purposes*, 6–9 May 2013, Madrid, Spain (pp. 789–893).

Bisignano, A., Trini Castelli, S., & Malguzzi, P. (2020). Development and verification of a new meteo-dispersive modelling system for accidental releases in the Italian territory: SMART. Air pollution modeling and its application XXVI. In C. Mensink, W. Gong, & A. Hakami (Eds.), *Springer proceedings in complexity* (pp. 77–81). Springer International Publishing Switzerland.

Malguzzi, P., Grossi, G., Buzzi, A., Ranzi, R., & Buizza, R. (2006). The 1966 "century" flood in Italy: A meteorological and hydrological revisitation. *Journal of Geophysical Research, 111*, D24106.

Tinarelli, G., Uboldi, F., & Carlino, G. (2018). Source term estimation using an adjoint model: A comparison of two different algorithms. *International Journal of Environment and Pollution, 64*(1/3).

Trini Castelli, S., Bisignano, A., Donateo, A., Landi, T. C., Martano, P., & Malguzzi, P. (2020). Evaluation of the turbulence parameterisation in the MOLOCH meteorological model. *Quarterly Journal of the Royal Meteorological Society, 146*, 124–140.

Questions and Answers

QUESTIONER: Clemens Mensink

QUESTION: Do you think your methodology could be used in a legal framework? (i.e. detecting sources or accidents?)

ANSWER: At present our methodology is still under development and needs further testing and validation. Yet, the final aim is in fact to support the Regional Environmental Agency in detecting possible sources of odor nuisance in the area, based on the NOSE App. This approach allows the Agency first-responders to intervene with prompt air sampling in restricted potential release areas, verifying in real time the data of the odorous compounds measured by the air quality network, and to contact the operators of the plants that might be identified as the possible sources.

QUESTIONER: Giovanni Bonafé

QUESTION: Are you satisfied with the current internal boundary layer parameterization used in your modelling system, given its effect for fumigation cases in coastal areas? Or do you think it needs to be improved?

ANSWER: The meteorological input to the Lagrangian particle dispersion model retroSPRAY is provided, through the ARAMIS code, by the atmospheric model MOLOCH, thus the boundary-layer parameterization is the one implemented in it. The turbulence scheme in MOLOCH model is based on a 1.5-order *E-l* closure, where the second-order turbulence quantities are parameterized in terms of the turbulent kinetic energy and the mixing length. Therefore, in SMART suite we are using a boundary-layer parameterization that is typical for regional models and that represents a consolidated and standard approach. At the considered model resolutions, the thermal internal boundary layer is thus expected to be resolved. Certainly, improvements are always possible and can be pursued. We have not (yet) tested the system specifically on cases that may produce fumigation in coastal areas, but such a case could indeed be an interesting challenge.

On the Utilization of Real-Time Activity and Air Quality Sensor Data in a Local-Scale Operative Dispersion Model in Helsinki

Lasse Johansson, Age Poom, Jarkko Niemi, Hilkka Timonen, and Ari Karppinen

Abstract We presented a prototype of an operative air quality modelling system for Helsinki region that utilizes measurement data of variable quality (stations and sensors). The system continuously taps into various sources of real-time activity data to support the modelling. Being an updated version of FMI-Enfuser and developed during UIA-HOPE project we refer to this new modelling system as HOPE-Enfuser. Using public cloud storage, the modelling output is freely accessible for 3rd party applications. This is also the first time when we provide modelled Lung Deposited Surface Area (LDSA), and black carbon (BC) concentration predictions with our modelling system. For shipping, we present a novel near-real-time approach for the dynamic modelling of shipping emissions. HOPE-Enfuser uses data fusion to adjust the modelling and learn from the measurement evidence. The Covid-19 pandemic caused disruptions in the behavior of people and in the patterns of local emission sources. In this paper we will also discuss how these disruptions were perceived and captured by the model.

Keywords Dispersion modelling · Sensor networks · Data fusion · Shipping emissions · Covid-19

1 Methods

The HOPE-Enfuser is an operative local scale air quality model (a combination of Gaussian Puff & Plume) used in the Helsinki Metropolitan area in Finland. The previous versions of the Enfuser model (Johansson et al., 2015; Mensink & Volker, 2021) has also been used in foreign installation sites such as Nanjing, China and

---

L. Johansson (✉) · H. Timonen · A. Karppinen
Finnish Meteorological Institute, P.O. Box 503, 00101 Helsinki, Finland
e-mail: lasse.johansson@fmi.fi

A. Poom
Helsinki Institute of Urban and Regional Studies (Urbaria), Helsinki, Finland

J. Niemi
Helsinki Region Environmental Authority HSY, Ilmalantori 1, 00240 Helsinki, Finland

C. Mensink and O. Jorba (eds.), *Air Pollution Modeling and its Application XXVIII*, Springer Proceedings in Complexity, https://doi.org/10.1007/978-3-031-12786-1_6

Delhi, India. The traditional modelled pollutant species are NO_2, NO, O_3, $PM_{2.5}$ and PM_{10} for which the model provides hourly average concentrations at breathing height of 2 m above ground. These model predictions are updated several times per day, each time including a "now-casting" period with measurements (up to 24 h in the past and a forecasting period to the future (up to 48 h) in a resolution of 13 $\times$ 13 m. The key source of information to support the modelling are (a), NWP meteorological data (HIRLAM, HARMONIE and GFS) (b), AQ measurement data for data fusion (c), GIS-data to describe the modelling environment (e.g., digital surface maps, land-use and road network (OpenStreetMap), Satellite data from Copernicus Sentinel-2, population data from Global Human Settlement) (d), regional scale AQ forecasts from FMI-SILAM (Sofiev et al., 2015) (e), local emission inventories and (f), supporting activity data. The most notable modelled emission source categories are (a), traffic (hourly emissions computed for each road separately addressing hourly flows of cars, heavy vehicles and buses (b), household combustion emissions (c), shipping emissions via FMI-STEAM (Johansson et al., 2017) (d), power plant emissions as elevated point sources and (f), the regional background.

Since the adoption, the model has been developed further during the on-going UIA-HOPE project in Helsinki, especially by connecting to new information sources to support the modelling. This extended version is also an operational service and provides online predictions to several use-cases. However, due to its novelty and the amount of new implemented features, the service is an unofficial one and is mainly used by internal partner of the UIA-HOPE project. There are several notable additions in the extended model version currently being developed. We have included LDSA, black carbon (BC) and CO to the list of modelled species. We also continuously extract several new types of supporting activity data (real-time shipping, traffic flow data and road weather measurements to name a few), which we focus more on Sect. 1.2. A real-time connection to a complementary AQ sensor network has also been added (Sect. 1.1).

Finally, as a technical novelty the modelling data is pushed to an Amazon S3 cloud storage. This storage a public access and a rolling archive of hourly modelling datasets is kept there for the past 2 weeks from where 3rd party application can access the model predictions. As an example of an application that utilizes the HOPE-Enfuser modelling data we present the Green Paths tool by University of Helsinki (Poom et al., 2020) (Fig. 1). This open-source tool assesses multiple environmental exposures (e.g., AQ and noise) for navigation paths and finds exposure-optimised routes for the users. For AQ related costs, the pathing algorithm uses Air Quality Index (AQI), that is provided by HOPE-Enfuser based on the pollutant species concentrations.

1.1 The Online Measurement Networks

HOPE-Enfuser utilizes data fusion with AQ measurements during its use. This includes online sensor data, being the first time where we have utilized such sensor data in an operational air quality modelling service. This can be regarded as a more

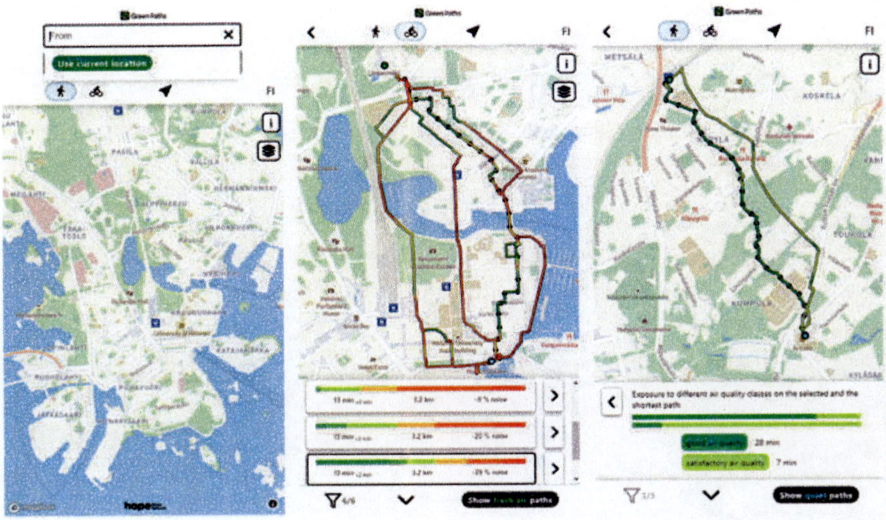

Fig. 1 Caption of the Green Paths—navigation application's user interface. The service can be accessed from: https://green-paths.web.app/

challenging approach than an offline-based demonstration, which the author of this study presented previously in Mensink and Volker (2021).

Briefly put, in the data fusion we apply adjustments to hourly emission factors to a set of emission source categories (e.g., traffic, households) to adapt the hourly modelling to a higher agreement with measurement evidence. For the background concentration a flat correction offset is also applied. In addition to a stronger hourly correction these adjustments also gradually contribute to a longer-term learning pattern, for which we provide a concrete example in Sect. 2. The longer-term corrections captured by the model can also adjust diurnal patterns for emission sources if needed.

The overall measurement network (as well as the modelling domain) configuration has been shown in Fig. 2. The base figure is a caption of monthly NO_2 average concentration (March 2021), automatically generated by the HOPE-Enfuser modelling service. Within the base figure we have highlighted three sensor focus areas of Pakila, Vallila and Jätkäsaari, in which most of the 25 AQ sensors (Vaisala AQT530) have been installed to. The official AQ measurement station network is more evenly distributed. It should be noted that for O_3 there are only 4 measurement sites, for BC there are 7 and for CO there are none. The network of LDSA-sensors (by Pegasor) is mostly concentrated on Pakila focus area (locations not shown in Fig. 1). The whole network is maintained by local authority (HSY), which has also provided emission inventories (households) for the Helsinki modelling area. Finally, it should be noted that during the data fusion we must address the measurement quality differences (i.e., preliminary weights) and we have assigned an order of magnitude lower quality estimate for the sensors with respect the refence stations.

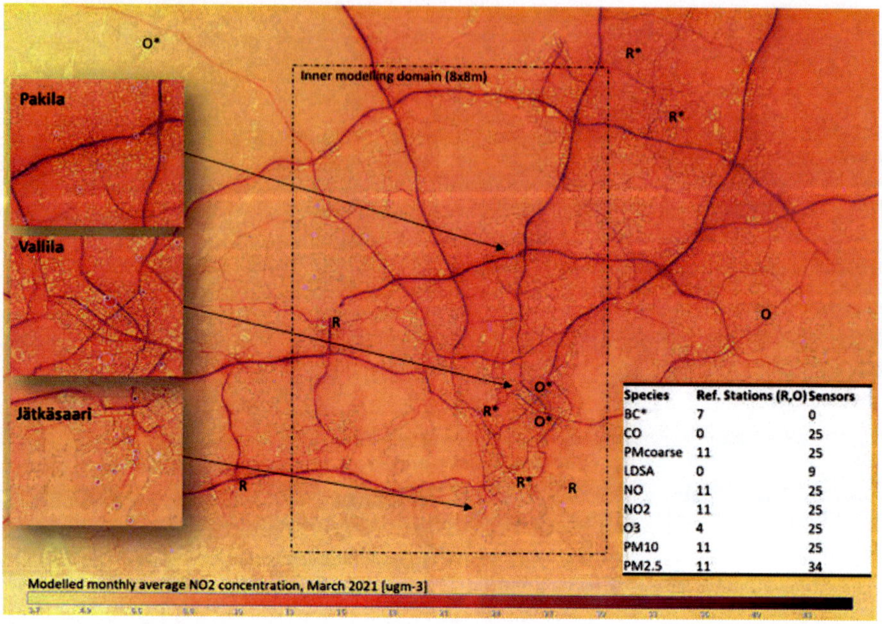

Fig. 2 Illustration of measurement network and modelling domain configuration, using a monthly average modelling result as an example. Coloured 'dots' correspond to observed monthly average NO_2 concentrations but more importantly shows the measurement location. For sensors the dots are smaller. * = station that also measures BC. O = station that measures O_3. All stations also measure $PM_{2.5}$, NO_2 and PM_{10}

This topic of measurement reliability and weighting is further discussed in Mensink and Volker (2021).

1.2 Real-Time Activity Data

For a reactive high-resolution AQ modelling system it is often not enough to rely on static emission inventories. This in mind HOPE-Enfuser continuously extracts information from online sources to get a more dynamic overview on the modelling area. As an example, we extract and use data from:

- Road weather measurements (more than 20 locations) to improve our road dust modelling capabilities. Source: DigiTraffic (https://tie.digitraffic.fi/en/).
- Traffic congestion data from Here.com. Main use-case is to address the effect of notable traffic jams and to analyse longer term traffic patterns in the area.
- Hourly traffic flow measurements from 55 monitoring sites, including a split between directions and vehicle categories. Source: DigiTraffic.

- Real-time shipping movement patterns via a local AIS-receiver coupled to the AISHub service (https://www.aishub.net/).

The most ambitious addition of these is the real-time shipping data and its derivation into continuous shipping emissions, which we focus on next. Helsinki centre has a busy passenger shipping port, which is also close to the Jätkäsaari focus area. To model the shipping emissions as they occur, we have adopted the following solution:

- We contribute with our own local AIS-receiver and connect to AISHub service to access real-time shipping activity data (AIS) globally but more importantly, for the Helsinki coastal area.
- On an hourly basis we use the archived recent AIS-data as input to the FMI-STEAM model, which creates a special kind of hourly emission inventory dataset. This is basically text-based line-data that describes emission release rates (g/s), time, location and an estimate for release height (m) for the emissions.
- Produced hourly emission inventory packages (the line-data text-files) by the STEAM-model are pushed to Amazon S3 cloud storage.
- Whenever HOPE-Enfuser modelling task occurs these fresh shipping emissions are accessed from the cloud storage (Fig. 3) and dispersion modelling occurs separately for each line of activity.

The special format of the emission inventory makes it possible to represent the shipping emissions at any chosen resolution by modelling the emissions as moving point sources. Due to technical limitations, we have settled in representing the shipping emissions as a gridded inventory (for Gaussian Puff releases) with a resolution of approximately 100 m and 5 min. For the forecasting period for the model, we simply use the shipping movements that occurred one week before.

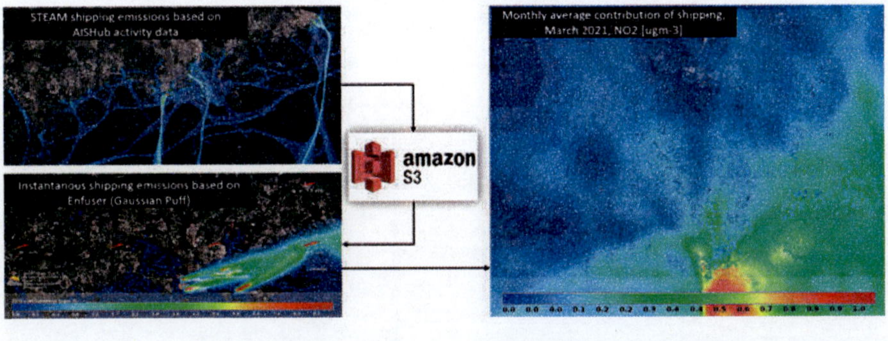

Fig. 3 The concept of near-real-time shipping emissions modelling in HOPE-Enfuser. AIS-activity data is used as input for FMI-STEAM that predicts shipping emissions. HOPE-Enfuser, while accessing AWS S3, uses this emission inventory data as input for dispersion modelling. In Right, an example of the shipping contribution to NO_2 concentrations near Helsinki is shown (March 2021, using the same geographic area as in Fig. 2)

2 Results and Discussion

The HOPE-Enfuser service has been operational during the Covid-19 epidemic, and this has provided the possibility to inspect the indirect effects of the epidemic to air quality in Helsinki. In Fig. 4 a comparison of diurnal average NO_2 concentration during Spring (March to May, 2018 and 2020) is shown for an urban traffic measurement site. The differences are drastic, and similar strong reductions in traffic-related concentrations are also visible in other urban measurement sites (not shown here). During early 2020, we also observe the traffic emission factors (being coupled to measurement evidence via the data fusion) started to decline gradually in HOPE-Enfuser. The hourly traffic flow measurement data, which we have formed into a generalized index (percent of average hourly patterns observed in 2017), also declined strongly for passenger cars at the same time. It should be noted that the decline in longer-term emission factor corrections is slower than the actual rapid decline in traffic flow counts, however the floating emission factor correction eventually stabilizes to a fixed level in July.

The modelled shipping emissions according to FMI-STEAM also reduced notably near the coastal areas of Helsinki, however the reductions occur later than with road traffic (after May 2020). For example, $PM_{2.5}$ shipping emissions in January-March 2020 near Helsinki are 120% with respect to the emissions in the previous year of 2019, but this relation is reduced to 73% during May–July respectively.

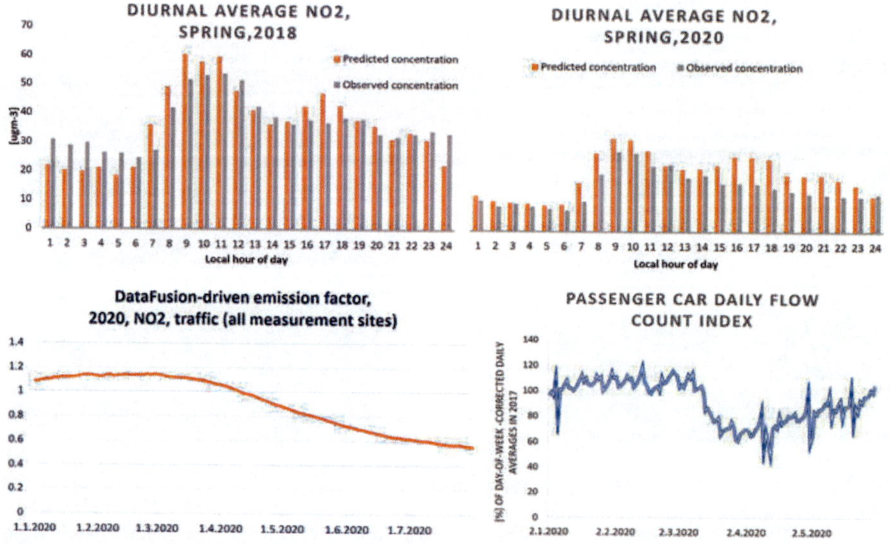

Fig. 4 Indirect effects of the pandemic during early 2020. Up, a comparison of diurnal average NO_2 concentrations at a selected urban traffic site (60.16964 N, 24.93924 E) is shown. Below, the data-fusion driven longer-term adjustment to traffic emissions is shown, as well as a derived passenger car flow count index based on 55 traffic flow count measurement sites in the area

As shown, the modelling system attempts to learn and adjust the modelling based on the measurement evidence. It has been designed to automatically adapt to changes in the modelling environment, with respect to local emission sources. The Covid-19 pandemic caused a notable disruption in the behavior of local emission sources in Helsinki and therefore provided a useful testing case for the performance of the modelling system. According to preliminary results, we have seen that the modelling system has been able to adapt to the changes in traffic patterns and to the changes in shipping emissions in the area. For traffic it should be noted that the corrections done to emission factors cannot yet account for geographical differences in the changes and the automatic adjustments are applied similarly to the whole modelling area. Interestingly, the traffic flow data shows a rapid recovery, while the AQ concentrations (and the traffic-related emission factors) remain low long afterwards with respect 2019, or 2018.

One of the key features of the modelling system is the simultaneous use of AQ data from both reference stations and a complementary network of sensors. This is a topic in which we concentrate more in further studies. Briefly put, there are 'pros and cons' related to the use of sensor data in such a way. Theoretically, an extended network of measurement devices makes the data fusion methodology more robust and facilitates more detailed input for the modelling system that attempts to learn via the measurement evidence. On the other hand, we have seen that when (a), the group of sensors is large and (b), their measurement biases are strongly correlated and (c), the group of reference stations is small, then the sensors can in fact harm the modelling performance. As an example, we currently observe this detrimental effect with our O_3 modelling results.

The evaluation of the modelling performance of our new pollutant species LDSA and BC will also be addressed in future papers. For LDSA there are a couple of challenges to solve: first, the LDSA sensors are located on suburban residential areas, some sensors being very close to households. This means that individual households may cause sudden peaks in LDSA concentrations, which are not possible to be captured by the model. Also, there is no LDSA background information available from the regional scale model we use; we simply proxy it from $PM_{2.5}$ and rely on data fusion -driven corrections. In our recent rolling 2-month assessment (Leave-one-out validation, July–August 2021) the hourly LDSA has an R2 of 0.49 against the measurements. There are similar issues with BC, however the rolling 2-month average R2 shows a higher value of 0.58. Again, we have no direct source of information for long range transportation of BC and we instead use elemental carbon mass (EC) from FMI-SILAM. The BC values are small (majority being lower than 1 ugm-3) in Helsinki up to the point that the sensitivity of the BC measurement stations may affect the correlation between model predictions.

The dynamic modelling of shipping emissions makes it possible to explain some of the high concentrations the stations and sensor measure, especially near the coastal regions. If done successfully then this in turn aids the data fusion method as the model can associate high measured concentrations to their proper source of origin. However, we also observe that the dynamic modelling of shipping emissions can cause large margin of error from time to time, as the dispersion of elevated strong emission plumes is highly challenging near the coast. For example, HOPE-Enfuser may model a false-positive strong shipping emission plume which are not captured by the nearby measurement network. The margin of error is especially noticeable for stationary ships at port, for which the modelling of auxiliary engine use is challenging.

References

Johansson, L., Epitropou, V., Karatzas, K., Karppinen, K., Wanner, L., Vrochidis, S., Bassoukos, A., Kukkonen, J., & Kompatsiaris, I. (2015). Fusion of meteorological and air quality data extracted from the web for personalized environmental information services. *Environmental Modelling & Software, 64*, 143–155. Elsevier.

Johansson, L., Jalkanen, J.-P., & Kukkonen, J. (2017). Global assessment of shipping emissions in 2015 on a high spatial and temporal resolution. *Atmospheric Environment, 167*, 403–415. https://doi.org/10.1016/j.atmosenv.2017.08.042

Mensink, C., & Volker, M. (2021). *Air pollution modeling and its application XXVII*. Springer International Publishing. https://doi.org/10.1007/978-3-662-63760-9

Poom, A., Helle, J., & Toivonen, T. (2020). Journey planners can promote active, healthy and sustainable urban travel. *Urbaria Summaries Series.*

Sofiev, M., Vira, J., Kouznetsov, R., Prank, M., Soares, J., Alves Antunes, J., & Genikhovich, E. (2015). Construction of the SILAM Eulerian atmospheric dispersion model based on the advection algorithm of Michael Galperin. *Geoscientific Model Development, 8*, 3497–3522. https://doi.org/10.5194/gmd-8-3497-2015

https://www.aishub.net/. Visited 15.4.2021.

3D Multi-scale Weather and Dispersion Models Applied to Assess the Impact of Industrial Plants on Human Health and the Environment

Patrick Armand and Christophe Duchenne

Abstract Routine operations of industrial facilities as accidents possibly affecting them often result in releases of gases or fine particles into the air. The evaluation of the space and time distribution of these substances is of utmost importance to verify the compliance of the facilities with the environmental regulation and their harmlessness to health in normal conditions or to estimate their health impact in adverse conditions. Most of the industrial sites encompass built areas and are located in complex environments characterized by a rugged terrain and a heterogeneous land use. While Gaussian-type models are not adapted to these configurations, there are still much used for impact evaluation. By contrast, this paper dedicates to exemplifying the implementation and use of a 3D modelling system. The approach consists in developing the 3D digital mock-up of a research center, performing a multi-scale meteorological study over a yearly period, and simulating the Atmospheric Transport and Dispersion (ATD) of gases emitted by the stacks of the center installations. The distribution of the gases is significantly influenced by the site effects and compares satisfactorily with measurements carried out in the environment. The results are definitely more realistic and more reliable than the ones of the usual Gaussian model and prove the interest of 3D modelling in the field of impact assessment.

1 Introduction

All over the world and notably in the European Union, the regulations require estimating the consequences on human health and the environment of the releases into the air due to either the authorized operations of the industrial facilities or accidents hypothetically affecting them. Impact assessment is preceded by Atmospheric Transport and Dispersion (ATD) computations with the aim to determine the space and time distribution of the released substances.

P. Armand (✉) · C. Duchenne
DAM, CEA, 91297 Arpajon, DIF, France
e-mail: patrick.armand@cea.fr

C. Mensink and O. Jorba (eds.), *Air Pollution Modeling and its Application XXVIII*, Springer Proceedings in Complexity, https://doi.org/10.1007/978-3-031-12786-1_7

Still today, Gaussian-type models are most commonly used to carry out the ATD computations. Yet, the strong assumptions underlying these models like a flat unobstructed terrain and a uniform wind field are far from being verified when real life is considered. Even if certain Gaussian models have been upgraded to account for isolated buildings or specific built-up configurations (like street canyons), these models cannot encompass the complexity and the diversity of most industrial sites and their environments. Thus, simple approaches may be inappropriate for the proper and rigorous impact evaluation of industrial facilities (Armand et al., 2015; Baumann Stanzer et al., 2015).

On the contrary, 3D models are not restricted to oversimplified configurations and benefit from thorough development and validation efforts in the last decades, and from increasingly affordable powerful computational resources. In this context, we have implemented a generic and flexible 3D modelling system cascading up and down from the meso-scale to the micro-scale and from outside to inside industrial plants or other premises. The system has been successfully demonstrated in the framework of urgency preparedness and response exercises, exemplified by the EMERGENCIES projects (Armand & Duchenne, 2019; Armand et al., 2017, 2021; Oldrini et al., 2021a, 2021b). Furthermore, it is exploited to assess the impact of installations under normal circumstances.

This paper shows an original application of our modelling system to evaluate the impact of the normal operation conditions at a research center. In the absence of full containment, some facilities release authorized amounts of gases into the atmosphere. The modelling system is used to simulate the spatial and temporal distribution of the gases for the whole 2018 year and *in fine* check that the activities of the center do not cause any impact on health and the environment.

Simulations are performed with PMSS, which is the parallel version of Micro-SWIFT-SPRAY, developed as a trade-off between a rigorous and quick solution of the flow and dispersion in built-up environments. PMSS combines the Micro-SWIFT 3D mass-consistent diagnostic flow model and the Micro-SPRAY 3D Lagrangian particle dispersion model (Tinarelli et al., 2013). The major features of PMSS are to deal with nested domains and to explicitly account for the influence of the obstacles (relief, buildings…) on the flow and dispersion. The parallelization of PMSS has proven to be very efficient to deal with extremely large computation domains (Oldrini et al., 2011, 2017, 2019). The greatest attention has been paid to PMSS validation against numerous experimental trials in wind tunnel and at full scale (Trini Castelli et al., 2016, 2018; Oldrini & Armand, 2019).

After this introduction, the paper has two parts dedicated to the meteorological study around the center and the 3D dispersion simulations of the authorized releases from the stacks of the facilities. Then, we conclude on the interest to apply 3D models in the field of impact assessment. While this paper focuses on dispersion computations, more details about the meteorological characterization of the site under consideration can be found in Armand and Duchenne (2021).

2 Meteorological Study and Flow Simulations

At first, a 3D digital mock-up was developed gathering the major features of the environment of the center, which has a quite large number of buildings and is located on a plateau surrounded by a rugged terrain with diverse land uses (forested areas, cultivated fields…).

The considered domain has horizontal dimensions of 13 km × 14 km. The numerical model of the terrain is provided at a resolution of 5 m by the French National Geographical Institute. The vegetation data are available at a resolution of 20 m on the site of the French Ecology Ministry.

2.1 Meteorological Considerations

A meteorological study was undertaken considering different kinds of data along the year 2018. On one hand, Fig. 1 shows the wind rose built using the data of the meteorological mast set up on the center measuring the wind speed and direction at 10 m and the air temperature at 2 m above the ground. On the other hand, AROME hourly data projected on a regular mesh with a resolution of 0.025° were compiled to establish wind roses in 5 velocity and 36 direction classes (like for the wind rose of the mast) at the 45 points of the AROME grid over the simulation domain. Figure 2 shows that the AROME wind roses present disparities due to orographic and land use effects, even on the relatively limited area considered. This justifies developing a 3D approach of the regional flow and dispersion simulations. Moreover, AROME data extracted at the coordinates of the mast are consistent with the wind data measured on this mast. This led us to use AROME as input data of the simulations performed at micro-scale with the PSWIFT flow model of the PMSS system.

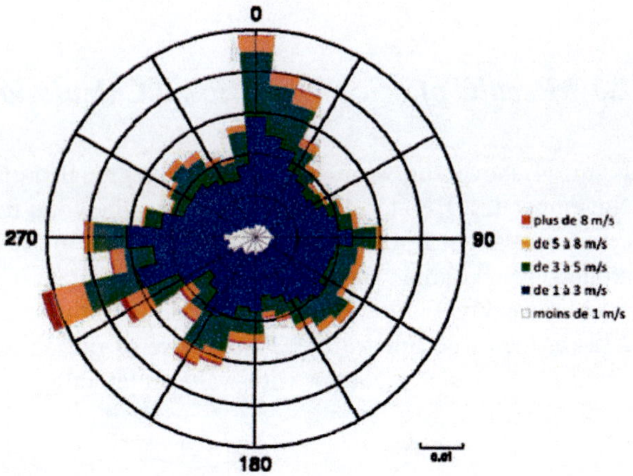

Fig. 1 Wind rose in 2018 at a height of 10 m built from data measured on the meteorological mast

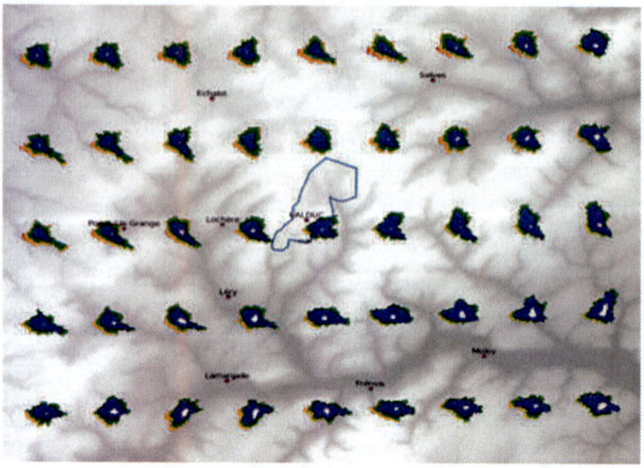

Fig. 2 Wind roses in 2018 at a height of 10 m built from AROME data

2.2 *Micro-scale Meteorological Simulations*

The resolution of the meso-scale flow computations (*circa* 2 km) is not enough to simulate the flow and the dispersion of releases around and in-between buildings. Therefore, AROME hourly data (wind, temperature, humidity…) were downscaled for the whole year 2018 using PMSS.

Simulations account for the topography and land use and integrate the influence of all buildings in the center and in villages surrounding the center where atmospheric sampling devices are set up. Figure 3 shows the extending of the computational domains. The regional domain has a resolution of 20 m and is divided into nine tiles so that the memory capacity of the computer nodes used for the calculations is not exceeded. Six local domains have a resolution of 2 m and cover the upper part and the lower part of the center, as well as four villages around the center. See also Table 1.

2.3 *Results of Micro-scale Flow Calculations*

Figure 4 shows the wind field over the regional domain at a height of 10 m, on November 1, 2018 at 9:00. Figure 5 shows the wind field over the upper part of the center at a height 2 m (at the same timeframe). Wind speed and direction are changed around each building. However, buildings are distant from each other and their wakes interact weakly. Figure 6 shows the wind field over village E domain at a height of 2 m (at the same timeframe). Houses are of old construction, close to each other, leading to a complex pattern with wakes interacting inside the streets of the village.

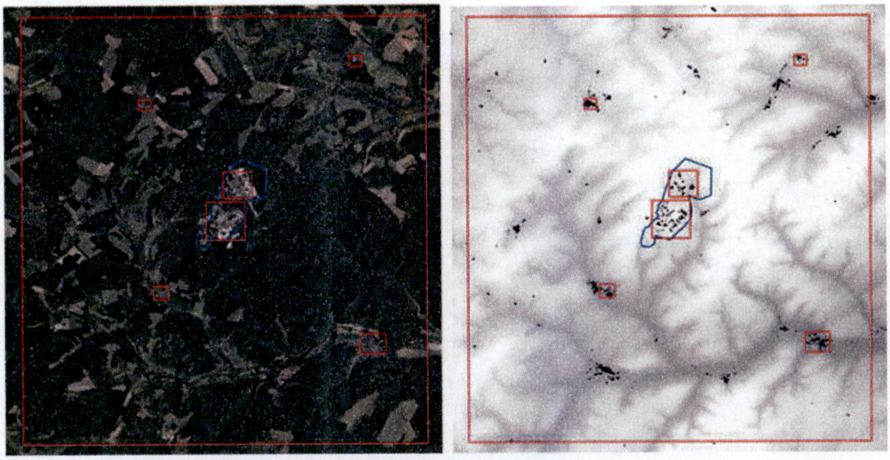

Fig. 3 Extending of the nested simulation domains

Table 1 Features of PSWIFT and PSPRAY computational grids

Domain	Resolution (in m)	Dimensions (in km × km)	Number of nodes	Division into tiles
Regional	20	12.98 × 13.98	650 × 700	Yes (9 tiles)
Center upper	2	1.24 × 1.22	621 × 611	Yes (4 tiles)
Center lower	2	0.92 × 0.86	461 × 431	No
Village L	2	0.50 × 0.45	251 × 226	No
Village M	2	0.80 × 0.65	401 × 326	No
Village E	2	0.40 × 0.35	201 × 176	No
Village S	2	0.40 × 0.35	201 × 176	No

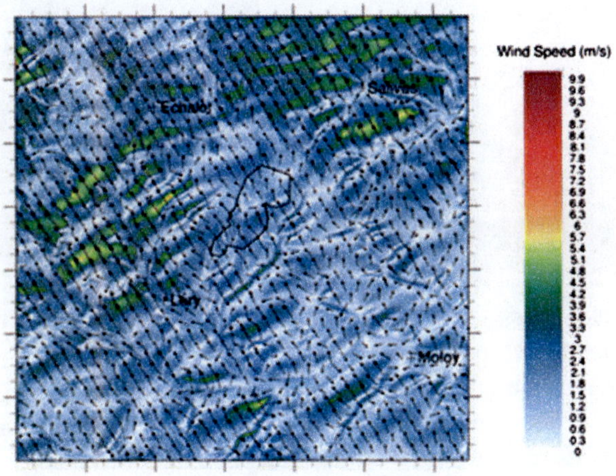

Fig. 4 Wind field at 10 m on the regional domain on November 1, 2018 at 9:00

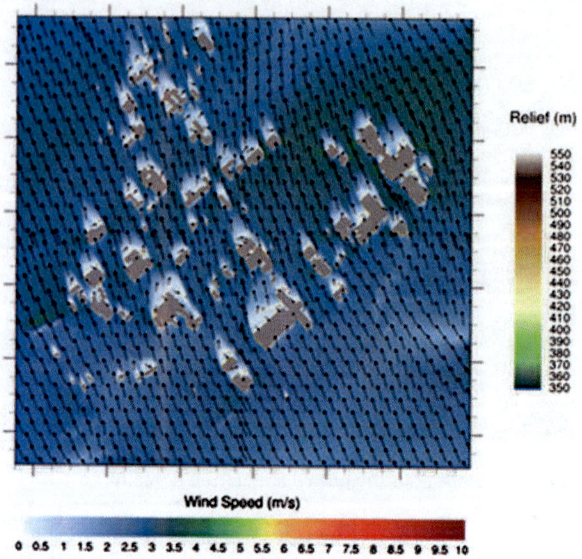

Fig. 5 Wind field at 2 m on the upper part of the center on November 1, 2018 at 9:00

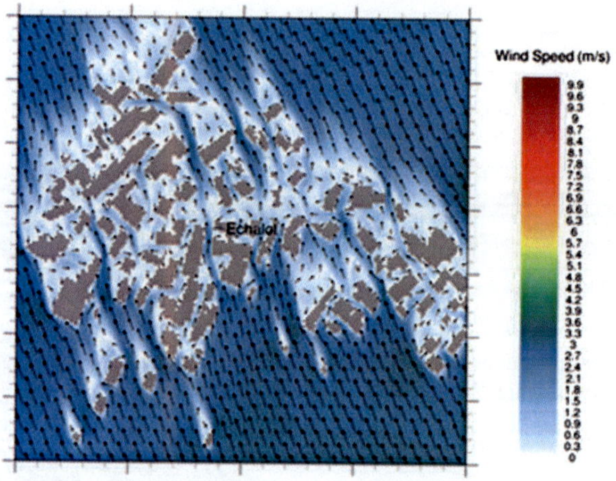

Fig. 6 Wind field at 2 m on the village E domain on November 1, 2018 at 9:00

3 Dispersion Simulations at Regional and Local Scales

The emissions from the center facilities are the subject of weekly measurements. The heights of the stacks and the gaseous release rates and temperatures are accounted for in the computations. Depending on the facilities, the releases are unequal in amplitude and temporal distribution, both daily and seasonally, in particular because of the activities that are carried out. Higher frequency measurements show that the actual timing of the releases is complex and has a number of peaks.

The atmospheric concentrations computed with PSPRAY are averaged and stored on an hourly basis throughout 2018. The calculations are linked in chronological order: the plume of particles obtained at the end of a calculation is taken to initiate the following calculation. In total, there are eight emission sources represented by numerical particle numbers proportional to the releases. For the whole of 2018, this corresponds to several hundred billion numerical particles emitted. Note that we take into account the dry deposition on the ground and washing by the rain of the gaseous emissions during their transport and dispersion.

Dispersion simulations were performed first without accounting for the buildings (considering only the 20 m resolution domain), then accounting for the buildings (considering the 20 m domain and the 6 calculation domains of 2 m resolution, the whole divided in 18 tiles) (*cf.* Table 1). Let us recall that the buildings are accounted for in the villages where sampling stations are set up with the aim to study their influence on the concentration measurements. The air intake on the roof of the sampling stations is located at the height of 3 m at which the concentrations results are shown.

3.1 Distribution of the Releases Without Taking into Account the Built Environment

Figure 7 shows the spatial distribution of the atmospheric concentration on November 1, 2018 averaged between 8:00 and 9:00 and between 18:00 and 19:00 at a height of approximately 3 m above the ground, following the releases of all the installations. The plumes are relatively straight, even when the wind is blowing at low speed around 19:00. The influence of the relief is visible in the valleys perpendicular to the axes of progression of the plumes. The low concentrations near the ground indicate that the plume passes mainly over the valley without descending the slopes.

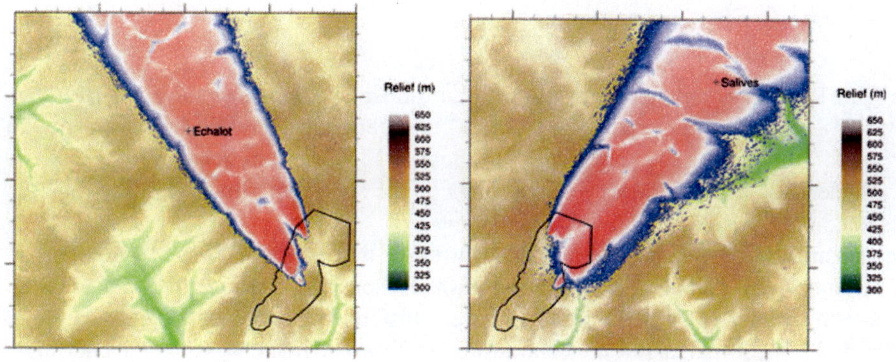

Fig. 7 Concentration fields at 3 m, on 01/11/2018 at 9:00 and 19:00, following the releases from all the installations

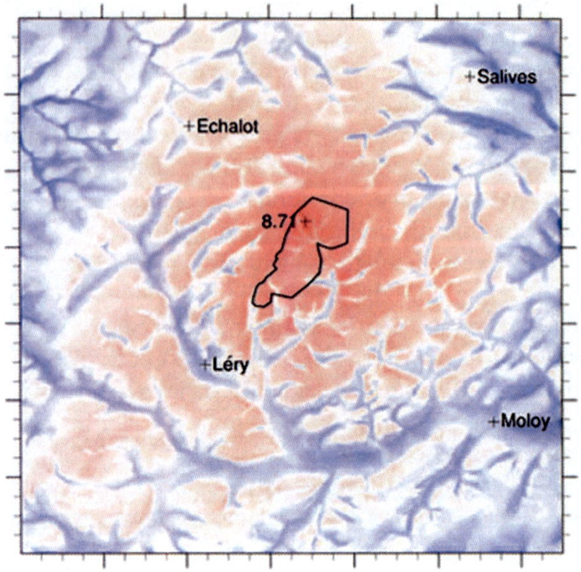

Fig. 8 Annual average concentration at 3 m for the releases of the year 2018

Figure 8 shows the annual average concentration computed for all releases for the year 2018. The point of the maximum annual average concentration is within the center. Outside the center, the annual average concentration is very low everywhere. The distribution near the ground appears to be influenced by the relief and land use. The concentrations evaluated inside the valleys, even the small ones, are much lower than the concentrations on the neighboring plateaus.

3.2 Distribution of the Releases Taking into Account the Built Environment

Figure 9 shows the spatial distribution of the atmospheric concentration on November 1, 2018 averaged between 8:00 and 9:00 and between 18:00 and 19:00, near the ground. The plumes are very similar to those in Fig. 7 and indicate that the explicit consideration of the influence of buildings on the spatial distribution of the releases is not significant outside of the center limits.

Figure 10 shows the average concentration field on the village S domain on November 1, 2018 between 18:00 and 19:00. Despite a pixilation effect linked to the Lagrangian model PSPRAY due to the very different volumes of the meshes in the 2 m and 20 m domains, low concentrations are observed in the wakes of the buildings. Consistently, the wakes extend downstream from the 2 m resolution domain into the 20 m resolution domain.

The respective contributions of the facilities to the annual average concentration were estimated at each point in the domain (results not presented in the article).

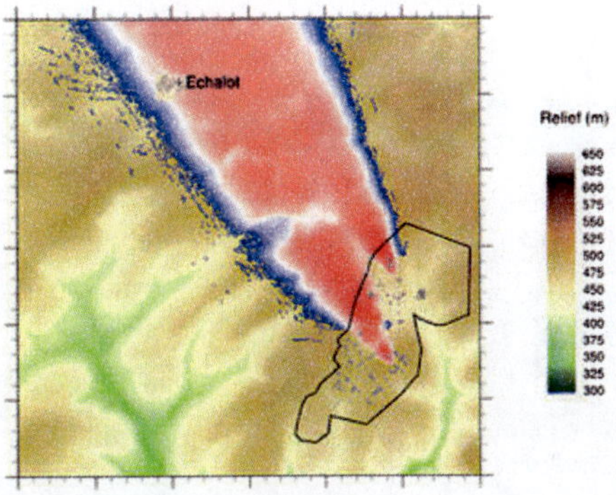

Fig. 9 Concentration fields at a 3 m, on 01/11/2018 at 9:00 and 19:00, following the releases from all the installations

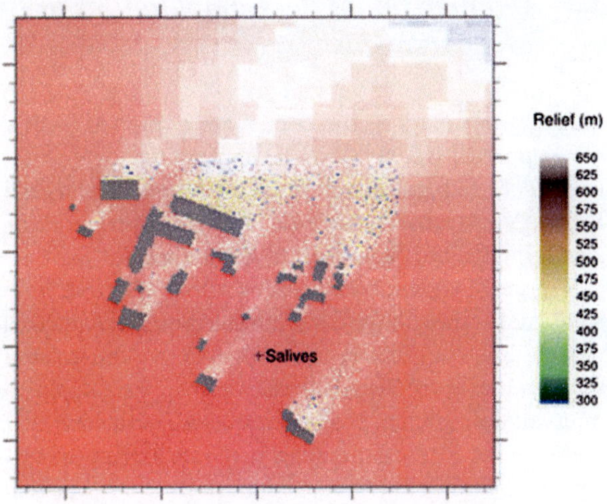

Fig. 10 Zoom on the village S of the concentration field, on 01/11/2018 at 19:00

We show that, for a given facility, the contributions are very disparate spatially (for example, to the south and north of the domain). In addition, depending on the point, the contributions of the facilities to the total concentration vary significantly and are not always in proportion to their respective source terms. This is felt for several kilometers from the center and explained by the different heights of the stacks and kinetics of the emissions linked to the activities of the facilities.

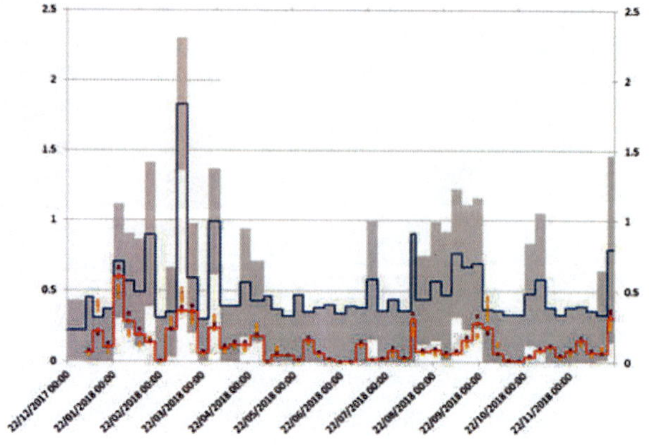

Fig. 11 Comparison of measurements at village S with the concentrations calculated by PSPRAY

3.3 Comparison of the Computations to Environmental Measurements

Villages surrounding the center are equipped with continuous sampling stations, which provide weekly measurements of the local atmospheric volume concentrations (after deferred analyses). These measurements as the concentrations computed by PSPRAY over the same time intervals show a strong spatial variability even by averaging the concentrations over one week.

Figure 11 shows, by way of example, the time series of the volume concentration measured at the village S station (blue curve) and computed by PSPRAY at the point of the same coordinates and at 3 m above the ground (red curve). Simulations and measurements are in agreement when the calculated concentrations are located in the intervals delimited by the uncertainty associated with the measurements (shaded gray area). Of the 51 measurements taken in 2018 at the village S station, 21 are above the detection limit. SPRAY concentrations are lower than the measurements. Still, for 10 of the 21 measures, they are within the confidence interval of the measurements.

Figure 12 shows an example of hourly evolution of the concentration computed by PSPRAY for the week of January 22 to 29, 2018. Take note that the average is close to the measured value.

4 Conclusion and Perspectives

This paper sums up the regional and local simulations on an hourly basis throughout 2018 of the wind flow and dispersion of gaseous releases emitted by the facilities at a research center located in a complex environment. The simulations are performed using a 3D digital mock-up and PMSS modelling system, explicitly accounting for the topography, the land use and the buildings both in the center and in the neighboring villages where environmental monitoring stations are set up. They are carried out at

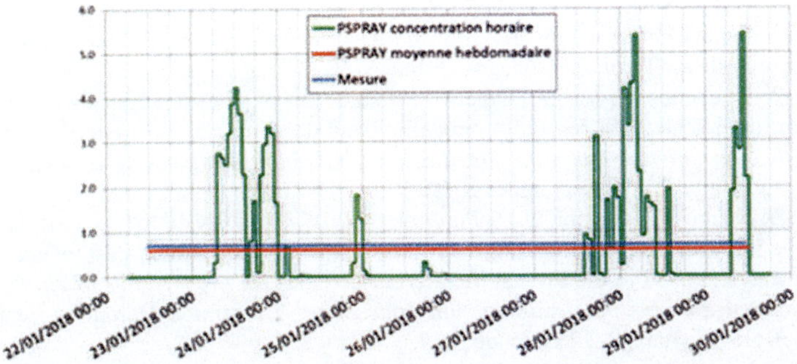

Fig. 12 Hourly evolution of the concentration computed at village S for a weekly period

spatial resolutions of 20 m in the regional domain of 13 × 14 km and 2 m in sub-domains encompassing the buildings. The simulations bring out the main following points:

- At the regional scale, the hourly as the annual concentrations clearly reflect the orographic effects. At the local scale, plumes sheltered areas in the buildings wakes are highlighted.
- Dispersion simulations cross-referenced with accessibility criteria make possible to identify areas favorable to detection and possibly decide on revising the sampling positioning.
- The comparison of the simulated concentrations against measurements over weekly periods is variously satisfactory depending on the stations. Computations are weaker than samples, may be due to the not precise enough knowledge of the weekly source terms.
- The hourly as the annual concentrations are very low all over the simulation domain. Thus, we check that the releases have no health consequences on the population around the center and lead to a minimal marking of the environment.

Ultimately, the simulations reveal the significant influence of the relief, land use and buildings on the flow and, subsequently, on the dispersion of releases into the atmosphere. Therefore, the use of 3D models should not be considered as superfluous, but they should be preferred when it comes to assessing the impact of gaseous or particulate releases on health and the environment.

References

Armand, P., et al. (2015). Best practice guidelines for the use of atmospheric dispersion models in emergency response tools at local-scale in case of hazmat releases into the air. *COST Action ES1006*.

Armand, P., & Duchenne, C. (2021). A new methodology based on 3D multi-scale modelling to assess the impact on health and the environment of industrial facilities. In *Proceedings of the*

20th International Conference on Harmonisation within Atmospheric Dispersion Modelling for Regulatory Purposes. Tartu, Estonia, 14–18 June 2021.
Armand, P., & Duchenne, C. (2019). Urgency management of adverse atmospheric releases with a forefront multi-scale high resolution modelling and decision-support system. In *Proceedings of the 37th International Technical Meeting on Air Pollution Modelling and its Application*. Hamburg, Germany, 23–27 September 2019
Armand, P., Duchenne, C., Oldrini, O., & Perdriel, S. (2017). EMERGENCIES Mediterranean—A prospective high-resolution modelling and decision-support system in case of adverse atmospheric releases applied to the French Mediterranean coast. In *Proceedings of the 18th International Conference on Harmonisation within Atmospheric Dispersion Modelling for Regulatory Purposes*. Bologna, Italy, 9–12 October 2017.
Armand, P., Oldrini, O., Duchenne, C., & Perdriel, S. (2021). Topical 3D modelling and simulation of air dispersion hazards as a new paradigm to support emergency preparedness and response. *Environmental Modelling and Software, 143*, 105129.
Baumann Stanzer, K., et al. (2015). Model evaluation case studies: Approach and results. *COST Action ES1006*, May 2015.
Oldrini, O., & Armand, P. (2019). Validation and sensitivity study of the PMSS modelling system for puff releases in the Joint Urban 2003 field experiment. *Boundary Layer Meteorology, 171*(3), 513–535.
Oldrini, O., Armand, P., Duchenne, C., Olry, C., & Tinarelli, G. (2017). Description and preliminary validation of the PMSS fast response parallel atmospheric flow and dispersion solver in complex built-up areas. *Journal of Environmental Fluid Mechanics, 17*(3), 1–18.
Oldrini, O., Armand, P., Duchenne, C., & Perdriel, S. (2019). Parallelization performances of PMSS flow and dispersion modelling system over a huge urban area. *Atmosphere, 10*(7), 404.
Oldrini, O., Armand, P., Duchenne, C., Perdriel, S., & Nibart, M. (2021a). Accelerated time and high-resolution 3D modeling of the flow and dispersion of noxious substances over a gigantic urban area—The EMERGENCIES project. *Atmosphere, 12*(5), 640.
Oldrini, O., Perdriel, S., Armand, P., & Duchenne, C. (2021b). High speed visualization of very large high-resolution simulations for air hazard transport and dispersion. *Atmosphere, 12*(7), 920.
Oldrini, O., Olry, C., Moussafir, J., Armand, P., & Duchenne, C. (2011). Development of PMSS, the parallel version of micro SWIFT SPRAY. In *Proceedings of the 14th International Conference on Harmonisation within Atmospheric Dispersion Modelling for Regulatory Purposes* (pp. 443–447). Kos, Greece, 2–6 October 2011.
Tinarelli, G., Mortarini, L., Trini-Castelli, S., Carlino, G., Moussafir, J., Olry, C., Armand, P. & Anfossi, D. (2013). Review and validation of Micro-Spray, a Lagrangian particle model of turbulent dispersion. In *Lagrangian modeling of the atmosphere* (Vol. 200, pp. 311–327); Geophysics. Monograph. American Geophysical Union.
Trini Castelli, S., et al. (2016). Evaluation of local-scale models for accidental releases in built environments—Results of the modelling exercises in COST Action ES1006. In *Air pollution modeling and its application XXIV* (pp. 497–502). Springer International Publishing Switzerland.
Trini Castelli, S., Armand, P., Tinarelli, G., Duchenne, C., & Nibart, M. (2018). Validation of a Lagrangian particle dispersion model with wind tunnel and field experiments in urban environment. *Journal of Atmospheric Environment, 193*, 273–289.

Questions and Answers

Questioner: Oriol Jorba Casellas

Question: Have you explored the use of GPUs to speed up your code considering the large computational demand of such an application at very high resolution?

Answer: To date, PMSS modeling system benefits from an effective parallelization that is both Eulerian, in timeframes and decomposition of the simulation domain, and Lagrangian, in numerical particles distributed to CPU computing cores. The system is relatively optimal for CPU cores. Using GPUs requires above all to study which parts of the code could benefit from them. It will also be necessary to identify if it is necessary to modify the code in depth or if external commands may be suitable. In any case, it is certain that the use of GPUs is envisaged to bring a gain in speed to PMSS simulations, as seems possible for other models of meteorology and atmospheric dispersion. This is a research question for years to come.

Calibration and Application of the Integrated Assessment Tool RIAT+ for Air Quality Planning in the Po Valley

Giovanni Bonafè, Roberta Amorati, Stefano Bande, Fabrizio Ferrari, Giuseppe Maffeis, and Michele Stortini

Abstract Aim of this work is to explore the uncertainty generated by using different source-receptor (S/R) functions in the air quality (AQ) planning, in terms of selecting the most cost-effective measures to improve AQ and reduce population exposure. Both technical and non-technical measures have been evaluated through the regional integrated assessment tool RIAT+. Applications of RIAT+ provides valuable information for drafting the local AQ plans. In Friuli-Venezia Giulia in order to reduce the population exposure to PM10 exceedances, priority should be given to measures concerning domestic heating, livestock, maritime and road traffic. Moreover, RIAT+ has been applied in the Po Valley to assess the impact on AQ of the COVID-19 lockdown.

G. Bonafè (✉)
ARPA-FVG, via Cairoli 14, 33057 Palmanova, Italy
e-mail: giovanni.bonafe@arpa.fvg.it

R. Amorati · M. Stortini
ARPAE, viale Silvani 6, 40122, Bologna, Italy
e-mail: ramorati@arpae.it

M. Stortini
e-mail: mstortini@arpae.it

S. Bande
ARPA Piemonte, via Pio VII 9, 10135 Torino, Italy
e-mail: stefano.bande@arpa.piemonte.it

F. Ferrari · G. Maffeis
TerrAria srl, via Melchiorre Gioia, 132, 20125 Milano, Italy
e-mail: f.ferrari@terraria.com

G. Maffeis
e-mail: g.maffeis@terraria.com

C. Mensink and O. Jorba (eds.), *Air Pollution Modeling and its Application XXVIII*, Springer Proceedings in Complexity, https://doi.org/10.1007/978-3-031-12786-1_8

1 Introduction

In the Po Valley, atmospheric pollution is a serious problem (Guerreiro et al. 2014). Relevant portions of the population are exposed to PM10 concentrations above the limits, in particular to more than 35 daily exceedances per year. The intervention policies for the mitigation of this problem require harmonized tools, common to all the regions involved. In the framework of the LIFE-PREPAIR project (Raffaelli et al. 2020), the RIAT+ tool (Carnevale et al. 2012) has been improved and calibrated to support local and regional planning in northern Italy, to identify the most effective measures.

2 Methods

RIAT+ is a free flexible integrated modelling environment that allows implementing both the scenario analysis and the optimization approaches, by means of S/R functions. Such functions approximate the response of a CTM to emission reductions, with dramatically shorter calculation times, thus allowing the quick analysis of different intervention options. In this work, the S/R functions are defined according to the SHERPA methodology (Thunis et al. 2016).

Two sets of S/R functions have been calibrated (Stortini et al. 2021) fitting the output of two CTMs, in order to approximate their response—in terms of PM10 and NO_2 annual averages—to variations in the precursors' emissions. Details of their configuration and calibration are described in Table 1.

3 Results

In order to verify whether the calibrated S/R functions are able to approximate the CTMs for annual averages estimates, 12 scenarios—independent from those used in the training phase—have been analyzed. Of these, 4 apply reductions over Lombardy and Emilia-Romagna, 4 over Piedmont and Veneto and 4 on some urban areas. Results of the validation were very satisfactory, as shown in Fig. 1.

Impacts of COVID-19 lockdown on air quality

RIAT+ has been used to estimate the impact on AQ of the epidemic containment measures implemented in Italy in spring 2020, as if they were applied homogeneously all year long. Five scenarios have been analyzed:

- three with PM and NO_x emissions reductions equal to those determined by the lockdown (*LDmin*, *LDmed*, *LDmax*, respectively corresponding to the minimum, average and maximum estimated reduction), changing emissions of road transport, home heating, airports, energy production and industrial activities;

Table 1 Setup of the two CTMs for RIAT+ calibration over the Po Valley

Element or feature	Value
CTM	*ninfa_er*: CHIMERE2017; *farm_pi*: FARM v4.13
Domain	Northern Italy
Resolution	dx=0.07°, dy=0.05°
Levels	*ninfa_er*: 9; *farm_pi*: 16
Meteorological input	COSMO-I7 (2018)
Boundary condition	CHIMERE (PrevAir, 2018)
Emissions	PREPAIR (2013 and 2025) Marongiu et al. (2019), ISPRA, TNO-MACC III
Training scenarios	5 scenarios with 50% reduction of a single precursor,
	1 scenario with 50% reduction of all precursors,
	7 scenarios with various combinations of base and MFR
Validation scenarios	8 regional scenarios, 4 local scenarios

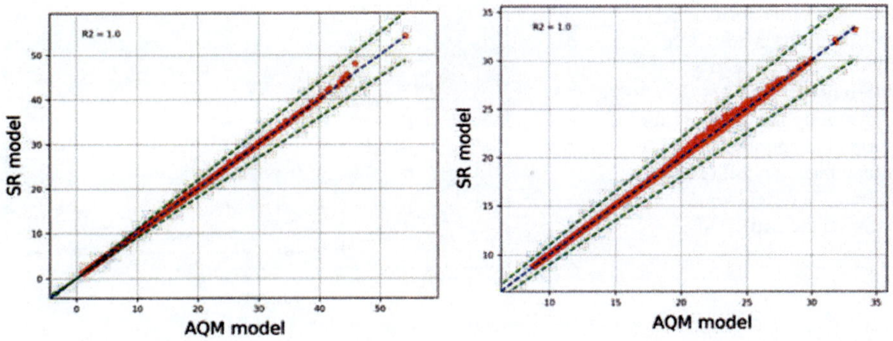

Fig. 1 Comparison of the PM10 concentration ($\mu g/m^3$) calculated by RIAT+ (y axis) with the CTMs' direct output (x axis): *farm_pi* scenario with local reductions (left) and *ninfa_er* scenario with reduction in Piedmont and Veneto (right). Each point represent a grid cell of the calculation domain; dashed lines represent linear fit with 90% confidence interval

- one (*LDmax+agr*) additionally including ammonia emission reduction in agriculture as planned for 2025;
- one (*plan2025*) considering the full application of current legislation, regional plans and additional planned commitments.

The three lockdown scenarios (Fig. 2) would lead to reductions in NO_2 concentrations ranging from −15 to −35% and PM10 reductions between −2 and −10%. With the *LDmax+agr* scenario a further decrease in PM10 concentrations is obtained, with higher relative reductions between the Piedmont plain, the Emilian plain, the eastern Lombard plain and the Venetian one (from −8 to −13%).

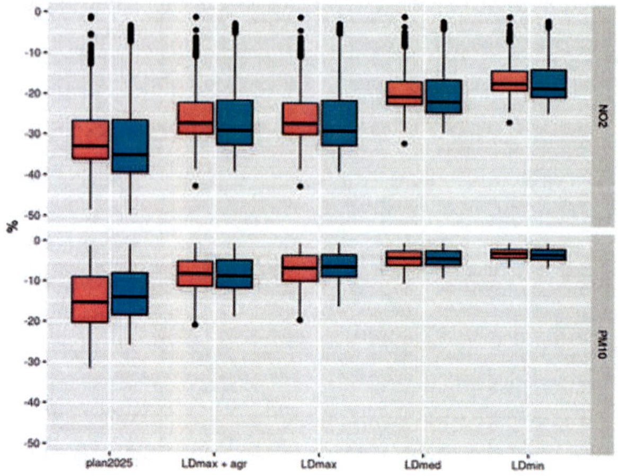

Fig. 2 Percentage reductions in the annual average concentrations of NO_2 (above) and PM10 (below) in the Po valley, obtained for the five scenarios analyzed with the two different RIAT+ calibrations (in each scenario *farm_pi* on the left, *ninfa_er* on the right)

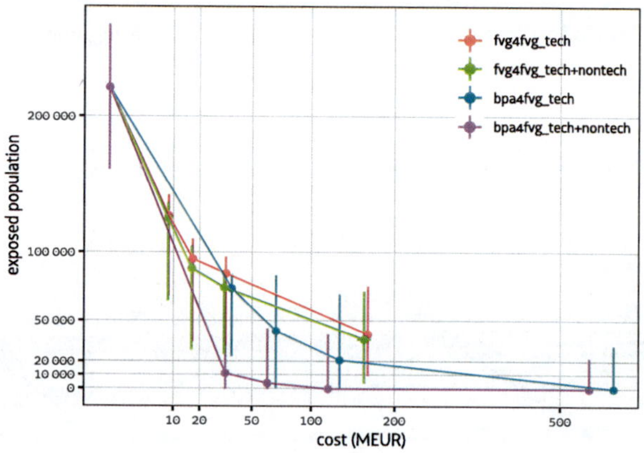

Fig. 3 Population exposed to more than 35 daily exceedances of PM10 in FVG, as a function of the total cost of the measures optimized for 2025. Each curve corresponds to a different strategy. Uncertainty bars account for (large) meteorological interannual variability and for (small) differences between the two CTMs

Support to regional AQ planning

The RIAT+ tool has been applied to the Friuli-Venezia Giulia (FVG) region in north-eastern Italy for AQ planning, for 2025. Some areas of FVG, close to Veneto region, experience more than 35 daily exceedances of PM10. In 2025, if additional local measures are not implemented, about 200,000 people will be exposed to concentrations above the limit. RIAT+ was used to identify the optimal set of measures to reduce the population-averaged annual mean PM10 concentration. Four different strategies have been analyzed: interventions implemented only on FVG (*fvg4fvg* in Figs. 3 and 4) or in the whole Po Valley (*bpa4fvg* in Fig. 3), with technical measures only (*tech*) or both with technical and non-technical measures (*tech+nontech*). For each strategy, the analyzes were conducted with the S/R functions calibrated by each of the two models. In order to reduce the impact of the models' bias, and to assess the interannual variability due to the varying meteorological conditions, the concentration estimated by RIAT+ has been corrected with a data fusion approach.

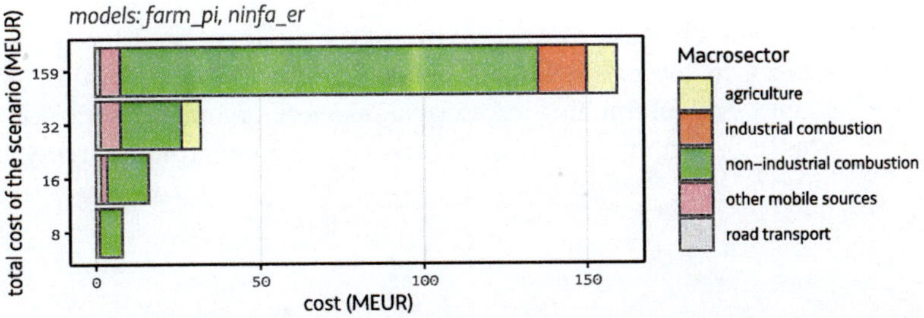

Fig. 4 Synthetic representation of the budget allocation as optimized by RIAT+ for year 2025 (target: PM10 over FVG region; intervention domain: FVG; measures: technical and non-technical)

The following results are obtained (Fig. 3): even applying the maximum feasible reductions (MFR), with a huge budget, does not completely avoid population exposure in FVG, if measures are limited to the FVG intervention area; nevertheless, with a budget of about 10 MEUR population exposure is halved; in addition, if the neighboring region Veneto is included in the intervention area, 30–50 MEUR may be enough to achieve the goal. RIAT+ identifies the most effective interventions (Fig. 4), in terms of costs and benefits, in wood-fired heating, livestock, maritime and road transport. Measures in the industrial sector have unfavorable cost/benefit ratios.

Acknowledgements This study was developed under the project LIFE-IP PREPAIR (Po Regions Engaged to Policies of AIR), which was co-funded by the European Union LIFE Program, Grant Number LIFE15 IPE/IT/000013. A special thanks is due to all the technicians of the environmental agencies who collaborated in the preparation of the emission dataset, in particular to ARPA Lombardia for the coordination. The update of RIAT+ was possible thanks to the database of the GAINS-Italy model D'Elia et al. (2018); Piersanti et al. (2021), provided by ENEA.

Questions and Answers

Question: One of the key emission sources identified by your system contributing most to air pollution in Po Valley is residential heating, what are the plans/actions in the region to replace current technology with more clean systems?

Answer: Through their air quality plans, the Italian regions included in the Po Valley are adopting a common system for defining the emission standards of wood combustion plants. Minimum standards are already adopted by some regions and mandatory for new plants. In some areas and in some critical periods there is a ban on the use of wood and derivatives for domestic heating. When critical levels of concentration of particulate matter in the atmosphere are exceeded, citizens are recommended to lower the temperature of their homes. Residential heating, and specifically wood burning stoves, are one of the pillars of the LIFE-PREPAIR project, involving the Italian regions included in the Po Valley. In this framework communication campaigns were organized to raise awareness of the importance of good stove management for air quality and to inform about best practices.

References

Carnevale, C., Finzi, G., Pisoni, E., Volta, M., Guariso, G., Gianfreda, R., Maffeis, G., Thunis, P., White, L., & Triacchini, G. (2012). An integrated assessment tool to define effective air quality policies at regional scale. *Environmental Modelling & Software, 38*, 306–315.

D'Elia, I., Piersanti, A., Briganti, G., Cappelletti, A., Ciancarella, L., & Peschi, E. (2018). Evaluation of mitigation measures for air quality in Italy in 2020 and 2030. *Atmospheric Pollution Research, 9*(6), 977–988.

Guerreiro, C. B., Foltescu, V., & De Leeuw, F. (2014). Air quality status and trends in Europe. *Atmospheric Environment, 98*, 376–384.

Marongiu, A., Angelino, E., Fossati, G., Moretti, M., Pantaleo, A., & Peroni, E. (2019). *Action A.1 Emissions data set*. Final report (Tech. Rep.). ARPA Lombardia.

Piersanti, A., D'Elia, I., Gualtieri, M., Briganti, G., Cappelletti, A., Zanini, G., & Ciancarella, L. (2021). The Italian national air pollution control programme: Air quality, health impact and cost assessment. *Atmosphere, 12*(2), 196.

Raffaelli, K., Deserti, M., Stortini, M., Amorati, R., Vasconi, M., & Giovannini, G. (2020). Improving air quality in the Po Valley, Italy: Some results by the LIFE-IP-PREPAIR project. *Atmosphere, 11*(4), 429.

Stortini, M., Amorati, R., & Bande, S. (2021). *Action C.3 implementing the integrated assessment model RIAT+*. Final report (Tech. Rep.).

Thunis, P., Degraeuwe, B., Pisoni, E., Ferrari, F., & Clappier, A. (2016). On the design and assessment of regional air quality plans: The SHERPA approach. *Journal of Environmental Management, 183*, 952–958.

Effect of Atmospheric Stability on Modeling Air Quality in and Around a Major Airport

Gavendra Pandey, Chowdhury Moniruzzaman, Akula Venkatram, and Saravanan Arunachalam

Abstract The impact of aircraft emissions on air quality is a major public health problem for communities near airports. It's difficult to quantify aircraft emissions and accurately model the air quality impacts of these in dispersion models. Airports located near a coast face additional challenges since meteorological conditions are not spatially homogeneous in and around the airport. The changes proposed in this study incorporates the features that define overall dispersion of aircraft emissions near a coastal region. To evaluate comparisons of these features with the AERMOD dispersion model, the SO_2 concentration measurements at a single site (CS) from the Los Angeles Air Quality Source Apportionment Study (LAX AQSAS) conducted during February 2012 have been utilized. As part of the LAX AQSAS, high frequency observations of NO_x, SO_2 and several other pollutants were collected at four core (AQ, CN, CS, and CE) locations during the study period at the Los Angeles airport. The qualitative performance of AERMOD incorporating atmospheric stability and roughness length changes is improved in terms of explaining the hourly diurnal pattern of observed concentration distribution at site CS. In addition, the quantitative analysis based on the fraction of model predictions within a factor of two of the observations (FAC2) improved from 0 to 27% and the ratio of medians of the observed to the predicted concentrations dropped from 4.2 to 2.8 at site CS.

Keywords AERMOD · Air quality · Atmospheric stability · LAX · Airport

G. Pandey · C. Moniruzzaman · S. Arunachalam (✉)
Institute for the Environment, University of North Carolina at Chapel Hill, Chapel Hill, NC, USA
e-mail: sarav@email.unc.edu

G. Pandey
e-mail: gavendra@email.unc.edu

A. Venkatram
University of California at Riverside, Riverside, CA, USA

C. Mensink and O. Jorba (eds.), *Air Pollution Modeling and its Application XXVIII*, Springer Proceedings in Complexity, https://doi.org/10.1007/978-3-031-12786-1_9

1 Introduction

Airports have several complex emission sources due to various aircraft-related activities and other non-aircraft sources such as traffic emissions. Various studies have explored impacts of aircraft and airport-related emissions on ambient air quality in and around the airports (Carslaw et al., 2008; Masiol & Harrison, 2014). The degradation of air quality in the vicinity of airports is indeed a threat to public health. It is necessary to determine the potential impacts of these airport-related emissions on the air quality-related health of nearby population. The dispersion modeling of aircraft emissions is a very challenging problem due to the transient nature of the aircraft emissions especially from the landing and takeoff modes and it becomes more difficult when the airport is situated in a coastal region due to heterogeneous meteorological conditions.

1.1 Methodology

The meteorological conditions of an airport situated near by a coastal region are never truly unstable or highly stable due to the cool wind from the ocean's water body over the airport region. Most Gaussian dispersion model's meteorological preprocessors like AERMET for AERMOD do not account for significant boundary layer features that occur along the coastlines (Yerramilli et al., 2009). The meteorological outputs from AERMET have been changed to account for these heterogeneous boundary layer features. A sensitivity analysis based on the following changes has been done.

- To account for the heterogeneous boundary layer features when the cold wind passes over the LAX airport region, all stable and unstable conditions in AERMET's output file are replaced by neutral conditions. For this, the Monin–Obukhov length is taken as 1000 m, and the friction velocity is calculated using the simplified neutral formulation (Venkatram & Schulte, 2018)

$$u_* = k\frac{U_r}{ln\left(\frac{z_r}{z_0}\right)}, \tag{1}$$

  where k is the von-Karman constant, U_r is the wind speed at z_r (reference height), and z_0 is the roughness length.
- When winds blew from the northeast quadrant, roughness length changed to indicate the flow going over the Los Angeles urban core having tall buildings.

AERMOD (v21112) dispersion model (Cimorelli et al., 2005) is used to simulate the SO_2 concentrations.

1.2 Description of LAX AQSAS

Los Angeles International Airport is one of the busiest airports in the USA which is located within the South Coast Air Basin of California. For illustration of model results, we have taken the SO_2 concentration measurements for February from the Los Angeles Source Apportionment Study (LAX AQSAS). This study was conducted at LAX in two different seasons (winter (01/31/2012 to 03/13/2012) and summer (07/18/2012 to 08/28/2012)) of 2012 for a total of 84 days. Several pollutants such as SO_2, NO_x, $PM_{2.5}$, CO, and BC were extensively measured especially at four core sites, the “Air Quality (AQ)”, “Community North (CN)”, “Community South (CS)”, and “Community East (CE)” during this campaign. There are four main runways: two each that are North and South of terminal area of the LAX airport (Fig. 1a) (Arunachalam et al., 2017, ACRP Report 179). Here, in this study, we have used the measurements of CS site only which was close to the South airfield and was largely affected by the aircraft operations of South runways as well as the urban core of the Los Angeles city. During February 2012, winds were westerly in most of the hours (Fig. 1b).

2 Results

AERMOD predicted results are presented as (1) diurnal concentrations averaged over a month, and (2) overall concentration analysis (Quantile–Quantile (QQ) plot) at site CS in step-by-step manner for three sets of meteorological inputs (a) original AERMET output (original meteorology), (b) updated AERMET output based on the stability (neutral meteorology) and (c) neutral meteorology with changes in roughness length. In addition, we have also evaluated the overall AERMOD model

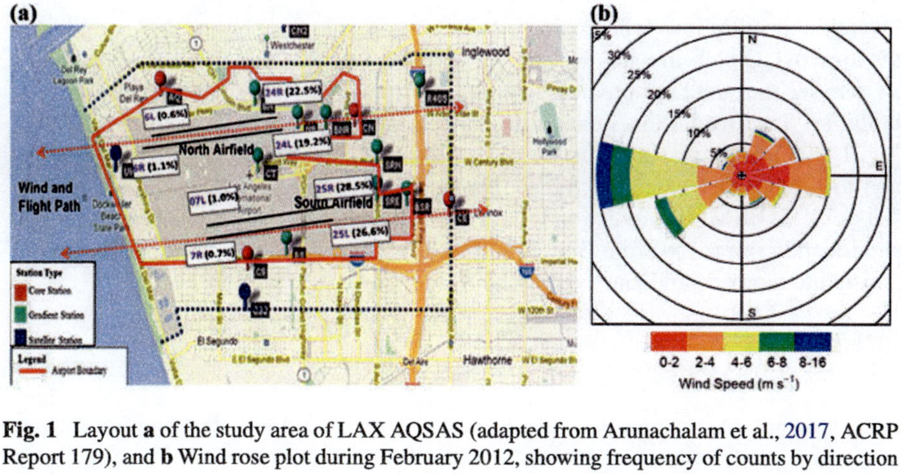

Fig. 1 Layout **a** of the study area of LAX AQSAS (adapted from Arunachalam et al., 2017, ACRP Report 179), and **b** Wind rose plot during February 2012, showing frequency of counts by direction

performance with the help of statistical measures such as Fractional Bias (FB) based on the top 26 robust highest concentrations (Cox & Tikvart, 1990) and factor of two to the observations (FAC2). The ideal values for these two statistical measures for best model performance are FB $= 0$ and FAC2 $= 100\%$. An over/under-estimate of the predicted concentration from the measurements is indicated by a negative/positive value of FB.

2.1 Diurnal Variation of Concentrations

The diurnal variation of concentrations averaged over the month of February at site CS is depicted in the form of diurnal line plots for both observed and AERMOD predicted concentrations (Fig. 2a–c). From Fig. 2a, we can clearly see that AERMOD is highly overpredicting the observed concentrations in morning and late evening with original meteorology. On the other hand, when we change all the stable and unstable conditions into neutral conditions, then concentrations are closer to the observed concentrations and slightly matching the diurnal characteristics during daytime (Fig. 2b). After changing roughness length, when the wind is blowing from Northeast direction, the high diurnal concentrations are reduced (Fig. 2c). We can further see that the morning and late evening peaks are now closer to observations.

In addition, the peak ratios (PR) of maximum predicted and observed concentrations (unpaired in time) are also shown on each figure (Fig. 2a–c). PR is 11.69 with original meteorology, but drops to 5.12 when using neutral meteorology, and further to 3.2 when using neutral meteorology and roughness length alteration.

2.2 Overall Concentration Analysis

Overall concentrations when using all three meteorological inputs are exhibited in the form of QQ plots in a step-by-step manner (Fig. 2d–f). From these three QQ plots, we can see that the high predicted concentrations values are closer to FAC2 $= 100$ or the 1:1 line. In addition, quantitative values of FB and FAC2 are also shown in the Fig. 2d–f. FB is decreasing from -1.62 to -1.19 from original to neutral meteorology and to -1.07 after changing the roughness length with neutral meteorology. The fraction of AERMOD prediction within a factor of two to the observation (FAC2) is significantly improved from 0 to 27%. However, slight over-prediction is still seen during the morning and late evening hours.

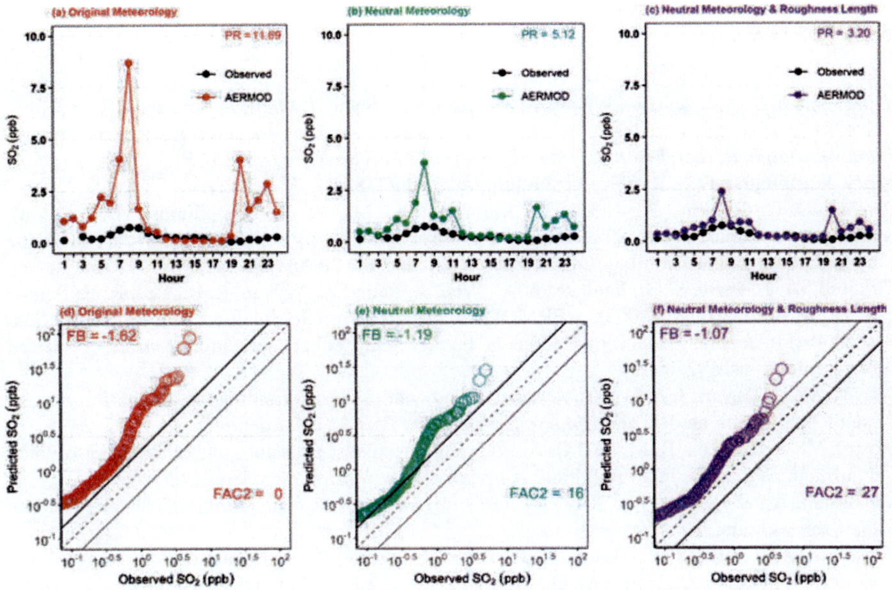

Fig. 2 **a–c** Diurnal variability in observed and AERMOD-predicted SO_2 concentrations, and **d–f** Quantile–Quantile concentration distribution of observed and AERMOD-predicted concentrations at site CS during February 2012. PR represents the peak ratio of maximum predicted to maximum observed concentration, FB is fractional bias, FAC2 = factor of two to the observations

3 Conclusion

In this study, AERMOD predicted concentrations using three different meteorological inputs are intercompared with the observations at CS site of the LAX Air Quality Source Apportionment Study (LAX-AQSAS) for February 2012. The qualitative and quantitative analysis suggest that AERMOD using the neutral meteorology with roughness length change predicts the observed concentrations significantly better when compared to using other inputs. We conclude that atmospheric stability matters to accurately assess the air quality in and around a major airport for regulatory applications, and airport modeling studies should pay attention to this issue especially around coastal regions.

Acknowledgements This research was funded by the U.S. Federal Aviation Administration Office of Environment and Energy through ASCENT, the FAA Center of Excellence for Alternative Jet Fuels and the Environment, project 19 through FAA Award Number 13-C-AJFE-UNC under the supervision of Jeetendra Upadhyay. ASCENT (Aviation Sustainability Center) (http://ascent.aero) is a US DOT-sponsored Center of Excellence.The authors gratefully acknowledge the Los Angeles World Airports (LAWA) for providing all data from the LAX AQSAS Study.

References

Arunachalam, S., Valencia, A., Woody, M., Snyder, M., Huang, J., Weil, J., Soucacos, P., & Webb, S. (2017). Dispersion modeling guidance for airports addressing local air quality concerns. In *Transportation Research Board Airport Cooperative Research Program (ACRP) Research Report 179*. Washington, D.C. Retrieved from http://nap.edu/24881

Carslaw, D. C., Ropkins, K., Laxen, D., Moorcroft, S., Marner, B., & Williams, M. L. (2008). Near-field commercial aircraft contribution to nitrogen oxides by engine aircraft type, and airline by individual plume sampling. *Environmental Science and Technology, 42*, 1871–1876.

Cimorelli, A. J., Perry, S. G., Venkatram, A., Weil, J., Paine, R., Wilson, R. B., Lee, R. F., Peters, E. D., & Brode, R. W. (2005). AERMOD: A dispersion model for industrial source applications. Part I: General model formulation and boundary layer characterization. *Journal of Applied Meteorology, 44*, 682–693.

Cox, W. M., & Tikvart, J. A. (1990). A statistical procedure for determining the best performing air quality simulation model. *Atmospheric Environment Part A General Topics, 24*, 2387–2395.

Masiol, M., & Harrison, R. M. (2014). Aircraft engine exhaust emissions and other airport-related contributions to ambient air pollution: A review. *Atmospheric Environment, 95*, 409–455.

Venkatram, A., & Schulte, N. (2018). *Urban transportation and air pollution*. Elsevier Science Publishing Co Inc.

Yerramilli, A., Challa, S. V., Dasari, H. P., Tuluri, F., White, L. D., Baham, J. M., Young, J. H., Hughes, R., Patrick, C., Hardy, M. G., & Swanier, S. J. (2009). Simulation of atmospheric dispersion of elevated releases from point sources in Mississippi gulf coast with different meteorological data. *International Journal of Environmental Research and Public Health, 6*, 1055–1074.

Modelling Street-Scale Resolution Air Quality for the West Midlands (UK) Using the ADMS-Urban RML System

Jian Zhong, Christina Hood, Kate Johnson, Jenny Stocker, Jonathan Handley, Mark Wolstencroft, Andrea Mazzeo, Xiaoming Cai, and William James Bloss

Abstract Air pollution is a major environmental concern in urban areas, causing substantial adverse effects on human health. This study simulated street-scale resolution air quality for the West Midlands region in the UK using a regional-to-local coupled system, which combined the regional CMAQ model with the local ADMS-Urban model, without double-counting emissions. CMAQ was used to represent dispersion on large temporal and spatial scales, while ADMS-Urban represents the local short-term dispersion from explicit point and road sources. Both models were evaluated against observations with reasonable performance, i.e. CMAQ captured measured air pollutant concentrations at background sites, while coupled ADMS-Urban RML (Regional Model Link) also captured air pollution concentrations at

J. Zhong (✉) · A. Mazzeo · X. Cai · W. J. Bloss
School of Geography, Earth & Environmental Sciences, University of Birmingham, Edgbaston, Birmingham B15 2TT, UK
e-mail: j.zhong.1@bham.ac.uk

A. Mazzeo
e-mail: a.mazzeo@bham.ac.uk

X. Cai
e-mail: x.cai@bham.ac.uk

W. J. Bloss
e-mail: w.j.bloss@bham.ac.uk

C. Hood · K. Johnson · J. Stocker · J. Handley
Cambridge Environmental Research Consultants, Cambridge CB2 1SJ, UK
e-mail: Christina.Hood@cerc.co.uk

K. Johnson
e-mail: Kate.Johnson@cerc.co.uk

J. Stocker
e-mail: Jenny.Stocker@cerc.co.uk

J. Handley
e-mail: jonathan.handley@cerc.co.uk

M. Wolstencroft
Birmingham City Council, Birmingham B5 5BD, UK
e-mail: mark.wolstencroft@birmingham.gov.uk

C. Mensink and O. Jorba (eds.), *Air Pollution Modeling and its Application XXVIII*, Springer Proceedings in Complexity, https://doi.org/10.1007/978-3-031-12786-1_10

roadside sites, where local effects were important. Street scale air quality maps were produced from the ADMS-Urban RML, which can be linked to health-related exposure studies. The coupled air quality modelling system for WM serves as an effective tool to evaluate potential regional and local air pollution mitigation policies.

Keywords Street-scale · Air quality · Regional-to-local coupled system · West Midlands · Road sources

1 Introduction

Air pollution is a major environmental concern in urban areas, causing adverse effects on human health. Air quality modelling is an important tool for the simulation of the combined effects of emissions, dispersion, physical and chemical processes in the atmosphere and for the quantification of air pollution concentrations at a variety of spatial and temporal scales (Hood et al., 2018). Air quality modelling can serve as an effective tool to evaluate potential air pollution mitigation policies. Owing to recent advances in High Performance Computing, air quality models have been widely used in many applications (Hood et al., 2018; Zhong et al., 2021). This study simulated street-scale resolution air quality for the West Midlands (WM) region (~902 km^2) in the UK using a multiscale air quality modelling system, i.e. the ADMS-Urban RML (Regional Model Link) system which couples the regional WRF-CMAQ (Community Multiscale Air Quality) model with the local ADMS-Urban model, without double-counting the local emissions.

2 Methodology

CMAQ model version 5.2.1 has been coupled with the meteorological WRF model version 3.9.1 for atmospheric chemistry simulations focused on the region of the West Midlands, UK. A system of 4 nested domains (Fig. 1a) has been set up to allow both models to simulate meteorological fields and air pollutant concentrations at spatial resolutions of 27, 9, 3 and 1 km. All domains account for 30 vertical levels above the ground, with 9 vertical levels below 1 km height and a surface layer depth of 20 m. Anthropogenic emissions for CMAQ simulations were processed from the National Atmospheric Emissions Inventory (NAEI, 2019) for the UK and from the CAMS-TNO emission inventory version 3.1 for north-west continental Europe (Granier et al., 2019). Both inventories' annual totals have been temporally distributed according to monthly, weekly, and hourly variations using time varying profiles (Simpson et al., 2012). The vertical distribution of anthropogenic emissions follows CAMS-TNO profiles. Biogenic emissions obtained from MEGAN model version 3.1 (Guenther et al., 2012) were also incorporated in CMAQ. The baseline year of simulation is 2016.

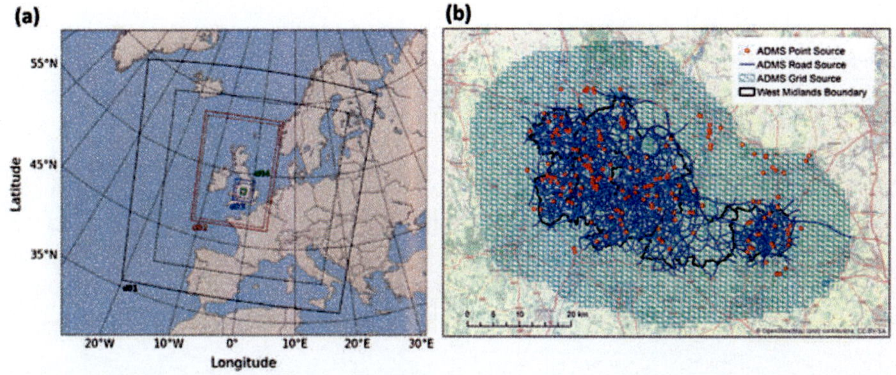

Fig. 1 **a** Domain setting in regional WRF (Solid lines) and CMAQ models. **b** Emission sources over West Midlands in the ADMS-Urban model

ADMS-Urban is a local scale air quality model which uses a quasi-Gaussian plume approach to represent the dispersion pattern of air pollution in the atmosphere. It can explicitly model full range of emission sources at high resolution and capture the effect of urban morphology on air pollution dispersion. The model can calculate street scale resolution air quality maps in urban areas. Modelling of the WM using ADMS-Urban as a stand-alone model with measured meteorological and background data was published in Zhong et al. (2021), local modelling in the coupled system was based on this configuration, with explicitly modelled emission sources shown in Fig. 1b.

The ADMS-Urban RML system (Hood et al., 2018) nested the local ADMS-Urban model within the regional CMAQ model over WM in this study. This coupled system combines both regional and local scale dispersion and chemical processes without double-counting local emissions, passing regional meteorology and background concentrations into local modelling for each regional model grid cell. This modelling system was installed and run in parallel on HPC clusters at the University of Birmingham.

3 Results and Discussion

Figure 2 plots CMAQ and ADMS-Urban RML model predictions against observations for annual mean NO_2 and $PM_{2.5}$. Both models had good performance in predicting NO_2 for airport and urban background sites. CMAQ had poor performance in capturing NO_2 at roadside sites, while the ADMS-Urban RML had the capability to capture street-scale NO_2 concentrations. For the limited number of $PM_{2.5}$ sites with available observation data in the WM region, both models underestimated measured $PM_{2.5}$ for both urban background and roadside sites. ADMS-Urban

RML predicted slightly higher $PM_{2.5}$ concentration for the single roadside site due to traffic emissions.

Figure 3 shows street scale annual average concentrations of NO_2 and $PM_{2.5}$ from the ADMS-Urban RML system. The dispersion of emissions from road sources was captured by the ADMS-Urban model, with higher concentrations predicted for both NO_2 and $PM_{2.5}$ near road sources, especially along motorways and busier major roads. Air quality maps are useful in identifying 'hot spots' where modelled air pollution concentrations are higher than the UK objective values. The street scale air quality maps can also be aggregated to health-related polygon layers such as Lower Layer Super Output Areas or Wards for the assessment of resulting health impacts.

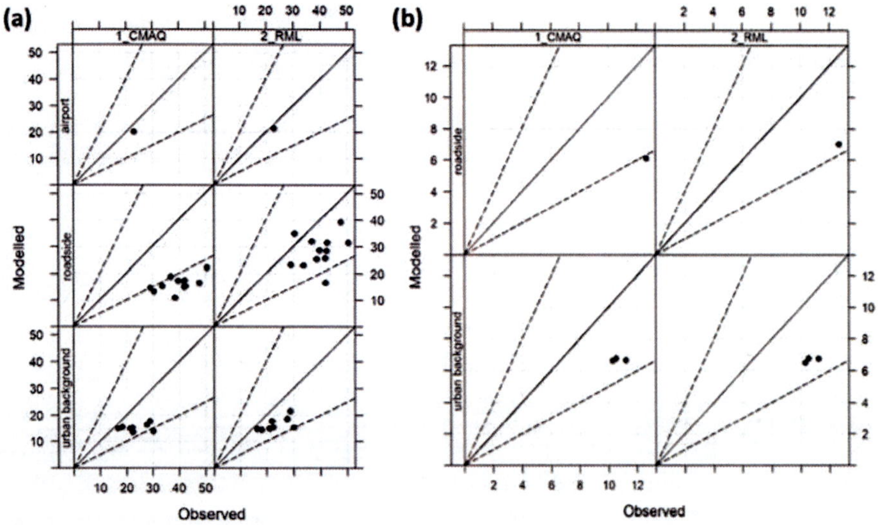

Fig. 2 Annual average concentrations from observations, CMAQ and ADMS-Urban RML models: **a** NO_2; **b** $PM_{2.5}$ (both in $\mu g\ m^{-3}$)

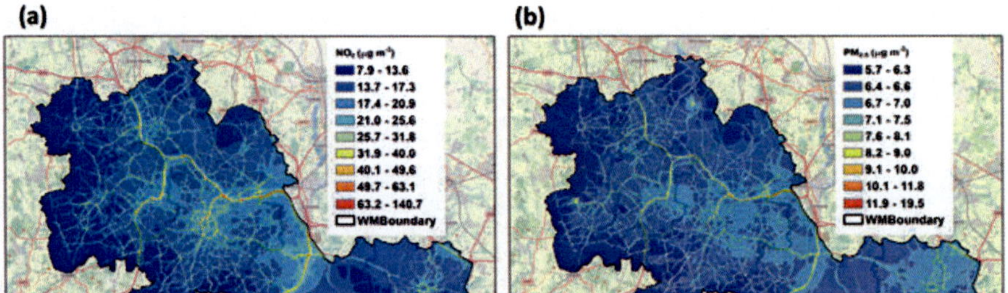

Fig. 3 Annual average concentrations of NO_2 and $PM_{2.5}$ at street scale (10 m × 10 m resolution) from the ADMS-Urban RML system

4 Conclusion

This study predicted air pollution concentrations at street-scale resolution for the WM using a regional-to-local coupled modelling system, which combined the regional CMAQ model with the local ADMS-Urban model without double-counting emissions. CMAQ was used to calculate dispersion on large spatial and temporal scales, while ADMS-Urban was used to represent the initial dispersion from explicit point and road sources. Both CMAQ and ADMS-Urban RML predictions were evaluated against observations with reasonable performance, i.e. CMAQ captured air pollution at the airport and urban background sites while the ADMS-Urban RML also captured air pollution at roadside sites. Street scale air quality maps from the ADMS-Urban RML were presented, which could be linked to health-related studies. The coupled air quality modelling system for WM in this study can be used as a baseline tool to assess the effects of specific interventions/policies within the region, such as clean air zones, lower traffic neighbourhoods, climate measures, traffic speed limit changes and adoption of electric vehicles etc., and to provide guidance to reduce air pollution exposure at population level, or for vulnerable sub-groups.

Acknowledgements This research was funded by the UK Natural Environment Research Council (NERC) project WM-Air, grant number NE/S003487/1.

References

Guenther, A. B., Jiang, X., Heald, C. L., Sakulyanontvittaya, T., Duhl, T., Emmons, L. K., & Wang, X. (2012). The model of emissions of gases and aerosols from nature version 2.1 (MEGAN2.1): An extended and updated framework for modeling biogenic emissions. *Geoscientific Model Developement, 5*, 1471–1492. https://doi.org/10.5194/gmd-5-1471-2012

Granier, C., Darras, S., Denier Van Der Gon, H., Jana, D., Elguindi, N., Bo, G., Michael, G., Marc, G., Jalkanen, J.-P., Kuenen, J., et al. (2019). *The Copernicus Atmosphere Monitoring Service global and regional emissions (April 2019 version)*; Copernicus Atmosphere Monitoring Service.

Hood, C., MacKenzie, I., Stocker, J., Johnson, K., Carruthers, D., Vieno, M., & Doherty, R. (2018). Air quality simulations for London using a coupled regional-to-local modelling system. *Atmospheric Chemistry and Physics, 18*, 11221–11245. https://doi.org/10.5194/acp-18-11221-2018

NAEI. (2019). Retrieved July 18, 2019, from http://naei.beis.gov.uk/data

Simpson, D., Benedictow, A., Berge, H., Bergström, R., Emberson, L. D., Fagerli, H., Flechard, C. R., Hayman, G. D., Gauss, M., Jonson, J. E., et al. (2012). The EMEP MSC-W chemical transport model—Technical description. *Atmospheric Chemistry and Physics, 12*, 7825–7865.

Zhong, J., Hood, C., Johnson, K., Stocker, J., Handley, J., Wolstencroft, M., Mazzeo, A., Cai, X., & Bloss, W. J. (2021). Using task farming to optimise a street-scale resolution air quality model of the West Midlands (UK). *Atmosphere, 12*, 983.

Questions and Answers

Question: How did the ADMS-Urban RML system avoid double-counting the local emissions?

Answer: The local ADMS-Urban model used concentrations from the regional CMAQ model as urban background conditions. These also included contributions from local emissions, so the concentration field due to initial dispersion from gridded local emissions in the ADMS-Urban model was removed before adding the contribution from the explicit street-scale local emissions.

Question: Can you explain $PM_{2.5}$ evaluation results from the RML system?

Answer: For urban background sites, the RML system predicted concentrations similar to those from the regional CMAQ model, as there was little influence from the explicit road sources. For the roadside site, RML predicted only slightly higher concentrations than the regional CMAQ model as the $PM_{2.5}$ contribution from long-range regional background remained dominant. There were a limited number of available sites for $PM_{2.5}$ within the region.

Question: How did the model capture the effect caused by buildings on air pollution dispersion?

Answer: The ADMS-Urban RML system has a street canyon module which calculates street canyon effects on air pollution dispersion from explicit road sources, and also an urban canopy module which represents the overall flow and turbulence effects caused by buildings within the grids.

Urban Air Pollution Reduction During the COVID Pandemia

Alba Badia, Veronica Vidal, Macià Mut Sbert, and Gara Villalba

Abstract Severe measures concerning the mobility of people have been implemented worldwide to contain the spread of COVID-19. This exceptional situation during COVID-19 can serve as a chemical experiment that expands our understanding of photochemical processes in the atmosphere. In this study, we analyze changes in the urban air quality during COVID lockdown restrictions, especially changes in O_3 concentrations due to drastic NO_2 reductions, in the Metropolitan Area of Barcelona. Model results are evaluated against ground monitoring stations available in this area.

Keywords Air pollution · Atmospheric modelling · Traffic reductions · WRF-Chem · COVID-19

A. Badia (✉) · V. Vidal · G. Villalba
Institute of Environmental Science and Technology (ICTA), Universitat Autònoma de Barcelona (UAB), Barcelona, Spain
e-mail: alba.badia@uab.cat

V. Vidal
e-mail: veronica.vidal@uab.cat

G. Villalba
e-mail: gara.villalba@uab.cat

V. Vidal
Departament d'Arquitectura de Computadors i Sistemes Operatius (CAOS), Escola d'Enginyeria, Universitat Autònoma de Barcelona (UAB), Barcelona, Spain

M. M. Sbert
Centre de Recerca Matemàtica, Faculty of Science, Universitat Autònoma de Barcelona (UAB), Barcelona, Spain
e-mail: macia.mut.sbert@gmail.com

G. Villalba
Department of Chemical, Biological and Environmental Engineering, Universitat Autònoma de Barcelona (UAB), Barcelona, Spain

C. Mensink and O. Jorba (eds.), *Air Pollution Modeling and its Application XXVIII*, Springer Proceedings in Complexity, https://doi.org/10.1007/978-3-031-12786-1_11

1 Introduction

The COVID-19 pandemic has spread rapidly around the world. To prevent the spread of the virus, drastic measures to limit mobility have been implemented on a global scale. Spain was one of the areas hardest hit by COVID-19 with mobility reductions reaching up to 84% (Google, 2020). Estimated averaged emissions reduction in Spain during the COVID-19 lockdown were reported with the road and air traffic reaching a drop of 80–90% (Guevara et al., 2021). Observations show nitrous oxide (NO_2) levels during the COVID-19 period were below the WHO and EU yearly average threshold, which had been exceeded at the same period in previous years in some measurement stations in Barcelona (Baldasano, 2020).

The goal of our study is to analyze changes in the air quality (NO_2 and ozone (O_3)) during this period in the Metropolitan Area of Barcelona (AMB). This study can help to quantify the pollution reduction during COVID lockdown restrictions and define future measures—for example promoting the reduction of people's mobility through teleworking (Badia et al., 2021) or applying emissions limitations—that could be implemented in the city of Barcelona to improve air quality.

2 Methodology

2.1 Study Case

The AMB is one of the largest cities in Europe with a population of more than 5 million people. Barcelona annually reports one of the highest air pollution levels Europe-wide, with the most problematic pollutants being PM2.5, PM10, and NO_2. Road traffic is the main source of NO_x (59%) while the seaport is the second largest emission source (16%) in the AMB (Schembari et al., 2014). For the model evaluation, we have selected the two weeks previous to the declaration of the emergency state in Spain (pre-covid). These two weeks together with the two weeks during the full lockdown period are analised (see Table 1).

Table 1 Different periods analised in this study

Week	Period
2–6 March	Pre-covid
9–13 March	Pre-covid
30 March–3 April	Full lockdown
6–10 April	Full lockdown

2.2 Model Description

In this study, we use the regional chemistry transport model WRF-Chem v4.1 (Grell et al., 2005). The model is defined with two domains covering the Iberian Peninsula (D1: 9 km × 9 km) and the region of Catalonia (D2: 3 km × 3 km) with 45 vertical layers up to 100 hPa. We are running two simulations for March-June 2020: (1) Busines As Usual (BAU) and 2) COVID. Here, we used a multi-layer layer urban canopy scheme, the Building Effect Parameterization (BEP) coupled with the Building Energy Model (BEP + BEM; Salamanca & Martilli, 2010) to represent the urban areas in our domain, that take into account the energy consumption of buildings and anthropogenic heat.

2.3 Data Sources and Speciation of Emissions

The meteorological and chemical initial and boundary conditions (IC/BCs) are from the ERA5 model (Hersbach et al., 2020) and WACCM (Marsh et al., 2013), respectively. The anthropogenic emissions files are from the CAMS-REG-APv3.1 database (Granier et al., 2019). To generate the emissions for the COVID run we use the emissions reduction factors provided by the Barcelona Supercomputing Center (Guevara et al., 2021).

3 Model Evaluation

BAU simulation has been compared with several urban traffic and urban background air quality monitoring stations from the network: Xarxa de Vigilància i Previsió de la Contaminació Atmosfèrica (XVPCA) for the pre-covid period (1–15 March 2020, see Table 1). Overall, the model shows a reasonable agreement with the observations during this period for NO_2 and O_3 concentrations, although specific periods are archiving significant negative and positive biases especially for the traffic stations for NO_2 and O_3, respectively (see Table 2). Part of the O_3 overestimation could be attributed to the boundary conditions used for this study (WACCM model). In addition, surface NOx concentrations are very sensitive to the emissions, therefore, model biases can be attributed to the low resolution in the emission inventory and our model runs.

Table 2 Model evaluation during the pre-covid period (1–15 March 2020)

Pollutant	Type (number of stations)	NMB (%)	RMSE (μg/m3)	R2 [0,1]
NO_2	Urban background (9)	−20	19.77	0.32
	Urban Traffic (2)	−35	24.46	0.26
O_3	Urban background (5)	−44	30.35	0.58
	Urban Traffic (2)	−60	34.28	0.5

4 Results

Here we present the modelled and observational data NO_2 and O_3 levels for the pre-covid and the full lockdown periods. For the full lockdown period, the two simulations (BAU and COVID) are presented. Only weekdays are considered in this study where the most significant changes are found. COVID-19 related restrictions during the lockdown lowered NO_2 concentrations (10–30%) compared with the pre-covid in AMB (Fig. 1) especially for the peaks concentrations from 45–57 $\mu g/m^3$ to 30–50 $\mu g/m^3$ for the urban stations. Despite the negative bias in modelled NO_2 concentrations, a clear reduction (25–40%) is also seen from 35–39 $\mu g/m^3$ to 20–28 $\mu g/m^3$ for the urban areas.

The highest reductions are seen for the traffic stations due to a significant drop in traffic emissions (e.g. NO). Consequently, NO reductions slow down O_3 consumption (titration of O_3, $NO + O_3 = NO_2 + O_2$) and cause an increase in O_3 concentrations. This increase is seen for observations where O_3 concentrations increase from 90 to 107 $\mu g/m^3$ in urban areas. Despite a positive bias for the O_3 levels (20 $\mu g/m^3$) during the pre-covid period, the model captures the daily pattern.

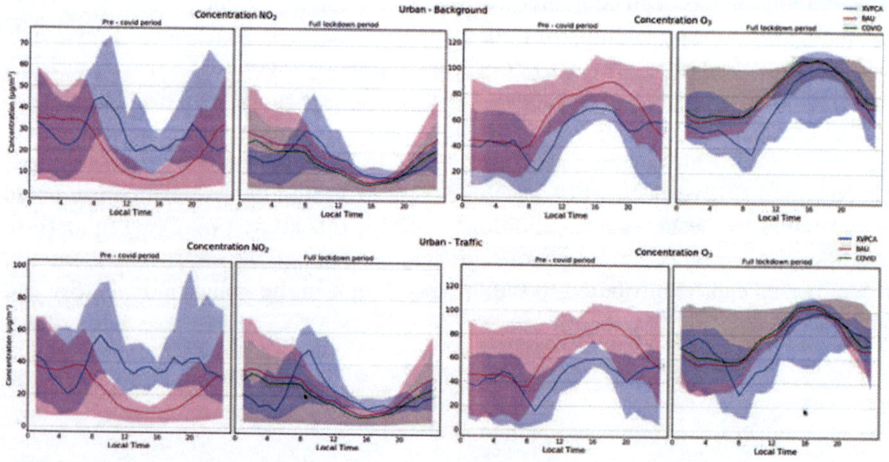

Fig. 1 Aggregated modelled and observational daily profiles of all the stations monitoring NO_2 and O_3 in the AMB

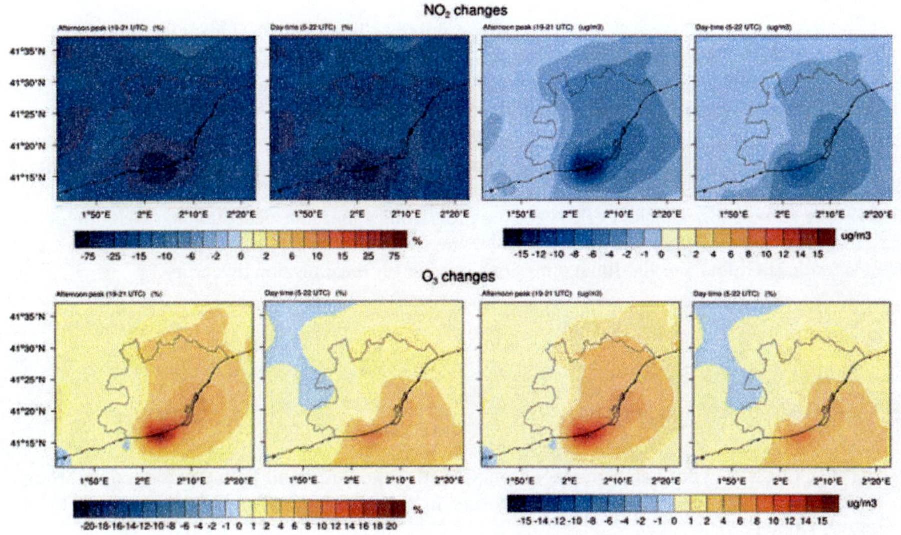

Fig. 2 NO_2 and O_3 changes for the BAU and COVID simulations for the lockdown period

A comparison between BAU and COVID simulations during the lockdown period illustrates significant changes in the concentration of NO_2, with a general decrease (up to -15 $\mu g/m^3$, -50%), and O_3, with a general increase (up to $+15$ $\mu g/m^3$, $+$ 20%), over the AMB (Fig. 2). Higher changes are seen around the airport area. Air traffic was drastically reduced (up to 90%) during the lockdown period (Guevara et al., 2021).

5 Conclusions

- Despite biases for the modelled NO_2 and O_3, the overall performance is captured by the model for the pre-covid and lockdown periods.
- Model uncertainties behind this biases are, (1) data for emissions inventory used here is for the year 2016, (2) chemical boundary conditions may affect long-living pollutants such as O_3, (3) the low resolution in our model simulations (3 km) can significantly underestimate primary pollutants, especially around the traffic areas.
- Model comparison during the lockdown period shows a general decrease of NO_2 (up to -50%) and an increase of O_3 (up to $+20\%$) levels between BAU and COVID simulations. Higher changes are seen around the airport area due to the drastic reduction of air traffic during the lockdown.
- Further steps include studying changes in the Barcelona urban plume which is transported inland by sea breezes to the Vic Plain, a rural area 40–65 km north of Barcelona, where every summer exceeded the O_3 limits.

Acknowledgements This work has been made possible thanks to the financial support of the ERC Consolidator Integrated System Analysis of Urban Vegetation and Agriculture (818002-URBAG) and the Spanish Ministry of Science, Innovation and Universities, through the "Maria de Maeztu" programme for Units of Excellence (CEX2019-000940-M). This research has been supported by MINECO-Spain (TIN2017-84553-C2-1-R), and by the Spanish government grant PRE2018-085425. The authors thankfully acknowledge the computer resources at PICASSO and XULA and the technical support provided by the Universidad de Málaga (RES-AECT-2020-2-0004). The authors also thank XVPCA for the provision of measurement stations. Also, thanks to the free use of HERMESv3_GR model and the developing team for their support. We also thank the Copernicus Global and Regional emissions service for the emission inventory.

References

Badia et al. (2021). A take-home message from COVID-19 on urban air pollution reduction through mobility limitations and teleworking. *npj Urban Sustainability*, *1*, 35. https://doi.org/10.1038/s42949-021-00037-7

Baldasano, J. M. (2020). COVID-19 lockdown effects on air quality by NO_2 in the cities of Barcelona and Madrid (Spain). *Science of the Total Environment*, *741*(2).

Google L. L. C. (2020). COVID-19 community mobility reports. https://www.google.com/covid19/mobility/

Granier, C., Darras, S., Denier van der Gon, H., Doubalova, J., Elguindi, N., Galle, B., Gauss, M., Guevara, M., Jalkanen, J.-P., Kuenen, J., Liousse, C., Quack, B., Simpson, D., & Sindelarova, K. (2019). The copernicus atmosphere monitoring service global and regional emissions (April 2019 version). Copernicus Atmosphere Monitoring Service (CAMS) report. https://doi.org/10.24380/d0bn-kx16

Grell, G. A., Peckham, S. E., Schmitz, R., McKeen, S. A., Frost, G., Skamarock, W. C., & Eder, B. (2005). Fully coupled "online" chemistry within the WRF model. *Atmospheric Environment*, *39*, 6957–6975.

Guevara, M., Jorba, O., Soret, A., Petetin, H., Bowdalo, D., Serradell, K., Tena, C., Denier van der Gon, H., Kuenen, J., Peuch, V.-H., & Pérez García-Pando, C. (2021). Time-resolved emission reductions for atmospheric chemistry modelling in Europe during the COVID-19 lockdowns. *Atmospheric Chemistry and Physics*, *21*, 773–797. https://doi.org/10.5194/acp-21-773-2021

Hersbach, H., Bell, B., Berrisford, P., et al. (2020). The ERA5 global reanalysis. *Quarterly Journal Royal Meteorological Society, 146*, 1999–2049. https://doi.org/10.1002/qj.3803

Marsh, D. R., Mills, M. J., Kinnison, D. E., Lamarque, J., Calvo, N., & Polvani, L. M. (2013). Climate change from 1850 to 2005 simulated in CESM1(WACCM). *Journal of Climate*, *26*(19), 7372–7391. Retrieved Feb 1, 2022, from https://journals.ametsoc.org/view/journals/clim/26/19/jcli-d-12-00558.1.xml

Salamanca, F., & Martilli, A. (2010). A new building energy model coupled with an urban canopy parameterization for urban climate simulations-part II. Validation with one dimension *off-line simulations. Theoretical and Applied Climatology*, *99*(3–4), 345–356.

Schembari, A., Nieuwenhuijsen, M. J., Salvador, J., de Nazelle, A., Cirach, M., Dadvand, P., Beelen, R., Hoek, G., Basagaña, X., & Vrijheid, M. (2014). Traffic-related air pollution and congenital anomalies in Barcelona. *Environmental Health Perspectives*, *122*, 317–323. https://doi.org/10.1289/ehp.1306802

Predicting Hourly Street-Scale NO_2 and $PM_{2.5}$ Concentrations Using Machine Learning at One of the Danish Traffic Hotspots

Jibran Khan, Thomas Ellermann, and Ole Hertel

Abstract Road transport is one of the significant sources of ambient air pollution. Machine learning approaches are becoming increasingly popular to predict and forecast air quality. In conjunction, this study aims to explore the potential of commonly used machine learning algorithms, Artificial Neural Networks (ANN), Random Forests (RF), and Support Vector Machines (SVM) to accurately predict hourly, street-scale NO_2 (ppb) and particulate matter ($PM_{2.5}$) ($\mu g/m^3$) at one of the Danish traffic hotspots, H. C. Andersens Boulevard (HCAB), in Copenhagen (2013–2015). Results suggest that the RF, overall, outperforms others [RF: RMSE = 11.2–13.5, R^2 = 0.73–0.81; ANN: RMSE = 13.5–14.1, R^2 = 0.69–0.75; SVM: RMSE = 14.7–15.1, R^2 = 0.64–0.65]. One strength of this study is that it addresses the research gaps concerning the unexplored potential of machine learning techniques for air quality prediction in Denmark. Future work will include more predictor variables (e.g. traffic attributes) and testing of other techniques (e.g. Extreme Gradient Boosting) (XGBoost).

Keywords Road traffic · Air pollution · Machine learning · Prediction exposure

1 Introduction

Air pollution continues to remain a significant environmental health risk factor. Both short-term and long-term exposure to the key air pollutants such as, Nitrogen dioxide (NO_2), particulate matter with an aerodynamic diameter of less than 10 μm (PM_{10}) and 2.5 μm ($PM_{2.5}$), has been associated with serious illnesses, e.g. cardiovascular

J. Khan (✉)
Department of Environmental Science, and Danish Big Data Centre for Environment and Health (BERTHA), Aarhus University, Roskilde, Denmark
e-mail: jibran@envs.au.dk

T. Ellermann
Department of Environmental Science, Aarhus University, Roskilde, Denmark

O. Hertel
Department of Ecoscience, Aarhus University, Roskilde, Denmark

C. Mensink and O. Jorba (eds.), *Air Pollution Modeling and its Application XXVIII*, Springer Proceedings in Complexity, https://doi.org/10.1007/978-3-031-12786-1_12

disease and respiratory infections (Meng et al., 2021; Tong et al., 2020). Therefore, assessing air pollution levels to study individuals' exposure is necessary.

Traditionally, physics and chemistry-based dispersion models are often used to estimate local and regional pollution levels. These models, however, often require a significant number of inputs and utilize sophisticated equations, making them computationally expensive and time-consuming (Elangasinghe et al., 2014). In addition, dispersion models include several assumptions and parametrizations, leading to exposure misclassifications (Tayarani & Rowangould, 2020).

Machine learning and data-driven approaches, on the other hand, apply several mathematical and statistical techniques to extract patterns and new insights from large data sets (Bergen et al., 2019). Furthermore, these approaches can model complex, non-linear relationships between observations and predictor variables via flexible and automated procedures (Reichstein et al., 2019). Predicting air pollution using machine learning is an emerging area of research; however, there are scant scientific articles on this subject in the Danish context.

Therefore, this paper explores the potential of commonly used machine learning algorithms to predict hourly NO_2 and $PM_{2.5}$ at one of the Danish traffic hotspots, H. C. Andersens Boulevard (HCAB) in Copenhagen.

2 Materials and Methods

2.1 Study Site and Data Collection

The study site, HCAB, is one of the busiest streets of the Danish Capital, having significant traffic round the clock (see https://bit.ly/3la2qf8 for HCAB details). The HCAB has a street air quality monitoring station, part of the Danish Air Quality Monitoring Programme (Ellermann et al., 2020). The street station has high quality, EU reference-grade equipment, measuring hourly air quality levels (NO, PM_{10}, $PM_{2.5}$, O_3, etc.). The quality assurance of the measured pollutants is regularly performed.

We acquired measured NO_2 and $PM_{2.5}$ (hourly resolution) data from the HCAB station from 2013 to 2015 ($N = 26{,}280$). Moreover, the hourly meteorological data, temperature, wind speed, wind direction, global radiation and relative humidity were also acquired. Furthermore, measured background pollution levels were also obtained from a background monitoring station located at the rooftop of the University of Copenhagen (hourly values, 2013–2015). Table 1 shows summary statistics of the measured data. A significant variation in street-level values of NO_2 (range = 1.8–119.9 ppb) and $PM_{2.5}$ (range = 2.3–71.3 $\mu g/m^3$) can clearly be observed.

Table 1 The summary statistics of the measured data

	Min	Q1	Med	Q3	Max	Var	SD	NAs
NO_2	1.8	30.8	47.9	67.0	119.9	634.9	25.1	415
NO_2 (B)	0.0	7.6	12.6	20.7	88.8	126.0	11.2	490
Temp	−6.3	5.3	10.0	15.2	30.1	42.6	6.5	0
WD	0.0	103.4	188.7	250.9	359.9	7893.0	88.9	791
WS	0.5	2.9	4.0	5.3	16.3	3.4	1.9	790
Glob Rad	0.03	0.06	2.9	146.3	913.7	38,041.5	195.1	790
Rel Hum	9.3	63.0	74.9	83.0	97.3	196.2	14.0	790
$PM_{2.5}$	2.3	10.5	14.7	21.0	71.3	89.9	9.5	11,184
$PM_{2.5}$ (B)	0.4	7.1	10.3	15.2	58.4	66.1	8.1	1656

Note: NO_2 (B) = Measured background NO_2 levels (ppb); Temp = Temperature (° C); WD = Wind direction (degrees); WS = Wind speed (m/s); Glob Rad = Global radiation (kWh/m^2); Rel Hum = Relative humidity (%), $PM_{2.5}$ (B) = Measured background $PM_{2.5}$ levels (μg/m^3); NO_2 = Measured street-level NO_2 (ppb); $PM_{2.5}$ = Measured street-level $PM_{2.5}$ (μg/m^3); NAs = Missing values; Q1 = 25th percentile; Q3 = 75th percentile, Var = Variance; SD = Standard deviation

2.2 Data Processing and Model Developments

In Table 1, meteorological parameters and background pollution levels are the predictor variables of respective street-level pollutants, NO_2 and $PM_{2.5}$, the 'target' variables. The pre-processing of the data is summarized as follows.

The data structure, e.g. variance, range, possible missing values, were analyzed (Table 1). Then, data imputation was performed to compensate for missing values, and potential outliers were removed. In addition, data transformation was carried out. Afterwards, machine learning models were developed and evaluated, summarized in the following sub-sections.

2.3 Model Validations and Evaluations

The first step of model development was to identify predictor and response variables. As stated above, meteorological parameters and background pollution levels were the predictors of street-level pollution. The data set was split into training and validation/test sets (75% training, 25% validation). Then, ANN, RF and SVM algorithms were used to develop models to predict local-scale air quality. These models were applied to predict hourly NO_2 and $PM_{2.5}$ levels at HCAB. In short, the different combinations of parameters commonly utilized in the literature were used to tune and optimize the model performance.

To evaluate model performance and accuracy, we used a combination of two frequently used statistical parameters, root-mean-squared error (RMSE) and the coefficient of determination (R^2), and compared model predictions with test/validation datasets (“unseen” data). All pre-processing, data exploration, model development, and evaluation analyses were performed in R software version 4.0.4.

3 Results and Discussions

Table 2 summarises training and validation RMSE, R^2 values for predicting NO_2 and $PM_{2.5}$ using ANN, RF and SVM, whereas Fig. 1 shows the scatterplot of observed and RF-predicted NO_2 at HCAB. For NO_2 predictions, ANN and SVM performed nearly the same, $R^2 = 0.69$ and 0.65, respectively. On the other hand, RF predicted NO_2 values in a relatively good agreement with the actual/observed data, RMSE = 13.5 ppb, $R^2 = 0.75$. One of the possible reasons is that RF and other tree-based algorithms generate several models, which collectively describes the relationship between predictor and response variables to optimize the model performance (Carslaw & Taylor, 2009).

Table 2 Summary of training and validation RMSE, R^2 values for predicting street-level, hourly NO_2 and $PM_{2.5}$ at HCAB, Copenhagen, Denmark

	NO_2				$PM_{2.5}$			
	Train		Val		Train		Val	
	RMSE	R^2	RMSE	R^2	RMSE	R^2	RMSE	R^2
ANN	14.5	0.66	14.1	0.69	13.0	0.78	13.5	0.75
RF	13.8	0.71	13.5	0.73	11.6	0.80	11.2	0.81
SVM	15.1	0.64	14.7	0.65	15.9	0.59	15.1	0.64

Train = Training data set, Val = Validation data set

Fig. 1 NO_2 prediction using random forests at HCAB (test/validation data)

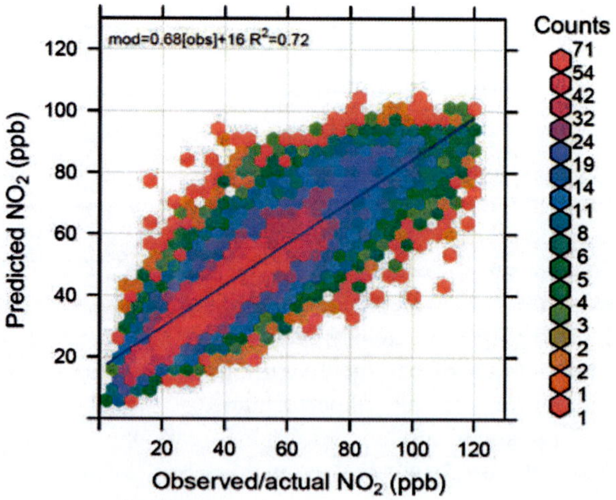

The same behaviour in the performance was observed, where RF performed better than ANN and SVM while predicting street-level $PM_{2.5}$. In addition, our finding based on RF's better performance was in line with several studies in the literature. For example, Xiao et al. (2018) reported that the RF algorithm was one of the two most accurate models predicting historic $PM_{2.5}$ (daily means) in China.

4 Strengths and Limitations

One main strength of this work is that it addresses the research gaps concerning the unexplored potential of machine learning algorithms for air quality prediction in Denmark. However, one main limitation is that the study mainly uses meteorological and background emissions data (limited time window) to predict street-level pollution, and traffic counts, one of the main predictors of street pollution, have not been used so far.

5 Conclusions and Outlook

This study explored the potential of commonly utilized machine learning algorithms, Artificial Neural Networks (ANN), Random Forests (RF) and Support Vector Machines (SVM) to predict hourly, street-level NO_2 and $PM_{2.5}$ concentrations. Overall, RF ($R^2 = 0.73$–0.81) performed better than the other algorithms (ANN: $R^2 = 0.69$–0.75, SVM: $R^2 = 0.64$–0.65) and showed relatively higher prediction accuracy.

The future work will include more predictor variables (e.g. traffic attributes) and test more machine learning algorithms, e.g. Light and Extreme Gradient Boosting (LGB, XGBoost), for a significantly large exposure period.

Acknowledgements Jibran's work is supported by the Danish Big Data Centre for Environment and Health (BER-THA), funded by the Novo Nordisk Foundation (NNF) Challenge Programme, grant number NNF17OC0027864.

Author Contributions Jibran Khan: Conceptualization, data acquisition and processing, model development and evaluation, visualization and analyses, software, original draft writing, revised writing; Thomas Ellermann and Ole Hertel: Reviewed the draft writing.

References

Bergen, K. J., Johnson, P. A., Maarten, V., & Beroza, G. C. (2019). Machine learning for data-driven discovery in solid Earth geoscience. *Science*, 363(6433).

Carslaw, D. C., & Taylor, P. J. (2009). Analysis of air pollution data at a mixed source location using boosted regression trees. *Atmospheric Environment, 43*(22–23), 3563–3570.

Elangasinghe, M. A., Singhal, N., Dirks, K. N., & Salmond, J. A. (2014). Development of an ANN–based air pollution forecasting system with explicit knowledge through sensitivity analysis. *Atmospheric Pollution Research, 5*(4), 696–708.

Ellermann, T., Nygaard, J., Nøjgaard, J. K., Nordstrøm, C., Brandt, J., Christensen, J., Ketzel, M., Massling, A., Bossi, R., Frohn, L. M., Geels, C. & Jensen, S. S. (2020). The Danish air quality monitoring programme. Annual Summary for 2018. Aarhus University, DCE—Danish Centre for Environment and Energy, 83 pp. Scientific Report No. 360. http://dce2.au.dk/pub/SR360.pdf.

Meng, X., Liu, C., Chen, R., Sera, F., Vicedo-Cabrera, A. M., Milojevic, A., & Kan, H. (2021). Short term associations of ambient nitrogen dioxide with daily total, cardiovascular, and respiratory mortality: Multilocation analysis in 398 cities. *BMJ*, 372.

Reichstein, M., Camps-Valls, G., Stevens, B., Jung, M., Denzler, J., & Carvalhais, N. (2019). Deep learning and process understanding for data-driven Earth system science. *Nature, 566*(7743), 195–204.

Tayarani, M., & Rowangould, G. (2020). Estimating exposure to fine particulate matter emissions from vehicle traffic: Exposure misclassification and daily activity patterns in a large, sprawling region. *Environmental Research, 182*, 108999.

Tong, R., Liu, J., Wang, W., & Fang, Y. (2020). Health effects of PM2. 5 emissions from on-road vehicles during weekdays and weekends in Beijing, China. *Atmospheric Environment*, *223*, 117258.

Xiao, Q., Chang, H. H., Geng, G., & Liu, Y. (2018). An ensemble machine-learning model to predict historical PM2.5 concentrations in China from satellite data. *Environmental Science & Technology*, *52*(22), 13260–13269.

Regional and Intercontinental Modelling

Sources and Processes Affecting Air Pollution in the Arctic and Northern High Latitudes—A Modelling Study

Wanmin Gong, Stephen Beagley, and Roya Ghahreman

Abstract Model simulations of atmospheric composition on a pan-Arctic domain at 15-km horizontal resolution were carried out for the year 2015 and for summer 2014. Through a series of sensitivity studies and analysis, we show that (1) biogenic emissions from northern boreal regions play an important role in controlling oxidation capacity over the northern high latitudes; (2) the Arctic is impacted by northern boreal fire emissions during the summer; while Eurasian (Siberian) fires mostly impact the European Arctic, the north American boreal fires dominate the impact over eastern and central Arctic; (3) deposition processes, both dry and wet deposition, are particularly important in affecting the air pollution levels in the Arctic. This study also includes a preliminary investigation on the impact of SO_2 emissions from natural fires, fueled by pyrites and organic materials, at the Smoking Hills in Northwest Territories.

Keywords Arctic atmospheric composition · Natural and anthropogenic sources of pollution · Aerosol dry and wet deposition

1 Introduction

Air pollution in the Arctic can have adverse effects on climate, ecosystems and health. A variety of sources contribute to air pollution in the Arctic and northern high latitudes. It is well established that during winter and spring the transport of anthropogenic emissions from Eurasia is a primary contributor to the Arctic haze

W. Gong (✉) · S. Beagley · R. Ghahreman
Air Quality Research Division, Science and Technology Branch, Environment and Climate Change Canada, Toronto, ON, Canada
e-mail: Wanmin.gong@ec.gc.ca

S. Beagley
e-mail: Stephen.beagley@ec.gc.ca

R. Ghahreman
e-mail: roya.ghahreman@ec.gc.ca

C. Mensink and O. Jorba (eds.), *Air Pollution Modeling and its Application XXVIII*, Springer Proceedings in Complexity, https://doi.org/10.1007/978-3-031-12786-1_13

events. During summer, the Arctic air pollution is generally much lower (particularly at low altitudes) due to the slower transport from the lower latitudes and more efficient wet deposition. However, there are episodic pollution events, some of which can be linked to transport of wildfire emissions from the northern boreal regions. Past studies on the Arctic air pollution have been much focused on characterizing the transport of anthropogenic pollutants from mid latitudes. However, it is becoming evident that local sources within the Arctic and northern high latitudes, both anthropogenic (e.g., oil and gas production, shipping, residential wood burning) and natural (biogenic, oceanic DMS and other biological precursors), can play an important role. The inadequate understanding and constraining of the local sources along with the deficiency in representing some of the processes during pollution transport contribute to the poor model performance in predicting air pollution in the Arctic.

In participating in a new AMAP assessment of SLCF (short lived climate forcers), model simulations of atmospheric composition on a pan-Arctic domain at 15-km horizontal resolution were carried out for the year 2015, using the Environment and Climate Change Canada's air quality prediction model GEM-MACH. Pan-Arctic simulations were also carried out for the 2014 NETCARE field campaign period. Comprehensive model evaluation against available observational data was conducted. This study explores the roles of several natural sources (at local and regional scale), such as wildfires in Eurasian and North American boreal regions, and sulfur emissions from natural fires at the Smoking Hills in the Northwest Territories. We also investigate the representation of biogenic emissions and particle dry and wet deposition in the model and their impact on modeled Arctic atmospheric composition. The investigation presented here focuses on summertime.

2 Model Setup for Pan-Arctic Simulations

GEM-MACH version 2.5 was used as the base model for the suite of simulations conducted in this study, with a 12-bin configuration for representing aerosol size distribution. The pan-Arctic limited area model domain is designed to cover north of 60°N extending to just south of the Canada-US border at 15-km resolution. Anthropogenic emissions are based on the 2016 US and 2015 Canadian national inventories, supplemented by the 2015 ECLIPSE global emissions (for outside North American portion of the domain). Shipping emissions over the Canadian waters are updated with the 2015 MEIT Canadian marine shipping emissions. Biogenic emissions are computed online based on BEIS v3.09 with BELD3 vegetation data (for North America) and GLC2000 land-use data (outside North America). The Canadian Forest Fire Emission Prediction System (CFFEPS, Chen et al., 2019) was used for producing the North American wildfire emissions while Eurasian wildfire emissions are processed from FINN v1.5 inventory (Wiedinmyer et al., 2011) using global satellite retrieval statistics for plume height estimate (Val Martin et al., 2018). Chemical lateral boundary conditions are processed from the MOZART4-GOES5 daily analysis. The 2015 annual simulations were conducted in 4 four-month segments (with one-month overlap as model spin-up); the 2014 summer simulations were conducted for a three-month period (June–August; covering the 2014 NETCARE summer campaign period).

3 Simulation Results and Discussion

3.1 Ozone—Impact of Biogenic Emission and Deposition Over the Ocean

The modelled ozone over the Arctic and northern regions was evaluated against available surface observational data obtained from a number of monitoring networks and database, including the Canadian Air and Precipitation Monitoring Network (CAPMoN), the National Air Pollution Surveillance network (NAPS), and EBAS (a database operated by the Norwegian Institute for Air Research). Figure 1 includes examples of the model-observation comparison at four selected Arctic and northern sites for the month of July. Overall, the model is able to capture the general trend of the observed ozone concentration at these sites. The model does not capture the full range of the variability in observations, which can be at least partly attributable to insufficient model resolution unable to resolve smaller scale processes. The model-observation comparison statistics show an overall model under-prediction of observed surface O_3 concentration at the northern sites. The causes for the model low bias are being investigated.

Biogenic emission update. Shown in Fig. 2a, b are the model simulated July-averaged surface ozone and isoprene. The swath of low ozone concentration coincides with the high concentration of isoprene over northern boreal region due to high

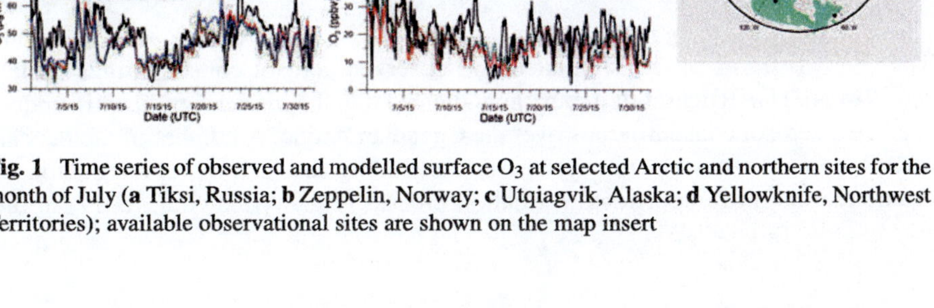

Fig. 1 Time series of observed and modelled surface O_3 at selected Arctic and northern sites for the month of July (**a** Tiksi, Russia; **b** Zeppelin, Norway; **c** Utqiagvik, Alaska; **d** Yellowknife, Northwest Territories); available observational sites are shown on the map insert

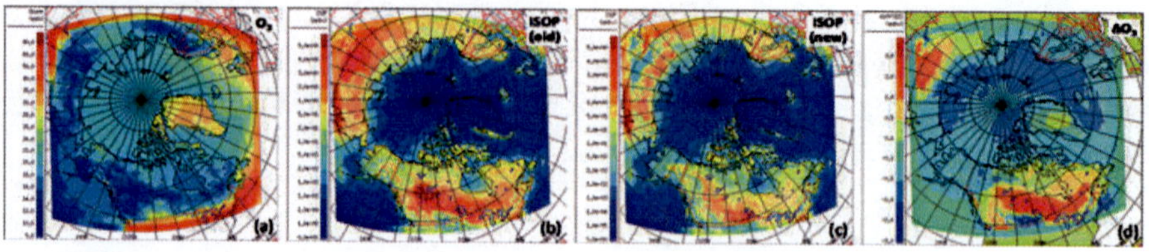

Fig. 2 Modelled (base case) July-averaged surface O_3 (**a**); modelled July-averaged surface isoprene concentrations, **b** base case and **c** with updated biogenic emissions; changes in modelled surface O_3 due to biogenic emission updates (**d**)

biogenic VOC emissions. Unlike populated (more polluted) areas where biogenic VOCs (particularly isoprene) contribute to O_3 formation, under low NO_x conditions (e.g., remote regions), biogenic VOCs contribute to O_3 loss. Inspecting the model simulated isoprene, it appears that biogenic emissions (based on BEIS3.09/BELD3) are over-estimated in the model. A new model simulation was then carried out after updating biogenic emissions (based on BEIS3.7/BELD4). The modelled isoprene concentrations are significantly reduced over the boreal region in the new run (Fig. 2c), which led to an increase in modelled O_3 concentrations over the northern boreal regions (Fig. 2d) and hence reducing the model bias.

O_3 deposition over the ocean. Dry deposition plays an important role in tropospheric O_3 budget (Pound et al., 2020). There is a large uncertainty in O_3 deposition velocity estimate over the water surfaces (Sarwar et al., 2015; Pound et al., 2020). In the GEM-MACH model, the dry deposition velocity for O_3 is set at a constant of 0.03 cm s^{-1} over both fresh and ocean waters (based on the recommendation of Wesley & Hicks, 2000), which is within the observed range 0.009–0.065 cm s^{-1} (Helmig et al., 2012). Sarwar et al. (2015) developed an updated parameterization for O_3 deposition velocity over the ocean, accounting for O_3 reaction with iodide at seawater surface, where a sea-surface-temperature (SST) dependence of surface-water iodide concentration is assumed. A sensitivity run was conducted replacing the constant O_3 deposition velocity with the Sarwar's parameterization. The impact on modelled O_3 concentration from the updated O_3 deposition velocity is limited to over the ocean and coastal regions, generally producing an increase over high latitudes due to colder SSTs and a reduction in O_3 dry deposition (Fig. 3).

3.2 *Aerosols—Impact of Particle Dry and Wet Deposition*

The comparison of model simulated and observed aerosol concentrations during the 2014 NETCARE aircraft campaign indicated that the model significantly under-predicted aerosol concentrations over the Canadian Arctic. A number of factors can contribute to the model under-prediction, including the lack of representation of Arctic marine biogenic aerosol formation process (e.g., Willis et al., 2017) in the

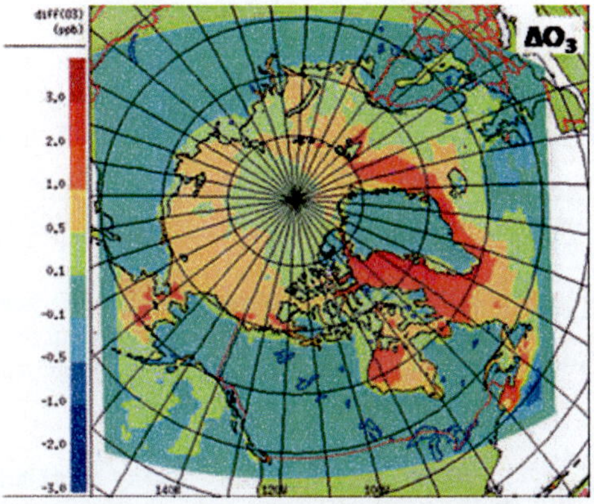

Fig. 3 Change in modelled July-averages surface O_3 concentration from revised O_3 dry deposition

model. There is also an indication that particle removal processes during long-range transport (e.g., via dry and wet deposition) may be over-estimated in the model.

Particle dry deposition. Emerson et al. (2020) examined the particle dry deposition scheme of Zhang et al. (2001), which is commonly used in air quality models (including the current GEM-MACH). Based on new measurements, they found that the original Zhang scheme significantly under represented the particle interception mechanism in dry deposition, and proposed a revised scheme, which shifts the minimum dry deposition to smaller particles over most of the vegetated land surfaces (Fig. 4, leftmost panel). This revised particle dry deposition scheme was tested in the model simulation of the 2014 NETCARE field campaign period. As expected the Emerson revised particle dry deposition scheme resulted in much reduced removal of the nucleation mode and the smaller end of accumulation mode aerosols and increased removal of larger particles, the latter in particular over the northern boreal regions, (Fig. 4c, d). Over the Arctic and northern regions, the net result is an increase of modelled $PM_{2.5}$ (~ 20% over the Arctic archipelago, Fig. 4a).

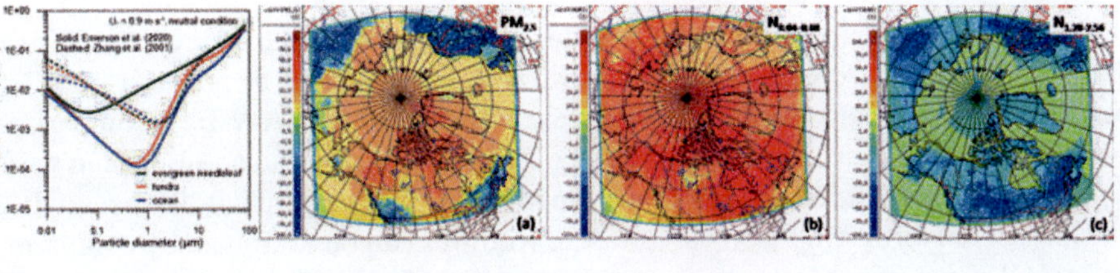

Fig. 4 Leftmost panel: comparison of revised particle dry deposition velocity based on Emerson et al. (2020) and the original Zhang et al. (2001) parameterization. **a** Relative change in modelled July-averaged $PM_{2.5}$ mass from the revised particle dry deposition; **b** and **c** relative change in modelled particle number concentration in bin 3 (0.04–0.08 μm) and bin 8 (1.28–2.56 μm), respectively

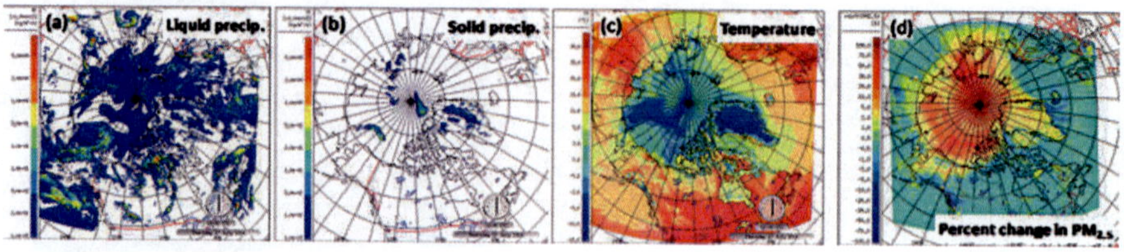

Fig. 5 **a–c** Snapshot example of modelled liquid and solid precipitation fluxes and temperature at surface level; **d** percent change in modelled July-averaged $PM_{2.5}$ when below-cloud scavenging is treated according to the modelled precipitation phase

Particle below-cloud scavenging. Wet scavenging can account for most of the loss of pollutants during transport to the Arctic during the summer season. The below-cloud scavenging of aerosols in GEM-MACH is currently set up to treat a single-phase (liquid or solid) precipitation at a given grid and time, i.e., the liquid and solid precipitation fluxes predicted from the meteorological module are added up to total precipitation and its phase decided by a temperature test: liquid, when $T > 0$ °C, and solid, when $T \leq 0$ °C. With this setup, the predicted liquid precipitation will be treated as solid (snow) precipitation if the environment temperature is below zero. An example is given in Fig. 5a–c. Below-cloud scavenging is generally more efficient for solid hydrometeors than liquid droplets for the same equivalent precipitation rate (due to differences in density and size of hydrometeors). As an alternative, modifications were made in the model to treat below-cloud scavenging according to the precipitation phases predicted by the meteorological module. The impact of this change in modelled $PM_{2.5}$ is shown in Fig. 5d. The modelled $PM_{2.5}$ over the Arctic is enhanced up to 50% as a result.

3.3 Impact of Northern Boreal Wildfire Emissions

To examine the impact of northern wildfire emissions on the Arctic, two additional model simulations were carried out (1) with wildfire emissions turned off ("noWF") and (2) including North American wildfire emissions only ("NA-WF"). Comparison between the base-case run, which includes both North American and Eurasian wildfire emissions, and the noWF run shows the impact of all northern wildfire fires, while the impact of North American vs. Eurasian wildfires can be examined by comparing the NA-WF run with the noWF run and comparing the base-case run with NA-WF run. Figure 6 illustrate these comparisons for the case of modelled July-averaged $PM_{2.5}$. It is shown that the northern boreal wildfire emissions have a strong impact on the Arctic, particularly the North American boreal wildfires. For the summer of 2014, the Eurasian wildfire mostly affected Eurasian side of the Arctic with some impact extending to Alaska, while North American wildfires has a profound impact on the Canadian Arctic and Greenland extending to central and European Arctic.

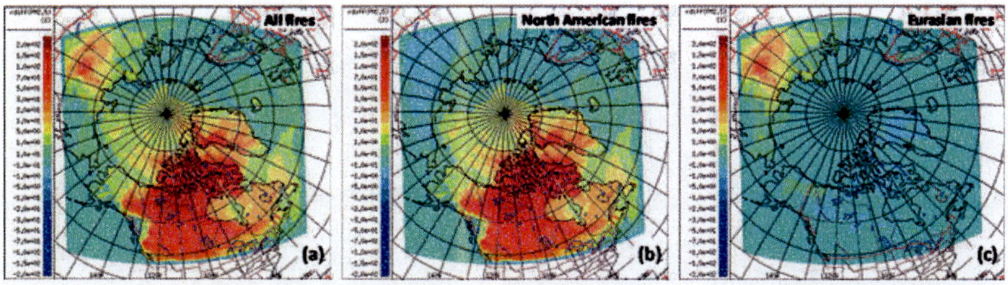

Fig. 6 Impact of northern boreal wildfire emissions on modelled ambient PM2.5 in the Arctic (July 2014). **a** impact of all fires; **b** impact of North American wildfires; **c** impact of Eurasian boreal fires

3.4 Sulfur Emissions from Natural Fires at Smoking Hills, NT

The Smoking Hills are an area of spontaneously burning sea cliffs facing on Franklin Bay at Cape Bathurst, NT (70° 14′ N, 127° 10′ W). The burning occurs at a number of sites over a distance of about 30 km along the 100 m high sea cliffs (Fig. 7). Based on airborne measurements, it was estimated that the fires at Smoking Hills put out emissions equivalent to 0.3 kg s^{-1} (10^4 T yr^{-1}) of SO_2 and 0.3 kg s^{-1} of particles (Radke & Hobbs, 1989). As a first attempt to assess the impact of this natural source of sulfur on the Arctic, two additional point sources were introduced to two model grids (located at Cape Bathurst), roughly 30 km apart, each assigned with an SO_2 emission rate of 150 g s^{-1} and stack parameters resembling wildfire (smoldering) emissions. Figure 7 show the impact of sulfur emissions from Smoking Hills on modelled ambient SO_2 and fine sulfate ($SU_{2.5}$) concentrations in the Arctic. The area of impact varies between seasons influenced by flow patterns, largely over the Beaufort Sea area. The predominantly stable boundary layer over the Arctic plays a role in limiting the plume influence to relatively local areas. As shown, the sulfur emissions from Smoking Hills contribute to sulfur deposition locally. There is also an enhancement in cloud condensation nuclei (CCN) from the sulfur emissions from the natural fires, which can modify cloud microphysical characteristics in the area.

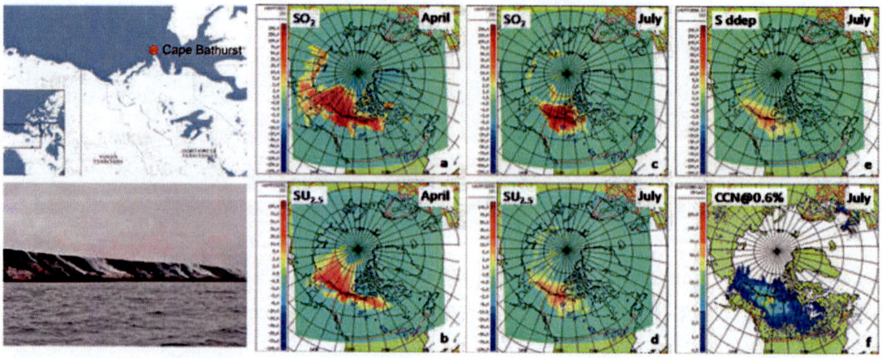

Fig. 7 Impact of sulfur emissions from Smoking Hills at Cape Bathurst; on modelled monthly averaged SO_2, April (**a**) and July (**c**); on modelled monthly averaged $SU_{2.5}$, April (**b**) and July (**d**); on modelled total sulfur dry deposition for July (**e**); on modelled July-averaged CCN (at 0.6% supersaturation)

References

Chen, J., et al. (2019). The FireWork v2.0 air quality forecast system with biomass burning emissions from the Canadian Forest Fire Emissions Prediction System v2.03. *Geoscientific Model Development, 12*, 3283–3310. https://doi.org/10.5194/gmd-12-3283-2019

Emerson, E. W., et al. (2020). Revisiting particle dry deposition and its role in radiative effect estimates. *PNAS, 117*(42), 26076–26082. https://doi.org/10.1073/pnas.2014761117

Helmig, D., et al. (2012). Atmosphere-ocean ozone fluxes during the TexAQS 2006, STRATUS 2006, GOMECC 2007, GasEx 2008, and AMMA 2008 cruises. *Journal of Geophysical Research, 117*, D04305. https://doi.org/10.1029/2011JD015955

Martin, V., et al. (2018). A global analysis of wildfire smoke injection heights derived from space-based multi-angle imaging. *Remote Sensing, 10*(10), 1609. https://doi.org/10.3390/rs10101609

Pound, R. J., et al. (2020). Influences of oceanic ozone deposition on tropospheric photochemistry. *Atmospheric Chemistry and Physics, 20*, 4227–4239. https://doi.org/10.5194/acp-20-4227-2020

Radke, L. F., & Hobbs, P. V. (1989). Arctic hazes in summer over Greenland and the North American Arctic. III: A contribution from the natural burning of carbonaceous materials and pyrites. *Journal of Atmospheric Chemistry, 9*, 161–167.

Sarwar, G., et al. (2015). Impact of enhanced ozone deposition and halogen chemistry on tropospheric ozone over the northern hemisphere. *Environmental Science and Technology, 49*(15), 9203–9211. https://doi.org/10.1021/acs.est.5b01657

Wesely, M. L., Hicks B. B. (2000) A review of the current status of knowledge on dry deposition. *Atmospheric Environment, 34*(12–14), 2261–2282. ISSN 1352-2310. https://doi.org/10.1016/S1352-2310(99)00467-7.

Wiedinmyer, C., et al., (2011). The Fire INventory from NCAR (FINN): a high resolution global model to estimate the emissions from open burning. *Geoscientific Model Development, 4*, 625–641. https://doi.org/10.5194/gmd-4-625-2011

Willis, M. D., et al. (2017). Evidence for marine biogenic influence on summertime arctic aerosol. *Geophysical Research Letters, 44*, 6460–6470. https://doi.org/10.1002/2017GL073359

Zhang, L. M., et al. (2001). A size-segregated particle dry deposition scheme for an atmospheric aerosol module. *Atmospheric Environment, 35*, 549–560.

Questions and Answers

Questioner: (not noted on the question sheet)

Question: How is the "Arctic death spiral" incorporated in the modelling as presented (in terms of land-use)?

Answer: The so-called "Arctic death spiral" depicts the drastic decreasing trend in the Arctic sea ice volume over the past 40 years. The current modelling study is a simulation of specific years (2014 and 2015) rather than a multi-year climate simulation. The sea ice cover used in the model simulation comes from the Canadian Meteorological Centre sea ice analysis for the simulation periods.

Contributions of Source Regions to Changes in Long-Range Transported Ozone to North America During 1990–2010: A Modeling Analysis

Rohit Mathur, Daiwen Kang, Sergey Napelenok, Christian Hogrefe, Golam Sarwar, Jia Xing, Syuichi Itahashi, and Barron Henderson

Abstract As measures to limit the impacts of domestic emissions on surface-level ozone (O_3) pollution are implemented, the increasing contributions of O_3 pollution transported from other regions become more influential in shaping local O_3 levels and confounds design of abatement measures to meet more stringent air quality standards. We investigate the contributions of emission changes from seven source regions, as well as the stratosphere, on the change in tropospheric ozone burden across the Northern Hemisphere and its influence on long-range transport (LRT) of O_3 to North America during the 1990–2010 period. Sensitivity calculations conducted with the Hemispheric Community Multiscale Air Quality (HCMAQ) model to estimate the response of O_3 to emissions from different source regions across the Northern Hemisphere are combined with simulated changes in O_3 due to emission changes during the multi-decadal analysis period to delineate impacts of different source regions in shaping the long-range transported O_3 to North America.

Keywords Hemispheric CMAQ · Long-range transport · Stratosphere-troposphere exchange · Air pollution trends

R. Mathur (✉) · D. Kang · S. Napelenok · C. Hogrefe · G. Sarwar
Office of Research and Development, U.S. Environmental Protection Agency, Research Triangle Park, NC, USA
e-mail: mathur.rohit@epa.gov

J. Xing
Tsinghua University, Beijing, China

S. Itahashi
Central Research Institute of Electric Power Industry, Tokyo, Japan

B. Henderson
Office of Air Quality Planning and Standards, U.S. Environmental Protection Agency, Research Triangle Park, NC, USA

C. Mensink and O. Jorba (eds.), *Air Pollution Modeling and its Application XXVIII*, Springer Proceedings in Complexity, https://doi.org/10.1007/978-3-031-12786-1_14

1 Introduction

Ambient ozone (O_3) levels in a region result from contributions from multiple sources including domestic and international anthropogenic precursor emissions as well as natural sources including stratosphere-troposphere exchange, biogenic, lightning, soil, and wildfire, emissions. Pollutants from continental source regions get lofted to the free troposphere and can be subjected to inter-continental transport (e.g., NRC, 2009). Several locations in non-compliance with the National Ambient Air Quality Standard for O_3 across the U.S. are thought to have non-negligible and increasing influences from pollution transported long distances and attributable to distant international emissions. Understanding the relative contributions of these international sources, the dominant natural contributions, and the changes in these contributions over time driven by global emission changes, is important for guiding local air quality management.

2 Model Configuration and Simulation Details

In this study we use simulations conducted with the Community Multiscale Air Quality (CMAQ) version 5.0 (https://doi.org/10.5281/zenodo.1079888) modeling system extended for hemispheric applications (HCMAQ) (Mathur et al., 2017). The process configuration options used in the model calculations follow those detailed in Xing et al. (2015) and Mathur et al. (2017). The horizontal domain extent covering the Northern Hemisphere set on a polar stereographic projection (see Fig. 1) is discretized with a 187 × 187 grid of 108 km grid spacing. A terrain-following vertical coordinate utilizing 44 layers of variable thickness is used to resolve the model vertical extent between the surface and 50 hPa. HCMAQ is driven by meteorological fields derived from the Weather Research and Forecasting (WRF) model simulations constrained by NCEP/NCAR Reanalysis. Year-specific emissions for 17 anthropogenic sectors were based on EDGAR (Emission Database for Global Atmospheric Research, version 4.2) while biogenic VOC and lightning NO_x emissions were based on GEIA (Global Emission Inventory Activity; http://www.geiacenter.org), as further detailed in Xing et al. (2015). The CB05 chemical mechanism was used to represent gas-phase photochemistry. In addition, O_3 mixing ratios in the top-most model layer (~50 hPa) were modulated based on scaling of the spatially and temporally varying potential vorticity (PV) fields to represent possible effects associated with stratosphere-troposphere exchange (Mathur et al., 2008).

To examine the impacts of emissions from different source region on tropospheric O_3 distributions, the Northern Hemisphere modeling domain was divided into seven source regions (depicted in Fig. 1) representing: (1) North America (NAM), (2) Europe and Russia (EUR + RUS), (3) East and Southeast Asia (EAS + SEA), (4) South Asia (SAS; the Indian subcontinent), (5) Northern Africa and the Middle East (NAF + MDE), (6) Central America (CAM), and (7) Other (the rest of the geographic

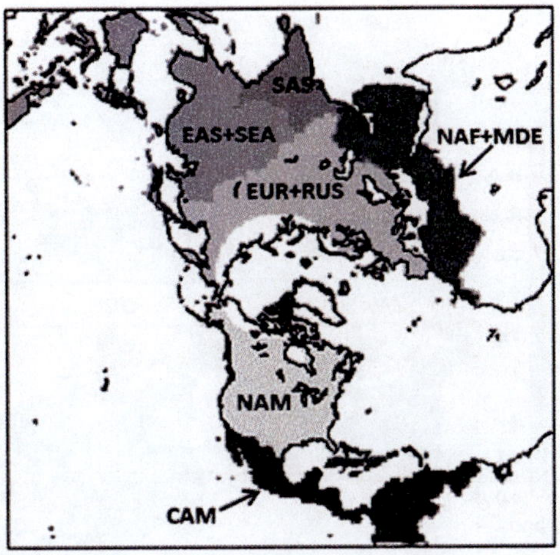

Fig. 1 The Northern Hemisphere modeling domain and source region definition (from Mathur et al., 2018)

domain consisting of the vast oceanic regions and portions of South America and Africa). We use the decoupled direct method in three-dimensions (DDM-3D; Hakami et al., 2003; Napelenok et al., 2008) implemented in CMAQ, to estimate the sensitivity of the modeled three-dimensional O_3 to emissions from each of the source regions. Specifically, we calculate the first-order sensitivity of modeled O_3 to NO_x and VOC emissions from each of the seven regions and also to the O_3 specified at the model top using the PV-parameterization to estimate the influence of stratosphere-troposphere exchange (STE) on the modeled O_3 fields. We also calculated second-order sensitivities only to global NO_x and VOC emissions. The DDM-3D CMAQ simulations with the above configuration then provided these emission sensitivities for the entire calendar year of 2006. In subsequent discussion we focus on seasonal- and 0–2 km-average values of these sensitivity coefficients and simulated mixing ratios.

3 Results and Discussion

Zero-out contributions, i.e., the difference between the base-case simulation and the concentration that would occur without any emissions from a specific source region, can be approximated as the sum of the semi-normalized first-order sensitivity coefficients for NO_x and VOC emissions for that region. Thus, these zero-out contributions for each source region other than NAM can be inferred as the long-range transported contribution from that source region to O_3 over the NAM. Figure 2 illustrates these estimated seasonal and 0–2 km average contributions from each of the source regions to the NAM domain for Spring. As can be seen by contrasting the different panels in Fig. 2, the contributions of each source region vary across continental North America. Emissions from Central America can contribute sizeable amounts of O_3

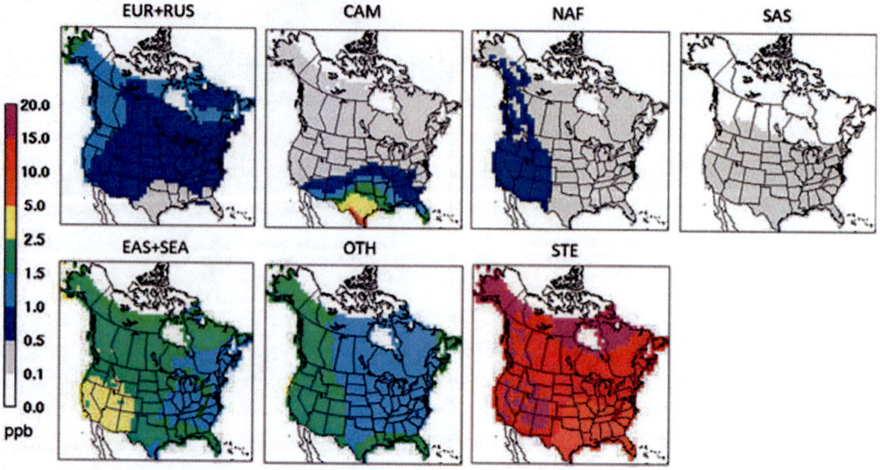

Fig. 2 Contributions of long-range transported O_3 from different source regions to 0–2 km average O_3 over North America during Spring 2006

to regions in the southwestern U.S. while cross-Pacific transport of both pollution from Asian emissions as well as emissions from the vast oceanic regions result in higher O_3 contributions in the western U.S. Also noticeable in Fig. 2 is the sizeable contribution of O_3 originating from the stratosphere to boundary-layer O_3 mixing ratios across the continental U.S., highlighting the need for consideration and accurate assessment of the impacts of stratosphere-troposphere exchange processes on regulating Springtime background O_3 levels.

To assess changes in long-range transported O_3 over time in response to changing global emission patterns, we combine these seasonal average sensitivities from DDM-3D with simulated multi-decadal changes in seasonal average O_3 distributions simulated by HCMAQ and emission changes across the Northern Hemisphere during 1990–2010 based on the model simulations described in Xing et al. (2015). Using the sensitivity coefficients for base year 2006, emission changes for each year relative to 2006, and the simulated seasonal average O_3 distributions for each year, one can estimate the long-range transported O_3 to NAM using a reduced form model approximation as follows:

$$C^N_{LRT,m} = \left(\sum\nolimits_{j \neq m} \left(\overline{S}_{NO_x,j} \Delta E^N_{NO_x,j} + \overline{S}_{VOC,j} \Delta E^N_{VOC,j}\right)\right)\left(\frac{C^N}{C^N_{recon}}\right)$$

where, $C^N_{LRT,m}$ is the estimated long-range transported O_3 to a region, m, for year N; $\overline{S}_{NO_x,j}$ and $\overline{S}_{VOC,j}$ are the first-order sensitivity coefficients to NOx and VOC emissions from region j; C^N is the simulated O_3 for year N, and C^N_{recon} is the reconstructed O_3 distribution based on a Taylor series approximation using the first- and second-order sensitivity coefficients described in Sect. 2 (Itahashi et al., 2020). Figure 3

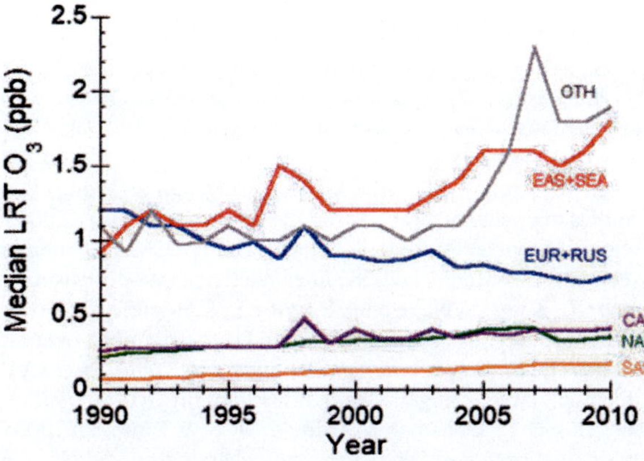

Fig. 3 Changes in median long-range transported O_3 to the North American domain during Spring from the different source regions shown in Fig. 1

illustrates the resulting estimates of changes in the contributions from each of the source regions to LRT O_3 to NAM during the 1990–2010 period.

Figure 3 indicates that while long-range transport contributions to NAM from Europe have steadily declined during 1990–2010, resulting from implementation of control measures and technological advances that have reduced emissions of O_3 precursors, contributions from other source regions across the Northern Hemisphere have steadily increased during this period. Transport from East Asia was a dominant and increasing contributor to long-range transported O_3 during this period. Also noticeable in Fig. 3 is the rising contribution of emissions attributable to the vast oceanic regions of the model domain (see Fig. 1) which are primarily associated with commercial shipping operations. Increasing international seaborne trade has likely increased commercial shipping activity and associated emissions, and point to the need for improved quantification of the magnitude and spatial and temporal variability in multipollutant emissions from this sector so that not only can their impacts on air quality in ports and near coastal areas be accurately assessed, but also their impact on hemispheric pollution and modulation of long-range transport be better quantified.

Disclaimer The views expressed in this paper are those of the authors and do not necessarily represent the view or policies of the U.S. Environmental Protection Agency.

References

Hakami, A., Odman, M. T., & Russell, A. G. (2003). High-order, direct sensitivity analysis of multidimensional air quality models. *Environmental Science and Technology, 37*, 2442–2452.

Itahashi, S., Mathur, R., Hogrefe, C., Napelenok, S. L., & Zhang, Y. (2020). Modeling stratospheric intrusion and trans-Pacific transport on tropospheric ozone using hemispheric CMAQ during April 2010—Part 2: Examination of emission impacts based on the higher-order decoupled direct method. *Atmospheric Chemistry and Physics, 20*, 3397–3413. https://doi.org/10.5194/acp-20-3397-2020

Mathur, R., Xing, J., Gilliam, R., Sarwar, G., Hogrefe, C., Pleim, J., Pouliot, G., Roselle, S., Spero, T. L., Wong, D. C., & Young, J. (2017). Extending the Community Multiscale Air Quality (CMAQ) modeling system to hemispheric scales: Overview of process considerations and initial applications. *Atmospheric Chemistry and Physics, 17*, 12449–12474. https://doi.org/10.5194/acp-17-12449-2017

Mathur, R., Lin, H. M., McKeen, S., Kang, D., & Wong, D. (2008). Three-dimensional model studies of exchange processes in the troposphere: Use of potential vorticity to specify aloft O_3 in regional models. Presented at the 7th Annual CMAS Conference, available at: https://www.cmascenter.org/conference/2008/slides/mathur_three-dimension_model_cmas08.ppt

Mathur R., Kang D., Napelenok S., Xing J., & Hogrefe C. (2018). A modeling study of the influence of hemispheric transport on trends in O_3 distributions over North America. In C. Mensink & G. Kallos (Eds.), *Air pollution modeling and its application XXV*. ITM 2016. Springer proceedings in complexity. Springer, Cham. https://doi.org/10.1007/978-3-319-57645-9_2

Napelenok, S. L., Cohan, D. S., Odman, M. T., & Tonse, S. (2008). Extension and evaluation of sensitivity analysis capabilities in a photochemical model. *Environmental Modelling and Software, 23*, 994–999.

NRC. (2009). Global sources of local pollution: An assessment of long-range transport of key air pollutants to and from the United States. National Academy Press. https://doi.org/10.17226/12743

Xing, J., Mathur, R., Pleim, J., Hogrefe, C., Gan, C., Wong, D.C., Wei, C., Gilliam, R., & Pouliot, G. (2015). Observations and modeling of air quality trends over 1990–2010 across the northern hemisphere: China, the United States and Europe. *Atmospheric Chemistry and Physics, 15*, 2227–2914. https://doi.org/10.5194/acp-15-2723-2015

Diagnostic Analysis of CMAQ Dry Deposition Fields in the Context of AQMEII4

Christian Hogrefe, Jesse Bash, Jonathan Pleim, Donna Schwede, Robert Gilliam, Kristen Foley, K. Wyat Appel, and Rohit Mathur

Abstract Phase 4 of the Air Quality Model Evaluation International Initiative (AQMEII4) is focused on the diagnostic intercomparison and evaluation of deposition simulated by regional-scale air quality modeling systems and employs both grid and box modeling techniques to accomplish this goal. This study presents an analysis of CMAQv5.3.1 simulations that were performed in the context of AQMEII4 over a North American modeling domain. Results show that variability in the estimated dry deposition sink affected model performance and resulted from both differences in model process formulation and the representation of the underlying land use (LU). To gain mechanistic insights into model behavior, the study employs diagnostic variables defined for the AQMEII4 project, i.e., grid scale and land-use specific effective conductances and deposition fluxes for the major dry deposition pathways (stomatal, cuticular, lower canopy, and soil). Analysis of these variables indicates that there generally is a greater contribution to ozone dry deposition from the stomatal and cuticular pathways in the CMAQ M3DRY dry deposition scheme compared

C. Hogrefe (✉) · J. Bash · J. Pleim · D. Schwede · R. Gilliam · K. Foley · K. W. Appel · R. Mathur
Center for Environmental Measurement and Modeling, US Environmental Protection Agency, 109 T. W. Alexander Dr., RTP, Durham, NC 27711, USA
e-mail: hogrefe.christian@epa.gov

J. Bash
e-mail: bash.jesse@epa.gov

J. Pleim
e-mail: pleim.jon@epa.gov

D. Schwede
e-mail: schwede.donna@epa.gov

R. Gilliam
e-mail: gilliam.robert@epa.gov

K. Foley
e-mail: foley.kristen@epa.gov

K. W. Appel
e-mail: appel.wyat@epa.gov

R. Mathur
e-mail: mathur.rohit@epa.gov

C. Mensink and O. Jorba (eds.), *Air Pollution Modeling and its Application XXVIII*, Springer Proceedings in Complexity, https://doi.org/10.1007/978-3-031-12786-1_15

to CMAQ STAGE dry deposition scheme and that differences in LU classification schemes can have a pronounced impact on simulated fluxes.

Keywords Dry deposition · Model evaluation · AQMEII · Model intercomparison · CMAQ

1 Introduction

As discussed in Galmarini et al. (2021), Phase 4 of the Air Quality Model Evaluation International Initiative (AQMEII4) is focused on the diagnostic intercomparison and evaluation of deposition simulated by regional-scale air quality modeling systems through both grid and box modeling techniques. The AQMEII4 grid model activity encompasses annual simulations for two years over North America and Europe performed by eight modeling groups and collects detailed dry deposition diagnostics for a range of trace gases. In this study, we provide examples of AQMEII4 concepts applied to simulations over North America performed with the Community Multiscale Air Quality (CMAQ) modeling system and focus on analyzing ozone simulations for May–September 2016.

2 Model Simulations and Observations

The 2016 annual CMAQ simulations analyzed in this study were performed at a horizontal grid spacing of 12 km and employed the two dry deposition schemes available in CMAQv5.3.1 (U.S. EPA, 2019), i.e., M3DRY and STAGE (Appel et al., 2021; Galmarini et al., 2021, and references therein). In the subsequent discussion, these two CMAQ simulations are simply referred to as M3DRY and STAGE. Meteorological fields for the CMAQ simulations were prepared with version 4.1.1 of the Weather Research and Forecasting (WRF) model using the Moderate Resolution Imaging Spectroradiometer (MODIS) land-use (LU) classification scheme. WRF outputs were processed through version 5.1 of the Meteorology—Chemistry Interface Processor (MCIP). The emissions and boundary condition inputs for the CMAQ simulations were common to all AQMEII4 modeling groups and were described in Galmarini et al. (2021). An additional annual M3DRY simulation was configured to use WRFv4.1.1 with the National Land Cover Dataset (NLCD40) LU classification scheme. For the analysis of LU-specific deposition fluxes, the MODIS and NLCD40 LU categories used in the three CMAQ simulations described above were mapped to the 16 common AQMEII4 categories defined in Galmarini et al. (2021). Observations of ozone were obtained from the Air Quality System (AQS), Clean Air Status and Trends Network (CASTNET), and National Air Pollution Surveillance (NAPS) databases.

3 Results

Figure 1 illustrates the effects of using the two CMAQ dry deposition schemes on simulated spatial fields of ozone deposition velocities (V_d), dry deposition fluxes (DDEP), and concentrations (CONC). These fields depict May–September averages for V_d and CONC and May–September totals for DDEP. As expected by the fundamental relationship DDEP = $V_d \times$ CONC and the regional-scale nature of ozone, regions with higher (lower) V_d in M3DRY than STAGE correspond to regions with higher (lower) DDEP. This increased (decreased) ozone loss through DDEP then yields corresponding regions of lower (higher) CONC. For example, much of the southeastern portion of the modeling domain exhibits higher V_d, higher DDEP, and correspondingly lower CONC for M3DRY than STAGE, while the reverse is true for much of the southwestern portion of the modeling domain.

To assess the extent to which the ozone CONC differences shown in Fig. 1 affect comparisons to available observations, Fig. 2 presents a time series of the bias (model—observations) of 2016 monthly means of daily maximum 8-h average ozone simulated by M3DRY and STAGE averaged over all AQS, CASTNET, and NAPS stations in the modeling domain. This analysis indicates that the dry deposition scheme selected in CMAQ can influence model performance but that the overall persistent domain-wide overestimation likely is driven by other processes and/or input fields. It should also be noted that the spatial heterogeneity of the CONC differences in Fig. 1 suggests that larger bias differences than those shown in Fig. 2 likely exist for specific regions and that the bias differences in Fig. 2 cannot readily be associated with the corresponding CONC differences in Fig. 1 due to the non-uniform spacing of monitoring sites.

Overall, the results in Figs. 1 and 2 show that the effects of using different dry deposition schemes in CMAQ varies seasonally and spatially, but they do not identify the processes causing these differences. To this end, we analyze some of the diagnostic variables generated as part of AQMEII4 and described in Galmarini et al. (2021). Figure 3 shows May–September ozone effective conductances for M3DRY and STAGE for the southwestern and southeastern portions of the modeling domain.

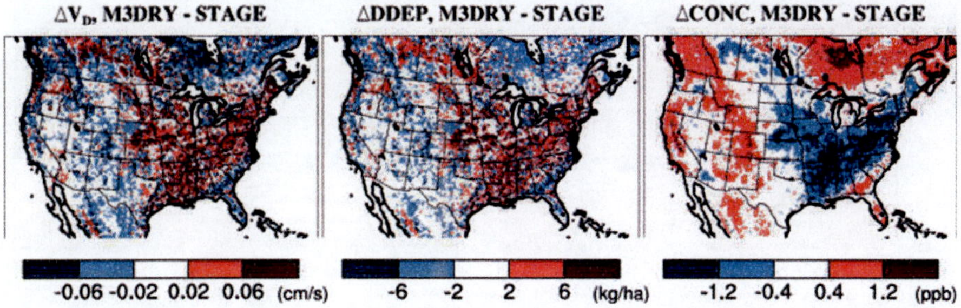

Fig. 1 M3DRY-STAGE differences for May–September average ozone V_d and CONC and May–September total ozone DDEP

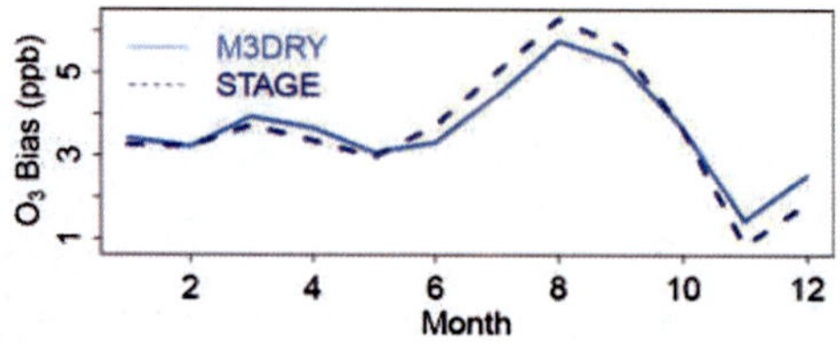

Fig. 2 Bias (model—observations) of 2016 monthly mean daily maximum 8-h average ozone simulated by M3DRY and STAGE averaged over all AQS, CASTNET, and NAPS stations in the modeling domain

As defined in Paulot et al. (2018) and Clifton et al. (2020), effective conductances quantify the contribution of the four major surface deposition pathways present in most dry deposition schemes (stomatal, cuticular, lower canopy, and soil) to V_d. For the M3DRY and STAGE deposition schemes, the "lower canopy" pathway corresponds to deposition to vegetated soil while the "soil" pathway corresponds to deposition to unvegetated soil.

Interpreting Fig. 3 in the context of Fig. 1 indicates that V_d in the southwestern region (where V_d is lower for M3DRY than STAGE) is dominated by deposition to vegetated and unvegetated soil (soil and lower canopy pathways), which have a larger contribution to STAGE V_d than M3DRY V_d in both absolute and relative terms. Conversely, V_d in the southeastern region where V_d is greater for M3DRY than STAGE is dominated by deposition to leaves (cuticular and stomatal pathways) which has a larger absolute and relative contribution to V_d in M3DRY.

As discussed in Galmarini et al. (2021), the concept of apportioning V_d using effective conductances can be extended to DDEP using the analogous concept of effective

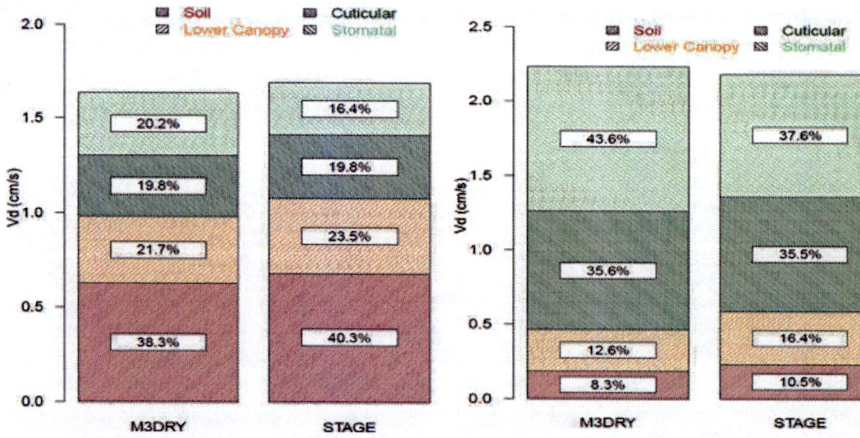

Fig. 3 Effective conductances M3DRY and STAGE for the southwestern (left pair) and southeastern (right pair) portions of the modeling domain. The sum of the effective conductances for the soil, lower canopy, cuticular, and stomatal pathways is equal to V_d

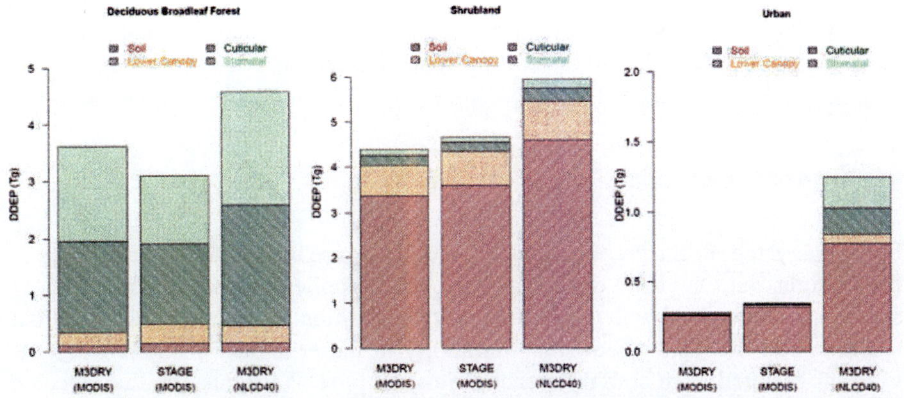

Fig. 4 Effective ozone dry deposition fluxes for the deciduous broadleaf forest (left), shrubland (center), and urban (right) land use categories. All values are totals for May–September 2016 across the entire modeling domain

fluxes, i.e., apportioning DDEP fluxes to the four pathways discussed above. Moreover, the AQMEII4 protocol calls for the calculation and reporting of LU-specific effective conductances and, by extension, effective fluxes to investigate the impacts of different deposition scheme formulations and different land use datasets on modeled deposition. Figure 4 presents three examples of LU-specific May–September average effective ozone fluxes for the base M3DRY and STAGE simulations using the MODIS LU scheme discussed thus far, as well as the M3DRY sensitivity simulation using the NLCD40 LU classification scheme.

Because the importance of each pathway varies by LU category, the relative differences in pathways between M3DRY and STAGE (generally higher for cuticular/stomatal in M3DRY and for soil/lower canopy in STAGE) are manifested differently in terms of deposition fluxes for different categories. For example, deposition via the cuticular and stomatal pathways dominates for deciduous broadleaf forest, and consequently, the M3DRY simulations show higher deposition fluxes than STAGE resulting from greater emphasis on these pathways. Conversely, deposition via the soil and lower canopy pathways dominates for shrubland, and consequently, the simulations with STAGE show higher deposition fluxes than M3DRY due to the greater importance of those pathways in that scheme.

Moreover, these examples also illustrate that the LU classification can affect dry deposition fluxes to a similar or even greater extent than the deposition scheme itself. Differences between the M3DRY MODIS and NLCD40 simulations for a given LU category are driven by differences in the spatial coverage and distribution for that LU category. For example, while less than 1% of the modeling domain is classified as urban for MODIS, 2.4% is classified as urban for NLCD40 due to its higher underlying spatial resolution and the inclusion of lower density developed areas with greater vegetation coverage in the urban category. This leads to a substantially higher amount of ozone dry deposition to urban LU in the M3DRY NLCD40 simulation, along with an increased role of deposition via the cuticular and stomatal pathways compared to the M3DRY and STAGE MODIS simulations.

4 Summary

In summary, the results presented here illustrate the use of AQMEII4 diagnostics to gain insights into modeled dry deposition. When applied to a set of three CMAQ simulations performed with two different dry deposition schemes and two different LU schemes, the results show that variability in the estimated ozone dry deposition sink resulted from both differences in model process formulation that affected the partitioning of dry deposition across pathways and the representation of the underlying LU.

Disclaimer The views expressed in this paper are those of the authors and do not necessarily represent the view or policies of the U.S. Environmental Protection Agency.

References

Appel, K. W., Bash, J. O., Fahey, K. M., Foley, K. M., Gilliam, R. C., Hogrefe, C., Hutzell, W. T., Kang, D., Mathur, R., Murphy, B. N., Napelenok, S. L., Nolte, C. G., Pleim, J. E., Pouliot, G. A., Pye, H. O. T., Ran, L., Roselle, S. J., Sarwar, G., Schwede, D. B., Sidi, F. I., Spero, T. L., & Wong, D. C. (2021). The community multiscale air quality model version 5.3 and 5.3.1: System updates and evaluation. *Geoscientific Model Development*. https://doi.org/10.5194/gmd-14-2867-2021

Clifton, O. E., Paulot, F., Fiore, A. M., Horowitz, L. W., Correa, G., Fares, S., Goded, I., Goldstein, A. H., Gruening, C., Hogg, A. J., Loubet, B., Mammarella, I., Munger, J. W., Neil, L., Stella, P., Uddling, J., Vesala, T., & Weng, E. (2020). Influence of dynamic ozone dry deposition on ozone pollution. *Journal of Geophysical Research: Atmospheres*. https://doi.org/10.1029/2020JD032398

Galmarini, S., Makar, P., Clifton, O., Hogrefe, C., Bash, J., Bianconi, R., Bellasio, R., Bieser, J., Butler, T., Ducker, J., Flemming, J., Hozdic, A., Holmes, C., Kioutsioukis, I., Kranenburg, R., Lupascu, A., Perez-Camanyo, J. L., Pleim, J., Ryu, Y.-H., San Jose, R., Schwede, D., Silva, S., Garcia Vivanco, M., & Wolke, R. (2021). *Technical note—AQMEII4 activity 1: Evaluation of wet and dry deposition schemes as an integral part of regional-scale air quality models. Atmosheric Chemistry and Physics, Discussion*. https://doi.org/10.5194/acp-2021-313

Paulot, F., Malyshev, S., Nguyen, T., Crounse, J. D., Shevliakova, E., & Horowitz, L. W. (2018). Representing sub-grid scale variations in nitrogen deposition associated with land use in a global Earth system model: Implications for present and future nitrogen deposition fluxes over North America. *Atmospheric Chemistry and Physics, 18*(24), 17963–17978.

United States Environmental Protection Agency. (2019). CMAQ (Version 5.3.1) [Software]. Available from https://doi.org/10.5281/zenodo.3585898

Forecasting Birch Pollen Levels in Belgium: First Analysis of the 2021 Season

Andy W. Delcloo, Willem W. Verstraeten, Rostislav Kouznetsov, Lucie Hoebelke, Nicolas Bruffaerts, and Mikhail Sofiev

Abstract Belgium is a highly industrialized and densely populated country with substantial air pollution and at least ~10% of the people develop allergic rhinitis symptoms due to birch tree pollen. The only information on airborne birch pollen is coming from five offline monitoring stations, lacking spatial and timely details of the pollen levels. In contrast, Chemistry Transport Models (CTM's) are able to quantify the spatial and temporal distributions of airborne birch pollen levels a few days ahead. Recently, a new dataset on the birch pollen emission sources has been developed for Belgium. Here we show the preliminary results of the forecast set-up for Belgium using the CTM SILAM (System for Integrated modelling of Atmospheric composition, http://silam.fmi.fi) driven by ECMWF meteorological fields and an updated map of birch pollen emission sources to produce the spatio-temporal distributions of airborne birch pollen levels for the current birch pollen season of 2021 four days ahead. We have evaluated the model performance by comparing the forecasted airborne birch pollen levels time series with four offline birch pollen monitoring stations in Belgium and with the CAMS forecast data.

A. W. Delcloo (✉) · W. W. Verstraeten
Royal Meteorological Institute of Belgium, Ukkel, Brussels, Belgium
e-mail: Andy.Delcloo@meteo.be

W. W. Verstraeten
e-mail: willem.verstraeten@meteo.be

A. W. Delcloo
Department of Physics and Astronomy, Ghent University, Ghent, Belgium

R. Kouznetsov · M. Sofiev
Finnish Meteorological Institute (FMI), Helsinki, Finland
e-mail: Rostislav.Kouznetsov@fmi.fi

M. Sofiev
e-mail: Mikhail.Sofiev@fmi.fi

L. Hoebelke · N. Bruffaerts
Belgian Scientific Institute of Public Health (SCIENSANO), Brussels, Belgium
e-mail: Lucie.Hoebeke@sciensano.be

N. Bruffaerts
e-mail: Nicolas.Bruffaerts@sciensano.be

C. Mensink and O. Jorba (eds.), *Air Pollution Modeling and its Application XXVIII*,
Springer Proceedings in Complexity, https://doi.org/10.1007/978-3-031-12786-1_16

Keywords Forecast · CTM · Birch pollen · SILAM

1 Introduction

Allergenic pollen grains affect the public health badly, especially in combination with long-term exposure by other air pollutants (Landrigan et al., 2017). More than 20% of the population in Europe suffers from pollinosis, while in Belgium it is estimated that ~10% is sensitive for birch pollen (Blomme et al., 2013). In order to take preventive measures for people suffering from birch pollen allergies, timely and spatially distributed information on current and forecasted birch pollen levels is essential. To date, the only birch pollen info that is available in Belgium is coming from five aerobiological stations which report offline daily concentrations with at least one-day lag. Chemistry Transport Models (CTM's), however, are able to quantify and forecasts the spatial and temporal distributions of airborne birch pollen levels a few days ahead. We apply the CTM SILAM (System for Integrated modelling of Atmospheric composition, http://silam.fmi.fi) (Sofiev et al., 2006) driven by ECMWF meteorological fields and an updated map of birch pollen emission sources for Belgium to produce the spatio-temporal distributions of airborne birch pollen levels for the birch pollen season of 2021 four days ahead in a bottom-up approach. The model performance is evaluated by comparing the forecasted airborne birch pollen levels time series with the current four offline birch pollen monitoring stations in Belgium that provide observations in 2021.

2 Methods and Data

Birch pollen are biogenic aerosols with a diameter of typically 22 μm, which is much larger than conventional atmospheric aerosols. CTM's such as SILAM are able to simulate the dispersion of aerosols such as pollen based on transport with air masses (wind advection), turbulent mixing, gravitational settling (dry deposition), and scavenging by precipitation (wet deposition) (Sofiev et al., 2006).

Here we use a bottom-up approach by inserting maps with birch pollen emission sources into the CTM. This requires the spatial distribution of birch trees. At the European scale such a map was first compiled by Sofiev et al. (2006). For Belgium, a new dataset on the birch pollen emission sources has been developed based on forest inventory data (Verstraeten et al., 2019) and applied in the forecast run of SILAM (Delcloo et al., 2020; Verstraeten et al., 2019). The emission timing and intensity in SILAM is based on the thermal time flowering model which assumes that the timing of birch flowering is mostly driven by accumulated ambient temperature during a certain time period (Linkosalo et al., 2010). The final amount of birch pollen in the air is determined by incorporating short-term meteorological conditions such as wind speed, relative humidity and precipitation rate.

3 Observational Data

ECMWF forecast meteorological data were used to drive the pollen transport model SILAM in an online mode. The simulated pollen levels are evaluated using observations of pollen data on four locations monitored by Sciensano (Hoebeke et al., 2018). For comparison, the ERA5 Reanalysis meteorological data set—available three months after the end of the birch pollen season—is also used to run SILAM. The updated birch pollen emission sources of Belgium were derived from Flemish and Walloon forest inventory data at each sampling plot (for details see Verstraeten et al., 2019) based on the diameter at breast height (DBH) of birch trees. The model set-up in the forecast mode was operational from the 1st April 2021 on, which is somewhat later than the general start of the birch pollen season in Belgium.

4 Results

The daily airborne birch pollen levels for the 2021 season (April–May) for two major monitoring sites (De Haan at the coast side and Brussels) are illustrated in Fig. 1. The birch pollen levels from monitoring sites (in black) are shown, as well as the pollen levels from SILAM simulations based on the ECMWF meteorological forecast data (in orange), the pollen levels from SILAM simulations based on ERA5 (in grey), and the difference of the modelled pollen between both simulations (in yellow). This difference indicates the effect of using different meteo input data on the pollen modelling. Overall, the mean difference for De Haan is daily −4 pollen/m^3, indicating a slight higher pollen amounts from ERA5, but it ranges between 193 and −485, showing that the simulation with ERA5 tends to produce higher peak values. For Brussels, the mean difference is +39 pollen/m^3, with a range of −942 and + 934 suggesting a higher pollen amount with the forecast simulation, and quit some differences in timing of the peaks between both simulations.

The comparison statistics for the four monitoring sites are shown in Table 1 for the forecast and the ERA5 runs.

5 Discussions and Conclusions

We have provided a preliminary evaluation of the forecasted birch pollen levels in Belgium for the current season 2021 using SILAM and evaluated the model performance with four birch pollen monitoring sites. Generally, the model overestimates the amount of birch pollen in the air. Only for the coast side site the R^2 between modelled and observed time series is very high (0.719). For the other sites the R^2 values are substantially lower. The observed birch pollen peak in Brussels at the start of the pollen season on 1st April—one of the highest daily pollen counts ever measured in

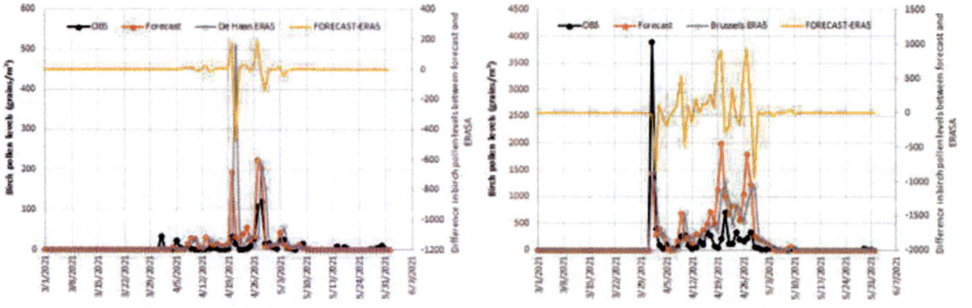

Fig. 1 Left panel, the March–May 2021 time series of daily observed birch pollen levels (in black), the corresponding SILAM simulated levels based on the ECMWF forecast meteo (in orange, April–May), the simulations based on ERA5 meteo (in grey, April–May) and the difference between both simulations (in yellow) for the monitoring site of De Haan (51.274 N, 3.022 E). Right panel, the same for the monitoring site of Brussels (50.825 N, 4.383 E)

Table 1 The model performance statistics for four birch pollen monitoring sites in Belgium (April–May 2021) based on daily pollen grain values. Left panel shows the model run with the forecast meteo of ECMWF, right panel is the model run driven by ERA5 meteo. The monitoring site of Genk is located at 50.965 N, 5.495 E, and Marche-en-Famenne at 50.200 N, 5.312 E

Season 2021	April–May	De Haan	Brussels	Genk	Marche
Model **FORECAST METEO**	Mean	19.04	304.31	600.62	353.40
	Stdev	44.79	464.05	1061.84	456.16
	Max	223	1986	5613	1798
Obs	Mean	9.78	148.33	298.34	139.95
	Stdev	21.54	504.78	572.00	234.87
	Max	120	3892	3809	1117
Model(y)-Obs(x)	Slope	1.805	0.423	0.513	0.942
	Intercept	1.754	241.511	447.480	221.621
	R^2	0.719	0.212	0.076	0.235

Season 2021	April–May	De Haan	Brussels	Genk	Marche
Model **ERA5 METEO**	Mean	21.39	269.87	592.24	284.84
	Stdev	71.46	394.65	1131.97	358.99
	Max	508	1472	6848	1344
Obs	Mean	9.78	148.33	298.34	139.95
	Stdev	21.54	504.78	572.00	234.87
	Max	120	3892	3809	1117
Model(y)-Obs(x)	Slope	1.141	0.447	0.594	0.716
	Intercept	10.972	207.939	424.839	189.338
	R^2	0.111	0.325	0.089	0.218

Belgium—were detected by the model, but at a much lower magnitude. We have also compared the simulated airborne birch pollen levels from the SILAM run driven by the forecast with that driven by ERA5 meteo data and observations. Generally, the model performance is much lower with the ERA5 run. A slightly better correlation between modelled and observed time series was only obtained at the site of Brussels. For the coast side, the ERA5 run degrades the model performance substantially, due to the fact that this meteo data set does not capture the sea breeze well. Overall, the pollen peaks simulated with ERA5 are 1–2 days later then simulated with the forecast data. For the coast side, the peaks from the forecast simulations are more in agreement with the observations, while for the Brussels case, the ERA5 simulations are closer to the observed peaks, but differ strongly in magnitude. Replacing the birch pollen emission map used in the simulations of season 2021 with maps that were produced for other seasons (Verstraeten et al., submitted) only have minor impacts on the model performance. A preliminary comparison with the 2021 CAMS dataset shows (see Table 2) that the correlations with observations are in general lower than the obtained values in Table 2. For some locations, some CAMS models might have a slightly better R^2.

All CAMS models show the same behaviour for the pollen monitoring site at the coast side where the daily pollen levels are much better modelled compared to the inland pollen monitoring stations. These models all have the same forecasting meteorology in common, suggesting that the sea breeze is better captured compared to the ERA5 meteorological data (see also Table 1).

Table 2 The R^2 statistics for four birch pollen monitoring sites in Belgium (April–May 2021) based on daily pollen grain values and the forecasts of the different CAMS models and the ensemble

MODEL R^2	De Haan	Brussels	Genk	Marche-en-Famenne
ENSEMBLE	0.755	0.010	0.100	0.217
SILAM_CAMS	0.663	0.014	0.127	0.294
CHIMERE	0.420	0.001	0.005	0.000
LOTOS	0.559	0.003	0.018	0.092
EMEP	0.229	0.187	0.088	0.142
MOCAGE	0.668	0.031	0.178	0.201
MATCH	0.390	0.000	0.020	0.245
GEMAQ	0.004	0.000	0.037	0.124
DEHM	0.559	0.060	0.184	0.162
EURADIM	0.583	0.032	0.138	0.326

References

Blomme, K., Tomassen, P., Lapeere, H., et al. (2013). Prevalence of allergic sensitization versus allergic rhinitis symptoms in an unselected population. *International Archives of Allergy and Immunology, 160*(2), 200–207.

Delcloo, A., Verstraeten, W. W., Dujardin, S., et al. (2020). Spatio-temporal monitoring and modelling of birch pollen in Belgium. *Air Pollution Modeling and Its Application XXVI*. https://doi.org/10.1007/978-3-030-22055-6_12

Hoebeke, L., Bruffaerts, N., Verstraeten, C., Delcloo, A., Desmedt, T., et al. (2018). Thirty-four years of pollen monitoring: An evaluation of the temporal variation of pollen seasons in Belgium. *Aerobiologia, 34*, 139. https://doi.org/10.1007/s10453-017-9503-5

Landrigan, P. J., et al. (2017). The lancet commission on pollution and health. Published online October 19, 2017. https://doi.org/10.1016/S0140-6736(17)32345-0

Linkosalo, T., Ranta, H., Oksanen, A., Siljamo, P., Luomajoki, A., Kukkonen, J., & Sofiev, M. (2010). A double-threshold temperature sum model for predicting the flowering duration and relative intensity of Betula pendula and B. pubescens. *Agricultural and Forest Meteorology, 150*(12), 6–11. https://doi.org/10.1016/j.agrformet.2010.08.007

Sofiev, M., Siljamo, P., Ranta, H., & Rantio-Lehtimäki, A. (2006). Towards numerical forecasting of long-range air transport of birch pollen: Theoretical considerations and a feasibility study. *International Journal of Biometeorology, 50*, 392–402. https://doi.org/10.1007/s00484-006-0027-x

Verstraeten, W. W., Dujardin, S., Hoebeke, L., Bruffaerts, N., Kouznetsov, R., Dendoncker, N., Hamdi, R., Linard, C., Hendrickx, M., Sofiev, M., & Delcloo, A. W. (2019). Spatio-temporal monitoring and modelling of birch pollen levels in Belgium. *Aerobiologia, 35*(4), 703–717. https://doi.org/10.1007/s10453-019-09607-w

Verstraeten, W. W. et al. (submitted to AFM in 2021). Under review.

QUESTIONER: Richard Kirchner

QUESTION: Have you considered in your model indexes of rain during the previous autumn or winter?

ANSWER: Since the chemical transport model SILAM is a conceptual model, this is not the case. However, the update of the pollen source emission map might be based using meteorological data from the seasons before. This was not done for this evaluation.

Air Quality Improvements During the COVID-19 Lockdown in Central Europe: Separation of Emission Reduction and Meteorological Impacts

Volker Matthias, Jan Arndt, Ronny Badeke, Lea Fink, Josefine Feldner, Markus Quante, Ronny Petrik, Daniel Schwarzkopf, and Ralf Wedemann

Abstract Corona lockdown measures caused unprecedented emission reductions in many parts of the world. However, this does not linearly translate into improved air quality, since meteorological conditions also have a significant impact on air pollutant concentration patterns. This study disentangles effects of emission reduction and the meteorological situation on the concentrations of air pollutants with a focus on elemental carbon (EC) in Central Europe during the first major lockdown in spring 2020. European emission data for 2016 was updated for 2020 by extrapolating previous emission trends for each country and sector for 2020. Lockdown emission reductions were approximated with daily adjustment factors for the sectors traffic (road, air, and ship), public power and industry and for almost all countries based on mobility data, energy consumption and industrial productivity data. Chemistry transport model results focus on traffic related emissions and show significant reductions between 20% and 55% for NO_2 concentrations during strongest lockdown measures between mid of March and mid of April. $PM_{2.5}$ concentration reductions were less strong compared to NO_2, only up to 15% in large parts of Europe. Elemental carbon (EC) was reduced by up to 20% in southern Germany and France. Exceptionally low EC concentrations, as observed in April in Northern Central Europe, were not an effect of lockdown measures. They were caused by advection of clean air from Scandinavia. To investigate this further, EC concentrations as well as lockdown emission reductions of EC were also calculated with meteorological data for the same spring period in 2016 and 2018. It could be seen that meteorological effects resulted in modified effects of the lockdown on EC concentration changes, particularly in Denmark and northern Germany in April 2020.

V. Matthias (✉) · J. Arndt · R. Badeke · L. Fink · J. Feldner · M. Quante · R. Petrik · D. Schwarzkopf · R. Wedemann
Institute of Coastal Environmental Chemistry, Helmholtz-Zentrum Hereon, Max-Planck-Strasse 1, 21502 Geesthacht, Germany
e-mail: volker.matthias@hereon.de

C. Mensink and O. Jorba (eds.), *Air Pollution Modeling and its Application XXVIII*, Springer Proceedings in Complexity, https://doi.org/10.1007/978-3-031-12786-1_17

1 Introduction

As a reaction to the spread of the Corona virus in Europe in spring 2020 European governments decided for significant lockdown measures in order to minimize personal contacts thereby avoid large infection rates in the European population. This led to an unprecedented traffic reduction in the period between end of February and end of June 2020, with strongest emission reductions in the second half of March and the first half of April. This affected mainly air traffic with more than 80% reduction for many weeks in most European countries. Road traffic was reduced by 30–80%, strongly depending on country. Part of the industrial production was also shut down, leading to additional emission reductions.

These extreme reductions had large effects on observed concentrations of pollutants like NO_2, $PM_{2.5}$, and PM components, like elemental carbon (EC). However, exceptional weather conditions in Central Europe with low rainfall, high solar radiation and advection of clean air from Scandinavia in North Central Europe also affected air quality during the lockdown. Emission modeling that considers lockdown effects in several sectors combined with air quality simulations was applied to quantify the effects of lockdown emission reductions on reduced pollutant concentrations. Through simulations with meteorological data from previous years, effects of the exceptional meteorological situation were additionally investigated.

2 Emission Data and Model Setup

Because emission inventories are usually not earlier available than a few years after the time under consideration, emission data for the year 2020 was constructed through extrapolation of CAMS-REGAP-EU version 3.1 data from 2016 (https://permalink.aeris-data.fr/CAMS-REG-AP, Granier et al., 2019). Individual extrapolation factors were derived for all sectors and countries based on recent emission trends (2014–2016) in EMEP emission data (https://www.ceip.at/webdab-emission-database/reported-emissiondata), however, extreme changes were avoided through limiting the maximum annual change to 10% for all sectors and countries. Subsequently, lockdown adjustment factors (LAFs) were derived for the sectors road traffic, aviation, shipping, industry and public power production. Publicly available statistical data like google mobility data and EUROSTAT data on industrial production were used to construct the LAFs. Reductions were strongest for road traffic and aviation. Details can be found in Matthias et al. (2021).

Gridded emission data was then produced with the HiMEMO emission modeling system developed at Hereon. Chemistry transport model runs for January–June 2020 were performed with CMAQ (v5.2), fed with meteorology data from COSMO-CLM (v16). The domains covered Europe on a $36 \times 36\ km^2$ grid and Central Europe on a $9 \times 9\ km^2$ grid, nested into the larger domain. Chemical boundary conditions for the outer domain were taken from IFS-CAMS (Inness et al., 2019) and did not include effects of lockdown measures outside Europe.

Two model runs were performed. One run considered full emissions for 2020 without lockdown measures (noCOV case), the other run used emissions modified with LAFs for the sectors given above (COV case). Differences between the model runs represent the effects of the lockdown measures. Additional runs were performed with the same emission data but with meteorological data for January–June 2016 and 2018.

3 Concentration Patterns

3.1 Lockdown Effects

In a previous analysis of the lockdown effects on NO_2 and O_3 (Matthias et al., 2021), it was demonstrated that NO_2 was strongly reduced in March and April 2020 due to lockdown measures. The reductions reached mere than 50% on a two weeks average in the second half of March and differed from country to country. On the other hand, O_3 was increased in the beginning of the lockdown period as a consequence of the reduced titration in the northern part of Central Europe and in highly polluted cities. The picture changed into a reduction of ozone concentrations by a few percent in late April and May, most likely because VOC emissions from natural sources increased and photochemical reactions produced more ozone with increased solar radiation levels.

The picture was different for particulate matter ($PM_{2.5}$) and its components. Here, we investigate the effects of the lockdown measures on elemental carbon (EC), a PM component that has a relatively high share in traffic emissions and that is chemically not reactive. Figure 1 (left) shows the average EC concentrations in the first half of April 2020 for the noCOV case and the relative reductions by the lockdown (Fig. 1, right). Highest concentrations between 0.5 and 1 $\mu g/m^3$ are found around harbour cities (Rotterdam and Antwerp) and in the English Channel. Also in the Po valley, high EC concentrations close to 1 $\mu g/m^3$ on average were simulated. Very low concentrations (<0.2 $\mu g/m^3$) prevail in the northeastern part of the Central European model domain. The lockdown measures reduced these concentrations by 10–20% in Germany, France, the Benelux countries, north Italy and the UK (Fig. 1, right). Lower reductions around 5% are found in Denmark, Czech Republic and in the North Sea area. Shipping contributed most to EC concentrations in northwestern Central Europe, but the reductions due to the lockdown were relatively small. Lower EC emissions from traffic led to much larger reductions of EC concentrations around Paris and in the southwestern part of Germany. The EC reductions are lower than those calculated for NO_2 in a previous study (Matthias et al., 2021), but higher than those for $PM_{2.5}$.

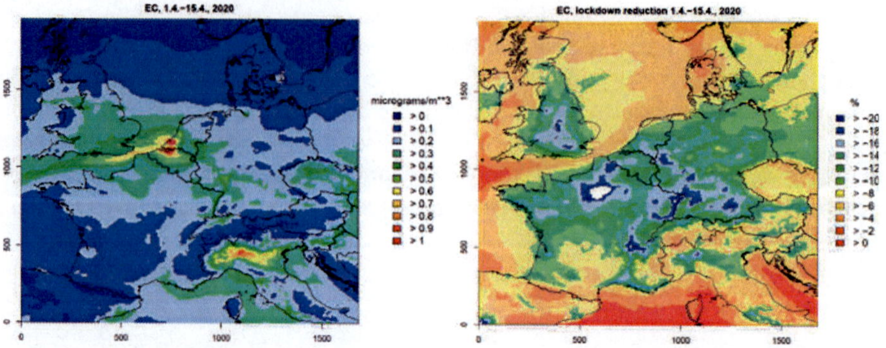

Fig. 1 Average EC concentrations between 1 and 15 April 2020 without lockdown (left) and relative reductions by the lockdown (right)

3.2 *Meteorological Effects*

Meteorological conditions have a large impact on atmospheric concentrations of pollutants. Lockdown effects on EC would be different when weather conditions as in 2016 or 2018 would have prevailed. Figure 2 shows the relative changes of EC reductions caused by the lockdown measures in 2020 compared to 2016 (left) and compared to 2018 (right). The reductions vary by $\pm 20\%$ in both directions in most areas of the model domain. Compared to 2018, reductions in EC in 2020 are relatively high in East Germany and Poland, central France, Switzerland and Austria. Interestingly, in the northeastern part of the domain, i.e. in Denmark, northern Germany and southern Sweden, EC lockdown reductions were much lower in 2020 compared to 2016 and 2018. This is connected with relatively clean air that was advected with north easterly winds from Scandinavia to the western Baltic Sea area. Figure 1a shows the very clean air with EC concentrations below 0.2 $\mu g/m^3$ on a two weeks average for the noCOV case. The model results demonstrate that low concentrations of air pollutants in central Europe during the lockdown period were not necessarily connected to reduced emissions. Meteorological conditions in the first two weeks of April led to very low concentrations of elemental carbon and also of other $PM_{2.5}$ components (see Matthias et al., 2021).

4 Summary and Outlook

Lockdown measures in Europe led to strong emission reductions in the transport sector as well as in industry and power production between March and June 2020. Accordingly, atmospheric concentrations of elemental carbon (EC) decreased by 10–20% in large parts of Central Europe. Air pollutant concentration changes depend on emissions and on meteorological conditions. This holds also for pollutants that

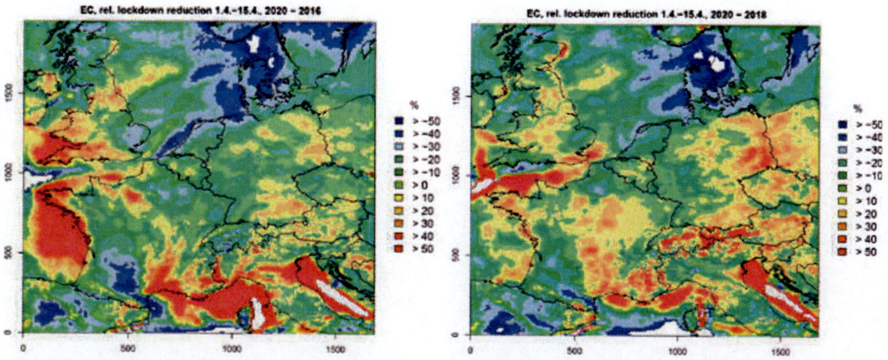

Fig. 2 Relative differences between lockdown reductions of EC concentrations for meteorological conditions in 2016 compared to 2020 (left) and 2018 compared to 2020 (right). Positive values denote higher lockdown reductions in 2020

are chemically inactive, like EC. EC concentrations were very low and lockdown effects very small in the first half of April in Denmark and northern Germany when compared to a situation with meteorological conditions as in 2016 or 2018.

The lockdown in Europe gives a unique opportunity to study effects of emission reductions on air quality. In addition, to gaseous pollutants like NO_2 and O_3, that were already extensively studied, chemical components of $PM_{2.5}$ should be taken into account. They may behave very differently because of their longer lifetime and strong dependence on meteorological conditions.

References

Granier, C., Darras, S., Gon, H. D. v. d., Doubalova, J., Elguindi, N., Galle, B., Gauss, M., Guevara, M., Jalkanen, J.-P., Kuenen, J., Liousse, C., Quack, B., Simpson, D., & Sindelarova, K. (2019). *The Copernicus Atmosphere Monitoring Service global and regional emissions* (April 2019 version), Copernicus Atmosphere Monitoring Service (CAMS) report, https://doi.org/10.24380/d0bn-kx16

Inness, A., Ades, M., Agustí-Panareda, A., Barré, J., Benedictow, A., Blechschmidt, A.-M., Dominguez, J. J., Engelen, R., Eskes, H., Flemming, J., Huijnen, V., Jones, L., Kipling, Z., Massart, S., Parrington, M., Peuch, V.-H., Razinger, M., Remy, S., Schulz, M., & Suttie, M.: The CAMS reanalysis of atmospheric composition. *Atmospheric Chemistry and Physics, 19*, 3515–3556. https://doi.org/10.5194/acp-19-3515-2019

Matthias, V., Quante, M., Arndt, J. A., Badeke, R., Fink, L., Petrik, R., Feldner, J., Schwarzkopf, D., Link, E.-M., Ramacher, M. O. P., & Wedemann, R. (2021). The role of emission reductions and the meteorological situation for air quality improvements during the COVID-19 lockdown period in central Europe. *Atmospheric Chemistry and Physics, 21*, 13931–13971. https://doi.org/10.5194/acp-21-13931-2021

Analysis of Emission-Driven Changes in the Oxidation Capacity of the Atmosphere in Major European Urban Cities

Daeun Jung, David de la Paz, Alberto Notario, and Rafael Borge

Abstract Air quality is highly affected by emissions generated by human activities. Anthropogenic emissions in Europe have gradually been reduced thanks to a combination of factors, including restrictive regulation and policy implementation, fuel switching, technological developments, and improved energy efficiencies. Many measures have been specifically introduced to meet the NO_2 limit values for the protection of human health, primarily targeting traffic emissions. Due to NO_X reduction policies in Europe, NO_2 levels have been generally reduced, but O_3 concentrations have been found to increase. This phenomenon would cause changes in the oxidant capacity of the atmosphere, promoting the formation of tropospheric oxidants. In order to observe the occurrence in Europe, the Community Multiscale Air Quality Modeling System (CMAQ) is applied in this study. We examine the variation for two months representative of winter and summer conditions (January and July) in the period of 2007–2015. In addition to observing the general situation, we selected 67 European large cities (those with more than 0.5 million inhabitants in 2015) to analyze the trends of pollutants along with NO_X emission changes by means of statistical methods. Additionally, we characterized the phenomenon of general changes in the oxidation capacity of the atmosphere dividing into five groups. As a general result, we found that NO_X emissions and NO_2 concentrations decreased in the period in both seasons. On the other hand, concentrations of O_3 and other main tropospheric oxidants (OH and NO_3) increased in winter but decreased in summer.

D. Jung (✉) · D. de la Paz · R. Borge
Laboratory of Environmental Modelling, Department of Chemical and Environmental Engineering, Escuela Técnica Superior de Ingenieros Industriales, Universidad Politécnica de Madrid (UPM), Madrid, Spain
e-mail: daeun.jung@upm.es

D. de la Paz
e-mail: david.delapaz@upm.es

R. Borge
e-mail: rafael.borge@upm.es

A. Notario
Physical Chemistry Department, Faculty of Chemical Science and Technologies, Universidad de Castilla-La Mancha, Ciudad Real, Spain
e-mail: alberto.notario@uclm.es

C. Mensink and O. Jorba (eds.), *Air Pollution Modeling and its Application XXVIII*, Springer Proceedings in Complexity, https://doi.org/10.1007/978-3-031-12786-1_18

However, the OH radical also shows an increase in urban areas in summer. Finally, HNO_3 shows a negative trend in general.

Keywords Oxidation capacity · CMAQ modeling · Tropospheric oxidants · Europe · Radicals

1 Introduction

Anthropogenic emissions in Europe have been gradually reduced thanks to a combination of factors, including restrictive regulation and policy implementation, fuel switching, technological developments, and improved energy efficiency (EEA, 2020). Many measures have been specifically introduced to meet the limit values of NO_2 for human health, primarily targeting traffic emissions (EEA, 2020). Due to this effort, emissions in Europe were reduced; specifically NO_X emissions decreased by 26% over the last decade (2009–2018) (EEA, 2020). As a result, annual mean concentrations of NO_2 decreased on average by 18–23% depending on the type of monitoring stations, while mean O_3 increased in most of the stations in the EU (EEA, 2020).

O_3 increases in Europe could be closely related to NO_X reduction policies (Henschel et al., 2016) that may have induced unanticipated changes in the oxidant capacity of the atmosphere in urban atmospheres. For example, Saiz-Lopez et al. (2017) found an increase in the formation of tropospheric oxidants (up to 70% and 90% of OH and NO_3, respectively) with a 11% decrease in nitric acid (HNO_3) in Madrid.

In this study, we would like to find out the phenomena in oxidation capacity caused by NO_X and ozone with the main tropospheric radicals such as OH and NO_3 in other large cities in Europe. To do this, we use a state-of-the-art mesoscale air quality modeling system to compare two simulations (2007–2015) across Europe, focusing on how the atmosphere has evolved in that period.

2 Methodology

In the first place, a simulation is performed throughout Europe using the Community Multiscale Air Quality (CMAQ) modeling system with 12 × 12 km resolution grid cells with 39 vertical layers covering all of Europe (Fig. 1). The Weather Research and Forecasting (WRFv 4.2.1) model is used to produce the meteorological inputs needed by the CMAQ model (Shamarock et al., 2008). With respect to the emission information, the Sparse Matrix Operator Kernel Emissions (SMOKEv 3.6.5) (Borge et al., 2008; UNC, 2015) is used during the period (2007–2015). The emission inventories in this study are based on CAMS-REG-APv 2.2.1 (Granier et al., 2019) that derives from TNO-MACC-II and TNO-MACC-III inventories (Kuenen et al., 2014).

The reference year for meteorology is 2015 for the two years to isolate the effects of emission changes and compare to the results of Saiz-Lopez et al. (2017). Besides, the meteorology of 2007 does not exist any significant differences with that of 2015 in average feature. Moreover, the observational data availability for 2015 is better than that for 2007 and therefore convenient for simulation assessment purposes. Since we chose 2015, the performance of the model is assessed for that year.

To focus on the largest agglomerations in Europe, we selected all 67 cities with more than 0.5 million inhabitants in 2015 according to Eurostat (available online at https://ec.europa.eu/eurostat/web/main/data/database, last access: 11 February 2021). Then these cities are divided into five groups according to k-mean clustering analysis with relevant variables, showing substantial differences. The input

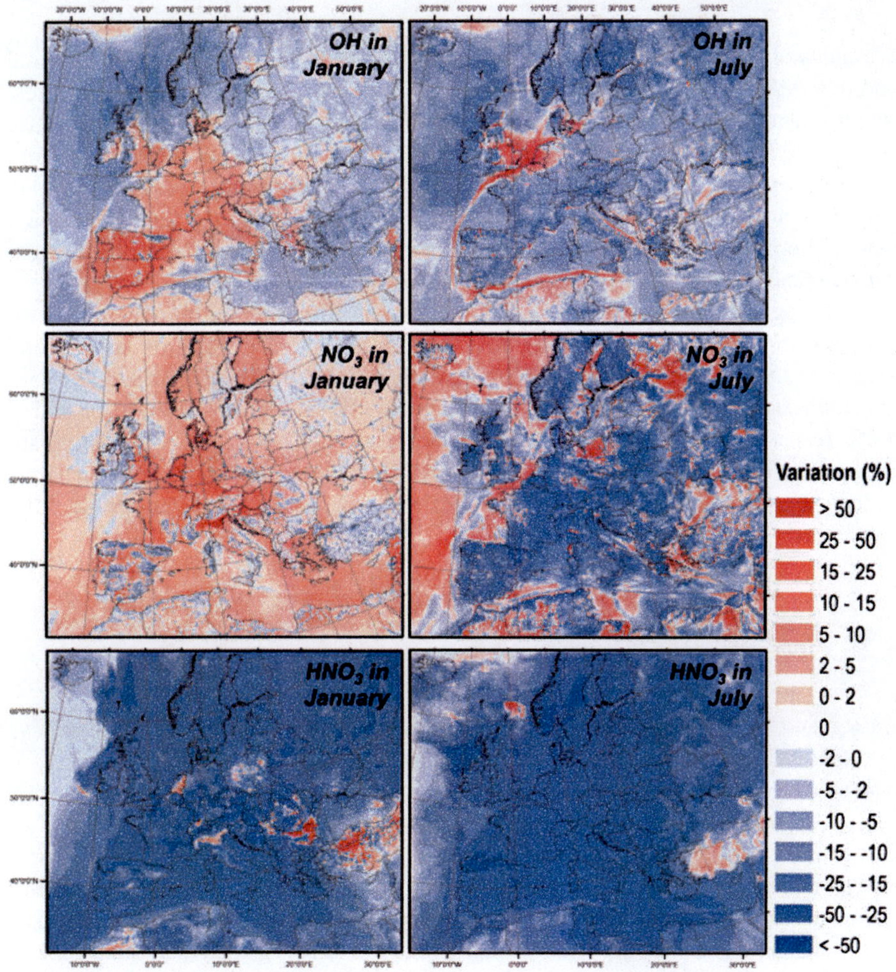

Fig. 1 Variations in atmospheric oxidants in January and July in the period 2007–2015

variables are the variations of pollutants that could have an impact on changes in the oxidation capacity of January and July between 2007 and 2015. The substances are as follows: NO_2, O_3, VOC, HO_2, HONO, H_2O_2, OH, NO_3, and HNO_3.

According to a Principal Component Analysis (PCA), the most relevant variables to segregate these groups are the variations of HO_2 in both months, NO_3 and OH in July and HNO_3 in January.

3 Results

3.1 NO_X Emission Trends

We analyze the variations in NO_X emissions throughout Europe in both the total and the traffic sectors. They mostly decrease but increase in some eastern countries. Regarding total emissions in January, the maximum decrease is 72%, but the increase is up to 207% at the grid cell level. In July, the emissions seem to decrease more intensely in most countries, up to 75%. Traffic emissions also decrease in many countries, and the decrease is relatively homogeneous compared to total NO_X emissions in both months. The country with the largest overall decrease in the 2007–2015 period is the Czech Republic (more than 50%).

Focusing on urban areas, total emissions are reduced on average by 23% in January and 26% in July. The maximum relative decrease is found in Paris in January and Prague in July (−45.1% and −42.3%, respectively). On the contrary, the cities with the smallest decrease are Hamburg in January and Stockholm in July (−8.9% and −8.4%, respectively). The cities with the greatest reduction in NO_X traffic emissions are Prague (−51.8% and −51.2%, in January and July, respectively), while Bucharest is the city with the smallest decrease (−5.8% and −5.1%, in January and July, respectively).

3.2 NO_2 and O_3 Changes

As a result of NO_X emission reductions, NO_2 concentrations generally decrease throughout Europe, noticeably in urban areas. However, O_3 levels clearly increase in January but tend to decrease in July. The increase occurs remarkably in the urban centers of the southern countries, as well as in entire countries in the west. However, the reduction can be found in most countries in July, but according to our simulations, a slight increase occurres from the south of the UK to the North Sea near the continent.

Observing cities specifically, the ambient concentration of NO_2 is reduced in all of them, most evidently in July. The average reduction is 16.3% in January and 27% in July. The maximum decrease can be found in Sheffield and Zagreb (−29.5% and −43.7%, in January and July, respectively) while the minimum corresponds to Krakow and Stockholm (−3.4% and −6.8%, in January and July, respectively). The smallest decrease in NO_2 concentrations in Stockholm in July is consistent with the NO_X emissions changes previously discussed.

Regarding O_3, levels increase in all studied cities in January, while they mostly decrease in July. The average increase in January is 11.8%, ranging from 0.1% in Sofia to 30.8% in Zagreb. In July, the increase is up to 20.2% in The Hague, but the decrease is up to 10% in Wroclaw, and the average change value is −2.9%.

3.3 Trends in Atmospheric Oxidants

Relative changes (in %) in the atomspheric oxidants with HNO_3 can be found in Fig. 1. An increase in OH radicals is shown in most European countries in January. However, a decrease can be observed in mountain areas such as the Pyrenees in the borders of France and Spain, and the Alpes between France and Italy and in northern countries. This radical forms by photolysis of O_3 and a subsequent reaction with water vapor. An additional factor that may have also affected net OH formation is decreases in NO_2 levels since the importance of the NO_2 + OH reaction (NO_2 + OH → HNO_3) must have also decreased reacting less with OH leading to its increase. On the other hand, this radical generally decreases in July, while increases are found along the main shipping routes and from the south of the UK and the north of Belgium and the Netherlands. In addition to those areas, the increase is also observed in cities. This phenomenon appears to be due to the decrease in the NO_2 + OH reaction with the deeper decrease in NO_2 reported, compared to January. It leads to a higher increase in OH since the reaction consumes even less OH. This radical increases on average by 1.4% and 6% in January and July, respectively, in the selected cities. Specifically, OH levels increase the most in Malaga (20.9%) and decrease in Stockholm (−11.5%) in January, while The Hague shows the greatest increases (76.8%) and Zagreb the largest reduction (−16.2%) in July.

A generalized increase is also found in the case of the NO_3 radical, the main nocturnal oxidant in January. The model shows an increase in most areas and is more noticeable in urban environments in January. This enhanced oxidation capacity in the nocturnal atmosphere has important implications for air quality due to NO_3-mediated oxidation of VOCs. However, concentrations tend to decrease in July except in the English Channel, the north of Poland, and some areas in Sweden. This reduction causes a sharp reduction in NO_2 levels in July and less availability of O_3. In urban centers, this radical mostly increase in January (10.9% on average) and Zagreb has the highest value (59.6%), while in Krakow it decreases the most (−12.2%). On the other hand, the radical generally decrease in July (−24% on average), ranging from −43.8% in Toulouse to 32.9% in The Hague.

HNO_3 is also a very important pollutant that has probably been affected by the reduction in ambient NO_2 concentration levels driven by NO_X emission reductions. HNO_3 can be favored by OH during day via NO_2 + OH → HNO_3, and by NO_3 during the night via NO_2 + NO_3 → N_2O_5, and heterogeneous reaction (N_2O_5 + H_2O →

Table 1 Urban cities of each cluster according to the K-mean cluster analysis

Cluster	Cities
1	Copenhagen, Lisbon, Valencia, London, Birmingham, Leeds, Glasgow, Bradford, Liverpool, Manchester, Sheffield
2	Naples, Amsterdam, Madrid, Barcelona, Helsinki, Thessaloniki, Athens, Rotterdam, Marseille, Genoa, Oslo, Sevilla, Zaragoza, Málaga
3	Frankfurt am Main, Stuttgart, Bremen, Rome, Milan, Turin, Munich, Cologne, Leipzig, Düsseldorf, Krakow, Wroclaw, Lodz, Poznan, Warsaw, Stockholm
4	Vienna, Brussels, Antwerp, Zagreb, Paris, Lyon, Toulouse, Lille, Nice, Hannover, Nuremberg, Budapest, Dublin, Palermo, Sofia, Praha, Bordeaux, Berlin, Hamburg, Essen, Dresden, Dortmund, Riga, Vilnius, Bucharest
5	The Hague

$2HNO_3$). Our results show a HNO_3 reduction in most of the European continent in both months, while a slight increase is shown in the north of Italy and the Netherlands in January. As a consequence of this reduction, the acidification of the atmosphere and the ammonium nitrate in $PM_{2.5}$ aerosols could have been significantly reduced in this period. With respect to urban areas, the greatest decrease is shown in Vienna and Prague in January and July (−67.7% and −45.8%, respectively).

3.4 Cluster Analysis

The resulting groups of the clustering procedure are shown in Table 1.

In cluster 1, there are 11 cities. It is clear that NO_2 and VOC concentrations decrease with a reduction in NO_X emissions. The O_3 concentration generally increase in both months, although some cities show a slight decrease in July. In addition, as O_3 increases, NO_3 also increases. It seems that OH is strongly related to HO_2 since they increase together in both months.

In cluster 2, there are 14 cities, and, in general, the O_3 concentration increases in all cities in January, but we find decreases in July. It seems that O_3 and OH have a strong correlation with NO_3 and HO_2, respectively. Furthermore, in July, VOC is related to HO_2 and OH, since when VOC decreases less, the other substances increase less or even decrease.

In cluster 3, 16 cities are included. NO_X emissions and NO_2 and VOC are very related. Specifically, VOC has a good correlation with NO_X traffic emissions. NO_2 is relevant to NO_X traffic emissions in January, but to the total in July. In this cluster, O_3 is strongly related to NO_3 and HNO_3 while OH to HO_2.

Cluster 4 is the largest group with 25 cities. Generally, the positive relationship between O_3 and NO_3 is evident due to the increase in NO_3 in January and the decrease in July with changes in O_3. On the other hand, it seems that HO_2 affects changes in NO_3 and OH in January, as the two substances show a similar tendency with changes in HO_2.

Lastly, cluster 5 has one city, which is The Hague. This is because the highest increases are found in atmospheric oxidants (OH, HO_2 and NO_3) with the greatest increase in O_3 in July among the other cities.

4 Conclusions

Generalized reductions in NO_2 concentration have occurred in Europe as a result of NO_X emissions abatements, causing a general increase as well in O_3 concentrations. As a consequence, the main atmospheric day oxidant, OH radical, generally increased during the period, especially in urban areas. The nighttime radical, NO_3, increased in winter, while decreased in summer. The other important pollutant, HNO_3, shows a reduction in both seasons, affected by the reduction in NO_X levels.

We examined the changes in the 67 largest European cities and divided them into five groups to understand how different urban environments may react to NO_X emission reductions. Division was performed according to k-mean cluster analysis with relevant variables that are variations of atmospheric oxidants.

In this study, we confirm that the oxidant capacity of different urban atmospheres in Europe has changed very differently across the continent in the 2007–2015 period, but more studies and more specific analyses are needed to better understand this phenomenon.

References

Borge, R., Lumbreras, J., & Rodríguez, E. (2008). Development of a high-resolution emission inventory for Spain using the SMOKE modelling system: A case study for the years 2000 and 2010. *Environmental Modelling and Software, 23*, 1026–1044. https://doi.org/10.1016/j.envsoft.2007.11.002

EEA. (2020). Air Quality e-Reporting (AQ e-Reporting) [WWW Document]. European Environment Agency. https://www.eea.europa.eu/data-and-maps/data/aqereporting-8

Granier, C., Darras, S., van der Gon, H.D., Jana, D., Elguindi, N., Bo, G., Michael, G., Marc, G., Jalkanen, J.-P., & Kuenen, J. (2019). The copernicus atmosphere monitoring service global and regional emissions (April 2019 version).

Henschel, S., Le Tertre, A., Atkinson, R. W., Querol, X., Pandolfi, M., Zeka, A., Haluza, D., Analitis, A., Katsouyanni, K., Bouland, C., Pascal, M., Medina, S., & Goodman, P. G. (2016). Trends of nitrogen oxides in ambient air in nine European cities between 1999 and 2010. *Atmospheric Environment, 117*, 234–241. https://doi.org/10.1016/j.atmosenv.2015.07.013

Kuenen, J. J. P., Visschedijk, A. J. H., Jozwicka, M., & Denier Van Der Gon, H. A. C. (2014). TNO-MACC_II emission inventory; a multi-year (2003–2009) consistent high-resolution European emission inventory for air quality modelling. *Atmospheric Chemistry and Physics, 14*, 10963–10976.

Saiz-Lopez, A., Borge, R., Notario, A., Adame, J. A., Paz, D. de la, Querol, X., Artíñano, B., Gómez-Moreno, F.J., & Cuevas, C. A. (2017). Unexpected increase in the oxidation capacity of the urban atmosphere of Madrid, Spain. *Scientific Reports, 7*, 45956. https://doi.org/10.1038/srep45956, https://www.nature.com/articles/srep45956#supplementary-information

Shamarock, W., Klemp, J., Dudhia, J., Gill, D., Barker, D., & Duda, M. (2008). A 497 description of the advanced research WRF version 3. NCAR technical note 498 NCAR/TN/u2013475.
University of North Carolina at Chapel Hill (UNC). (2015). SMOKE's v365 user's manual.

Questions and Answers

Questioner 1: Oriol Jorba.

Question: What is the sensitivity of your results to different meteorological years?

Answer: Meteorology is very important for ambient concentrations of pollutants due to influences on chemical reactions and vertical mixing. As we focused on the meteorology of 2015 to isolate the effect of the emission changes in this study, the potential influence of meteorological factors was not studied. According to the previous study (Saiz-Lopez et al., 2017), the meteorological factors between 2007 and 2014 were statistically insignificant. Therefore, we thought that the changes during the study period in meteorological conditions were supposed to be small. However, we would like to include this factor in further studies to confirm specific and possible differences.

Questioner 2: Joachim Fallmann.

Question: How are urban areas represented/parametrized? Canopy Model? How are urban areas resolved considering coarse resolution? I guess sometimes only ~1 grid cell.

Answer: In this study, we used CMAQ model with a 12 × 12 km resolution and the domain covers the whole European continent. As a result, urban areas are represented as one grid cell of the resolution. Due to this coarse resolution, pollutant concentrations are supposed to be represented as urban background sites, which are also used for the model assessment in the study. According to the assessment of the model, the CMAQ model can reproduce the concentrations of atmospheric chemical species in Europe, including urban areas with reasonable correlation coefficients for NO_2 and O_3 (0.60 and 0.44 for NO_2 in January and July, respectively, and 0.59 and 0.68 for O_3 in January and July, respectively). This enables us to investigate changes in the oxidation capacity associated with other relevant chemical species.

Attribution of the Californian Fire Emissions to the Surface Pollutant Levels in Sweden

Ana Cristina Carvalho, Lennart Robertson, and Manu Anna Thomas

Abstract In early September 2020, California experienced a combination of meteorological conditions leading to an "explosive fire growth" period. According to the estimates from ECMWF, the carbon levels from forest fires over California and Oregon peaked on the 16th of September. The amount of burned forest gave a distinct signal of the emitted pollutants that could be traced across the Atlantic reaching Europe within a few days, whilst undergoing chemical transformations. During the same period, Sweden experienced high hourly concentrations of PM10 in some air quality stations with values exceeding 150 $\mu g \cdot m^{-3}$. To understand the role of the contributing sources to these concentrations, a study was designed using the in-house chemical transport model MATCH. Two domains of simulations were set, one over the Northern Hemisphere and the other over Europe, both forced by the IFS-ECMWF meteorology. The European domain was defined according to the one used on the Copernicus Atmospheric Monitoring Service regional (CAMS regional) daily production. To trace the particulate matter from the fires transported from outside the domain a MATCH hemispheric run was set. The hemispheric run was restricted to cover PM fire emissions (from GFAS) and added to the boundaries of the European domain. This way the transported particulate matter from the fires occurring elsewhere to the boundaries of the European domain are uniquely tagged making it possible to follow their transport over Europe and to account for their contribution to the measured concentrations. The MATCH model results show a signal of the transported aerosol emitted by the fires over California when compared to the CALIPSO satellite data, in the free troposphere. Still it remains to conclude on the in-situ measurements.

A. C. Carvalho (✉) · L. Robertson · M. A. Thomas
Swedish Meteorological and Hydrological Institute, SMHI FoUmmk, 60176 Norrköping, Sweden
e-mail: ana.carvalho@smhi.se

L. Robertson
e-mail: lennart.robertson@smhi.se

M. A. Thomas
e-mail: manu.thomas@smhi.se

C. Mensink and O. Jorba (eds.), *Air Pollution Modeling and its Application XXVIII*, Springer Proceedings in Complexity, https://doi.org/10.1007/978-3-031-12786-1_19

1 Introduction

During the month of September 2020, several outliers in the PM10 measurements were detected over the Scandinavian countries, with concentrations higher than 150 $\mu g \cdot m^{-3}$ in least at three different locations on different days, between the 11th and the 19th of September. Given the emissions and chemical boundaries used for operational purposes, MATCH did not reproduce this elevated PM10 levels at high latitudes. However, this period coincides with the arrival of the air masses transported from North American subcontinent carrying the plumes of the fires occurred in California, as it was well monitored by the CAMS-Copernicus service (URL 1).

In order to attribute the high-altitude long-range transport of pollutants from across the Atlantic to near surface elevated concentrations at the Scandinavian in situ stations, one should consider the vertical exchange of air masses between the lower stratosphere and lower troposphere. One such example is the ozone transport which is mainly explained by different processes associated with low pressure systems and cut-off lows (Johnson & Viezee, 1981; Davies & Schuepbach, 1994). Several studies and review papers associate the folding of these weather systems with ozone concentration measured at surface, both in mountains as in other air quality stations (as examples: Monks, 2000; Elbern et al., 1998; San José et al., 2005, and recently Akritidis et al., 2018). However, physical processes also affect the distribution of other components in the free troposphere. Birmili et al. (2010) and the references therein suggested that the main mechanism by which the North American polluted air masses influence the air quality measurements over Europe is by lifting and long-range transport via warm conveyor belts. A statistical analysis of the aerosol vertical distribution from Calipso passes over Sweden indicated the presence of the aerosols in the free troposphere, above 3 km during the winter half of the year (Thomas et al., 2013). It was also observed that weather conditions such as north westerly winds and an enhanced positive phase of north Atlantic oscillation favour the long-range transatlantic transport of pollutants into the Nordic countries (Thomas & Devasthale, 2014).

In the light of the aforementioned studies, in this study, we investigate the attribution of the different processes that led to the observed elevated surface PM10 concentrations over the Scandinavian countries. We hypothesize that these high surface concentrations are as a result of the long-range transport of pollutants from the explosive fires in California that emitted the pollutants to high altitudes and a subsidence to the surface over the northern part of Europe via a deep fold that occurred during that time.

2 Methods

The preferred tool used is the chemistry transport model MATCH (Langner et al., 1998; Robertson et al., 1999), which is part of the ensemble of the Copernicus

CAMS regional services (URL 2 and URL 3), where it is validated both inside the consortium of models but also independently through other CAMS services and other model applications (Plu et al., 2021).

To look into the sources and mechanisms responsible for the high PM10 concentrations mentioned in the introductory part, the MATCH runs were perform using two domains: The European domain and a North Hemispheric one. The European domain has the same horizontal and vertical extension to the one used for the daily Copernicus Atmospheric Monitoring Service regional (CAMS regional) daily production; the chemical boundary conditions imposed are taken from the CAMS global archive, and the emissions treated according to their sources: the anthropogenic are defined by CAMS-REG-AP version 3.1, monoterpene, sesquiterpene and isoprene are calculated on-line, and the fire pollutant fluxes are given by the hourly Global Fire Assimilation System—GFAS product; The hemispheric run was restricted to cover PM fire emissions (from GFAS) represented as fine and coarse mode and added to the boundaries of the European domain as tagged species to enable distinct contribution to the measured concentrations; The period of simulation spans the interval between the 5th and 19th of September 2020, in order to assure a proper spin-up of the MATCH model.

3 Results

Hourly real time preliminary data of the PM10 surface measurements showed very high values in different days during the first three weeks of the September 2020, at least. The high concentrations values where measured over station in central and southern Sweden always during the morning (between 7 and 10 UTC), ranging from 153 up to 546 $\mu g \cdot m^{-3}$. However, the MATCH model was not able reproduce the measured peak concentrations, they were in fact one order of magnitude lower.

To understand if the MATCH model could reproduce the signal of the dynamic process behind the long-range transport, the simulated PM10 concentrations were qualitatively compared with the aerosol subtype classification applied to the measured data (URL 4) on the sensor on board of the CALIPSO satellite. As an example, we present the results for September 17th. The MATCH results' cross sections with hourly mean results of PM10 concentration, and the species tagged to follow the fire particulate matter (CAL_PM) entering from the boundaries through the simulation of the MATCH hemispheric run were plotted (Fig. 2) over the CALIPSO track under the red box depicted in Fig. 1.

In spite of the very low concentrations, the CAL_PM cross section helps to understand that there are possible two regions influencing the PM10 concentration fields, the fires occurring in the E and SE part of Europe, which are transported at lower altitudes, and the transport of the fires occurring over North America that travels at higher altitudes. The vertical transport to lower altitudes is achieved in regions where the vertical wind is downwards, which is consistent with the weather system over

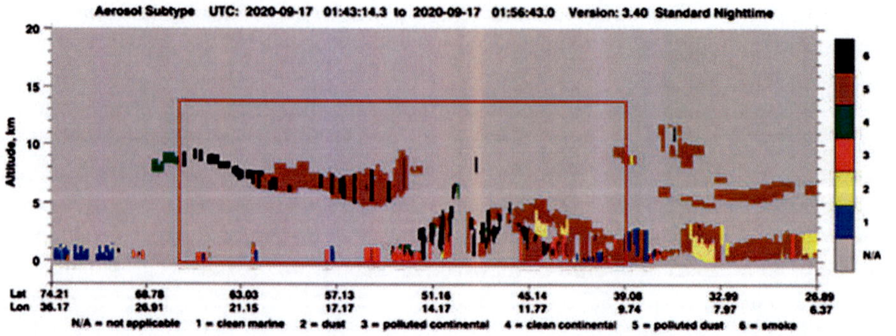

Fig. 1 Aerosol subtype classified over the track of lidar sensor on board CALIPSO (URL 4) passing near Sweden in the 17th of September 2020 (the red box represents the area the cross section of the MATCH model results in Fig. 2)

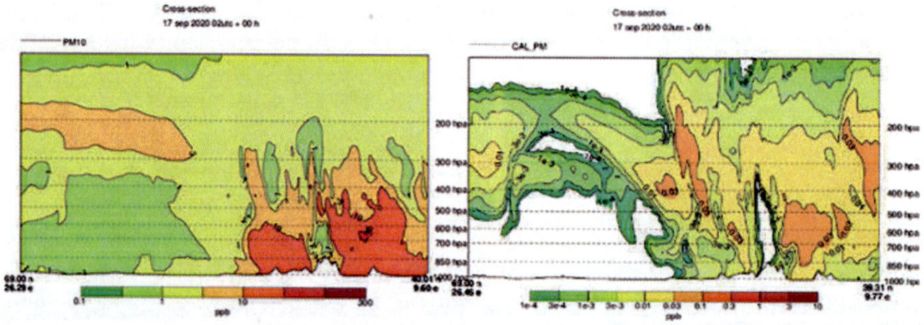

Fig. 2 Cross-sections of the PM10 and tagged specie (CAL_PM) of the fires entering and transported through the European MATCH domain

the region, a low-pressure system over Eastern Finland and a high-pressure located in the coast of Norway.

4 Final Remarks

Tagging species in limited area air quality models is a valuable tool when the processes leading to high concentration over are complex and need accountability to explain how surface concentrations can be attained. The work shows that the dynamical processes can be reproduced, however more effort is needed to accurately simulate the observed concentrations.

Acknowledgements The present work was funded by the Forte (2019-01550) for the Belmont Forum ACRoBEAR Project.

References

Akritidis, D., Katragkou, E., Zanis, P., Pytharoulis, I., Melas, D., Flemming, J., Inness, A., Clark, H., Plu, M., & Eskes, H. (2018). A deep stratosphere-to-troposphere ozone transport event over Europe simulated in CAMS global and regional forecast systems: Analysis and evaluation. *Atmospheric Chemistry and Physics, 18*, 15515–15534. https://doi.org/10.5194/acp-18-15515-2018

Birmili, W., Goebel, T., Sonntag, A., Ries, L., Sohmer, R., Gilge, S., Levin, I., & Stohl, A. (2010). A case of transatlantic aerosol transport detected at the Schneefernerhaus Observatory (2650 m) on the northern edge of the Alps. *Meteorologische Zeitschrift, 19*(6), 591.

Davies, T. D., & Schuepbach, E. (1994). Episodes of high ozone concentrations at the earth´s surface resulting from transport down from the upper troposphere/lower stratosphere: a review and case studies. *Atmospheric Environment, 28*(1), 53–68.

Elbern, H., Hendricks, J., & Ebel, A. (1998). A climatology of tropopause folds by global analysis. *Theoretical and Applied Climatology, 59*, 181–200.

Johnson, W. B., & Viezee, W. (1981). Estratosferic ozone in the lower troposphere—I. Presentation and interpretation of aircraft measurements. *Atmospheric Environment, 15*(7), 1309–1323.

Langner, J., Bergström, R., & Pleijel, K. (1998). *European scale modeling of sulphur, oxidized nitrogen and photochemical oxidants. Model dependent development av evaluation for the 1994 growing season.* SMHI report, RMK No. 82, Swedish Meteorological and Hydrological Institute, Norrköping, Sweden.

Monks, P. S. (2000). A review of the observations and origins of the spring ozone maximum. *Atmospheric Environment, 34*, 3545–3561.

Plu, M., Scherllin-Pirscher, B., Arnold Arias, D., Baro, R., Bigeard, G., Bugliaro, L., Carvalho, A., El Amraoui, L., Eschbacher, K., Hirtl, M., Maurer, C., Mulder, M., Piontek, D., Robertson, L., Rokitansky, C.-H., Zobl, F., & Zopp, R. (2021). A tailored multi-model ensemble for air traffic management: Demonstration and evaluation for the Eyjafjallajökull eruption in May 2010. Natural Hazards and Earth System Sciences Discussions. https://doi.org/10.5194/nhess-2021-96

Robertson, L., Langner, J., & Engardt, M. (1999). A Eulerian limited-area atmospheric transport model. *Journal of Applied Meteorology, 38*, 190–210.

San José, R., Stohl, A., Karatzas, K., Bohler, T., James, P., & Pérez, J. L. (2005). A modelling study of an extraordinary night time ozone episode over Madrid domain. *Environmental Modelling and Software, 20*, 587–593.

Thomas, M., Devasthale, A., & Kahnert, M. (2013). Exploiting the favourable alignment of CALIPSO's descending orbital tracks over Sweden to study aerosol characteristics. *Tellus B: Chemical and Physical Meteorology, 65*, 1. https://doi.org/10.3402/tellusb.v65i0.21155

Thomas, M. A., & Devasthale, A. (2014). Sensitivity of free tropospheric carbon monoxide to atmospheric weather states and their persistency: An observational assessment over the Nordic countries. *Atmospheric Chemistry and Physics, 14*, 11545–11555. https://doi.org/10.5194/acp-14-11545-2014

URL 1: https://atmosphere.copernicus.eu/cams-monitors-smoke-release-devastating-us-wildfires

URL 2: https://atmosphere.copernicus.eu/sites/default/files/2018-02/CAMS50_factsheet_201610_v2.pdf

URL 3: https://regional.atmosphere.copernicus.eu/index.php?category=ensemble&subensemble=hourly_ensemble&date=LAST&calculation-model=ENSEMBLE&species=o3&level=SFC&offset=000

URL 4: https://www-calipso.larc.nasa.gov/products/lidar/browse_images/

Wildfire Emissions and Atmospheric Dispersion

Karen De Causmaecker, Alexander Mangold, and Andy W. Delcloo

Abstract The austral summer of 2019–2020 was marked by a severe wildfire season. Continuous wildfires including many pyrocumulonimbus events were observed in Australia. On a global level climate change is expected to lead to a change in intensity, extent and location of wildfires. Intense wildfires inject a huge amount of pollutants in the air which can reach the upper troposphere and sometimes even the stratosphere. The smoke affects the nearby environment but can also indirectly affect the climate in various ways. Therefore, it is important to understand fire emissions and their atmospheric transport well. We use fire emissions from the CAMS Global Fire Assimilation System (GFAS) to run the Lagrangian particle model Flexpart and we investigate how well this model setup performs. We compare our results directly with observations from ground-based measurement stations as well as with observations from the CALIOP instrument on board of the CALIPSO satellite.

Keywords Wildfires Australia · Fire emissions · Atmospheric dispersion · Flexpart · CAMS GFAS · Lidar measurements · CALIOP

1 Introduction

The consequences of climate change are becoming more and more visible. Among them is the change in intensity, extent and location of forest fires (Zhai et al., 2018). It was found (Max et al., 2020) that the fire frequency could increase over 37.8%

K. De Causmaecker (✉) · A. Mangold
Royal Meteorological Institute of Belgium (KMI), Ukkel, Belgium
e-mail: karen.decausmaecker@meteo.be

A. Mangold
e-mail: alexander.mangold@meteo.be

A. W. Delcloo
Royal Meteorological Institute of Belgium (KMI), Ukkel, Belgium and Department of Physics and Astronomy, Ghent University, Ghent, Belgium
e-mail: andy.delcloo@meteo.be

C. Mensink and O. Jorba (eds.), *Air Pollution Modeling and its Application XXVIII*, Springer Proceedings in Complexity, https://doi.org/10.1007/978-3-031-12786-1_20

and decrease over 8.1% of the global land area during 2010–2039. This tendency seems to come true. In recent years, the number of wildfires in the Arctic tundra is increasing, the fire seasons in western USA are becoming stronger (Zhai et al., 2018) and also the Australian summer of 2019–2020 was marked by continuous fires including many pyrocumulonimbus events (as can be inferred from satellite observations, for example MODIS, 2020, VIIRS, 2020, OMPS, 2020). Wildfires such as these have a large impact on society and the world as a whole and therefore it is important to understand their behavior and consequences well.

In this work, we study the aerosol plumes from the Australian wildfires in the austral summer of 2019–2020 and investigate the consistency between observations and model results. We use emissions from the CAMS Global Fire Assimilation System (GFAS) (Kaiser et al., 2012) to run the Lagrangian particle dispersion model Flexpart (Stohl et al., 2005) and compare the results spatially as well as temporally with observations.

2 Methodology

The CAMS Global Fire Assimilation System (GFAS) v1.2 (Kaiser et al., 2012) provides daily estimates of the emissions from biomass burning. Among other information, they provide information on the location, time and mass flux of the release as well as the altitude of the top and bottom of the plume (ECMWF confluence, 2019). The information is available for many different kinds of smoke constituents with a spatial resolution of 0.1°. In this work, we select the emissions of particulate matter with a diameter $<2.5\ \mu m$, PM2.5. We select GFAS emissions of at least 10 kg per $0.1° \times 0.1°$ grid cell on a daily basis.

We use the Lagrangian particle dispersion model Flexpart 9.02 in the forward mode to calculate the dispersion of the emitted PM2.5 through the atmosphere. To do this, we need to provide the atmospheric wind fields and we rely on three-hourly meteo data from ECMWF ERA5 with 0.5° resolution (Hersbach et al., 2020). Flexpart gives an hourly output of the concentrations. We allow for convection and the particles are kept in the simulation unless they undergo wet or dry deposition.

3 Results and Discussion

3.1 Comparison with Satellite Data

The CALIPSO satellite (CALIPSO, 2020) and in particular the CALIPSO lidar, CALIOP, provides profile observations of clouds and aerosols. We use the CALIPSO Lidar Level 2 Virtual Feature Mask Product, Version 4.20. Since we work with

PM2.5 coming from the bush fires, we only use the relevant aerosol classifications of CALIOP (CALIPSO, 2020).

We focus on GFAS PM2.5 emissions due to fires in the whole southern hemisphere during the period from 27 December 2019 to 7 January 2020. In this period the wildfires in Australia were very intense. We then run Flexpart from 27 December 2019 to 31 January 2020 using the ERA5 ensemble of meteo data.

To compare the Flexpart results with the observations from CALIOP, we select the Flexpart output file which is closest in time to the CALIOP observations of interest. The location and time of the selected measurements is shown in Fig. 1. Then we make a cross section in space of the Flexpart output corresponding to the trajectory of the CALIPSO satellite and compare both profiles with each other. A selection of the results is shown in Fig. 2. We see that the location and altitude of regions with PM2.5 overlap quite well. The agreement is however not perfect.

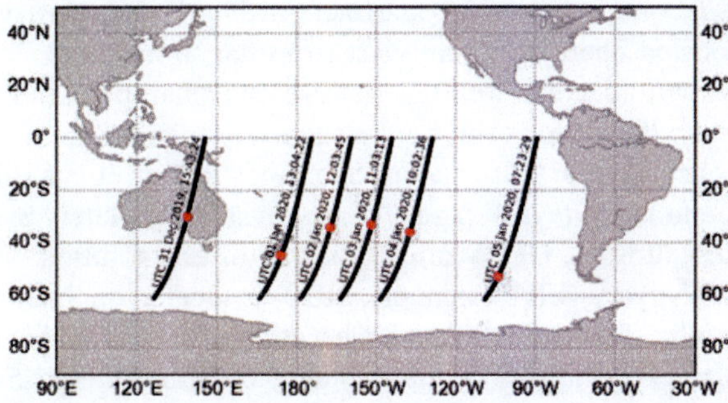

Fig. 1 Trajectories from CALIPSO satellite corresponding to the CALIOP data that are used to compare with the Flexpart results (adapted from CALIPSO, 2020). The date and time of the trajectories are indicated

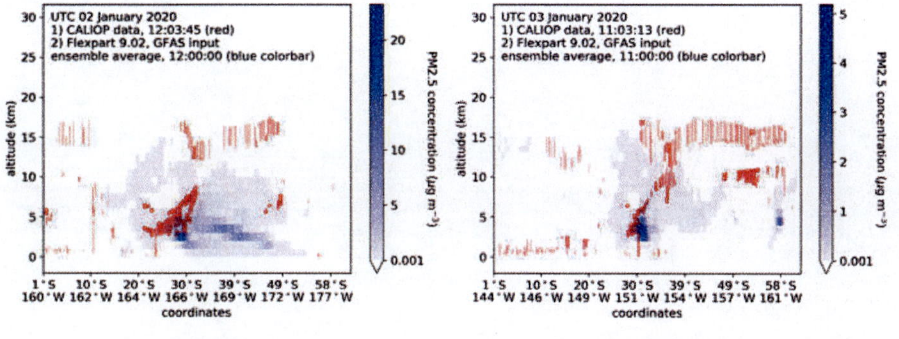

Fig. 2 Comparison between the CALIOP data (red) and the results from the Flexpart–GFAS analysis (blue). The CALIOP data show where the aerosol classifications we selected to correspond with PM2.5 are observed. The Flexpart result shows the PM2.5 concentration. The precise time and position can be read from the figures. *Contains modified Copernicus Atmosphere Monitoring Service Information [2020]*

3.2 Comparison with Air Quality Measurement Stations

A worldwide network of ground-based measurement stations is monitoring the air quality and many of them measure the concentration of PM2.5. These stations measure air pollution in the lowest levels of the atmosphere and also the Flexpart result for these stations is calculated at the ground level. Since the smoke quickly moves up in the atmosphere, we select stations that are located close to the fires. We selected four measurement stations in Australia and two in New-Zealand and compared their measurements of the PM2.5 concentration with the outcome of our analysis.

Figure 3 shows the results for three measurement stations in Australia. The station in Moranbah (Utah Dr) (Queensland Department of Environment and Science, 2020) is located roughly 1300 km to the northeast of the most intense fires close to Sydney. The PM2.5 concentration measured by the station is low and most likely comes from other sources which are not included in our analysis.

The station in Tibooburra (NSW Department of Planning, Industry and Environment, 2020) is located about 1000 km west from the strong fires close to Sydney. In the GFAS–Flexpart analysis, we only observe one small peak in PM2.5 concentration on 6 January 2020. However, on the same day the measurements do show a peak in PM2.5 concentration that is about five times higher. It is unlikely that such high PM2.5 concentrations (>100 $\mu g\ m^{-3}$) are caused completely by sources other than wildfires. Rescaling the GFAS emissions may offer a solution.

The station in Wynyard (EPA Tasmania, 2020) is located on Tasmania, an island southeast of Australia. Tasmania did not have a strong fire season, but the smoke did move from mainland Australia to Tasmania and some peaks in PM2.5 concentration were observed. The little peaks around 30 December 2019 and 6 January 2020 also come back in our analysis and the height of the peaks is in relatively good agreement with the measurements. On 3 January 2020 the station measured a large peak whereas in the analysis only a little peak is seen, about five times lower than the observation.

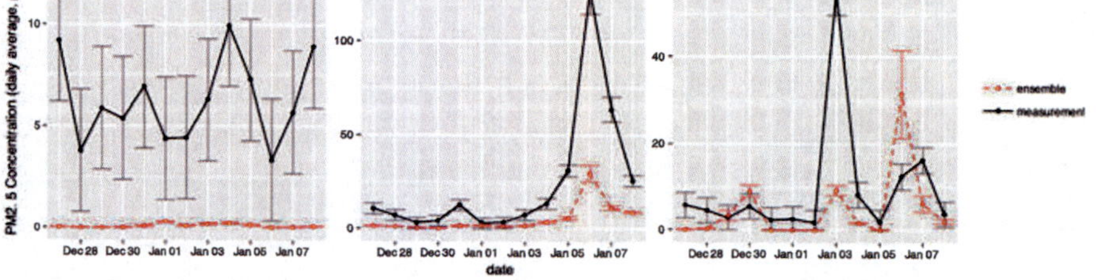

Fig. 3 Daily average of PM2.5 concentrations at three measurement stations in Australia. The black solid lines show the PM2.5 concentrations obtained from the observations, the red dashed lines show the results of the Flexpart analysis. *Contains modified Copernicus Atmosphere Monitoring Service Information [2020]*

4 Discussion and Conclusion

The comparison between the GFAS–Flexpart analysis and the CALIOP results is mostly qualitative and we observe a good overall agreement in height and shape of the plumes, up to 20 km. The plumes higher up in the atmosphere are more difficult to catch by the Flexpart simulations. The peaks in PM2.5 concentration observed by the air quality measurement stations, that are clearly due to the wildfires, are also observed in the GFAS–Flexpart analysis.

However, for the exact height of the peaks, model and observations agree less well. We observe that most of the time the height of the peaks is underestimated but there are also peaks for which our GFAS–Flexpart analysis resulted in an overestimation. This is partially due to other sources of PM2.5 that were not covered in the analysis. We plan to study this more in the future. We also plan to investigate whether rescaling the GFAS emissions could result in a better comparison.

Questions and Answers

Questioner: Patrick Armand.

Question: You carried out direct dispersion simulations. Why not try doing inverse modeling by using the ground and satellite measurements that you have at disposal?

Answer: Thank you for the suggestion. Inverse modeling could definitely be useful to identify the source regions of the measurements and it would aid in identifying the origin of any mismatch observed between the measurements and our model results. We did not do this kind of simulations so far, but if that kind of information would turn out to provide valuable additional insights, we will definitely include them as part of our future research.

References

CALIPSO. (2020). Calipso lidar data base. Received September 30, 2020, from https://www-calipso.larc.nasa.gov/products/lidar

ECMWF Confluence (2019). Global fire assimilation system (gfas). Received October 1, 2020, from, https://confluence.ecmwf.int/pages/viewpage.action?pageId=158643360

EPA Tasmania. (2020). Air. Received September 4, 2020, from https://epa.tas.gov.au/epa/air

Hersbach, H., et al. (2020). The era5 global reanalysis. *Q. J. R. Meteorol. Soc., 146*(730), 1999–2049.

IPCC. (2018). Global warming of 1.5 C. In V. Masson-Delmotte, P. Zhai, H. O. Pörtner, D. Roberts, J. Skea, P. R. Shukla, A. Pirani, W. Moufouma-Okia, C. Péan, R. Pidcock, S. Connors, J. B. R. Matthews, Y. Chen, X. Zhou, M. I. Gomis, E. Lonnoy, T. Maycock, M. Tignor, T. Waterfield

(Eds.), *An IPCC Special Report on the impacts of global warming of 1.5 C above pre-industrial levels and related global greenhouse gas emission pathways, in the context of strengthening the global response to the threat of climate change, sustainable development, and efforts to eradicate poverty*. In Press.
Kaiser, J. W., Heil, A., Andreae, M. O., Benedetti, A., Chubarova, N., Jones, L., et al. (2012). Biomass burning emissions estimated with a global fire assimilation system based on observed fire radiative power. *Biogeosciences, 9*(1), 527–554.
MODIS. (2020). Fire and thermal anomalies, cloud top height and true color corrected reflectance, 2020. NASA Worldview application, part of the NASA Earth Observing System Data and Information System (EOSDIS). Retrieved October 1, 2020, from https://go.nasa.gov/3igYcP9
Moritz, M. A., Parisien, M.-A., Batllori, E., Krawchuk, M. A., Van Dorn, J., Ganz, D. J., & Hayhoe, K. (2012). Climate change and disruptions to global fire activity. *Ecosphere*, *3*(6), art49.
NSW Department of Planning, Industry and Environment. Received September 4, 2020, from (2020). Current and forecast air quality. https://www.dpie.nsw.gov.au/air-quality
OMPS. (2020). Aerosol index (pyrocumulonimbus). NASA Worldview application, part of the NASA Earth Observing System Data and Information System (EOSDIS). Received October 1, 2020, from, https://go.nasa.gov/3jkXX74
Queensland Department of Environment and Science. (2020). Air quality. Received September 4, 2020, from https://apps.des.qld.gov.au/air-quality/
Stohl, A., Forster, C., Frank, A., Seibert, P., & Wotawa, G.(2005). Technical note: The lagrangian particle dispersion model flexpart version 6.2. *Atmospheric Chemistry and Physics*, *5*(9), 2461–2474.
VIIRS. (2020). Fire and thermal anomalies, cloud top height and true color corrected reflectance, 2020. NASA Worldview application, part of the NASA Earth Observing System Data and Information System (EOSDIS). Retrieved October 1, 2020, from https://go.nasa.gov/347v9bK

Ozone Episodes in Northern and Western Iberia: Understanding and Quantifying the Sources and Transport Mechanisms by Integrated Process Analysis

Eduardo Torre-Pascual, Joana Ferreira, Estibaliz Sáez de Cámara, Carla Gama, and Gotzon Gangoiti

Abstract Ozone (O_3) polluted air masses can impact regions of the Iberian Peninsula not traditionally studied, such as Northern and Western Iberia. As a contribution to a better understanding of O_3 episodes' origin, sources, and patterns, we simulated and analysed an O_3 episode recorded in July 2016 all along Northern Iberia and Western Iberia, which was concurrent with a blocking anticyclone over the British Isles. We used WRF-CAMx models and the Integrated Process Rate analysis tool to determine and quantify the most important contributions to the O_3 levels registered in rural areas. Preliminary results show that the high O_3 levels registered at rural areas were caused by the entry, with sea-breezes, of aged pollutants from the coastline.

Keywords Ozone · Northern Iberia · Western Iberia · WRF-CAMx · IPR · Regional transport

1 Introduction

O_3 EU thresholds and target value for protection of human health and ecosystems are exceeded in Northern Iberia (NI)-Spain and Western Iberia (WI)-Portugal when synoptic situations are characterised by blocking anticyclones located over NW

E. Torre-Pascual (✉) · E. S. de Cámara · G. Gangoiti
Faculty of Engineering Bilbao, University of the Basque Country (UPV/EHU), Plaza Ingeniero Torres Quevedo 1, 48013 Bilbao, Spain
e-mail: eduardo.delatorre@ehu.eus

E. S. de Cámara
e-mail: estibaliz.saezdecamara@ehu.eus

G. Gangoiti
e-mail: g.gangoiti@ehu.eus

J. Ferreira · C. Gama
CESAM & Department of Environment and Planning, University of Aveiro, Aveiro, Portugal
e-mail: jferreira@ua.pt

C. Gama
e-mail: carlagama@ua.pt

C. Mensink and O. Jorba (eds.), *Air Pollution Modeling and its Application XXVIII*, Springer Proceedings in Complexity, https://doi.org/10.1007/978-3-031-12786-1_21

France-S British Isles, with special relevance in the mountainous inland areas of both regions (de Blas et al., 2019; Monteiro et al., 2012). In NI, earlier papers have reported the transport mechanisms and pathways behind the build-up of O_3 episodes in this region: O_3 and its precursors are typically transported from the French Atlantic Coast and the English Channel to NI (Gangoiti et al., 2006; Valdenebro et al., 2011; Sáez de Cámara et al. 2018). In WI, previous photochemical simulations have shown that background and boundary conditions are the major contributors for the observed O_3 concentrations (Borrego et al., 2016). Transport mechanisms behind this kind of episodes, the interaction between both regions at different scales, and the quantification of the transported O_3 from other regions are still in the process of being further documented. In view of the above, we have used WRF-CAMx together with the Integrated Process Analysis (IPR) tool to analyse an O_3 episode occurred in July 2016 to perform a comprehensive O_3 transport process characterisation and quantification.

2 Methodology

A summer episode (15–19 July 2016) was selected as a case study. Our analysis is based on the use of CAMx version 6.5 (Comprehensive Air Quality Model with extensions) driven by WRF version 3.9.1.1. (Weather Research and Forecasting Model). The model domain includes the whole Europe and Northern Africa (with a resolution of 27 km, not shown), a first nested domain (d02) with a 9 km horizontal resolution over the Iberian Peninsula (IP), and a second nested domain (d03) over Northern Iberia with a 3 km horizontal resolution (Fig. 1). We used 6 h ECMWF's ERA-Interim reanalysis data and NOAA's OISST to provide initial and boundary conditions to WRF, whereas initial and boundary conditions for CAMx were extracted from CESM2.1: CAM-Chem. The photolysis rates were calculated using the TUV photolysis preprocessor developed by Dr. Madronich (NCAR, 2011) and, O_3 column densities were extracted from TOMS.

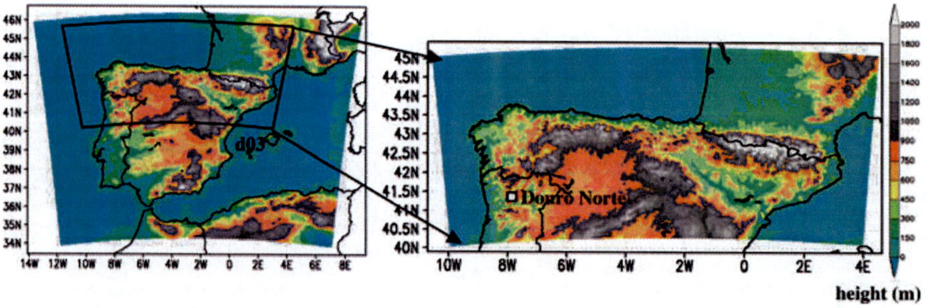

Fig. 1 Topographic map of the domains (d02-left and d03-right) used by WRF and CAMx and the location of the Douro Norte rural air quality monitoring station

Anthropogenic emissions were based on EDGAR 4.3.2 emission inventory (https://jeodpp.jrc.ec.europa.eu/ftp/jrc-opendata/EDGAR/datasets/v432_AP/). The data consisted of monthly average emissions for 2010 of CH_4, CO, NO_x, SO_2, NH_3, NMVOC, for IPCC sectors, on a $0.1° \times 0.1°$ grid. Biogenic emissions were calculated using MEGAN 3.0: this model used hourly meteorological data from WRF output, the adapted Spanish National Forest Inventory according to the method described in Torre-Pascual et al. (2021) and the reprocessed MODIS LAI (Yuan et al., 2011).

The meteorological model validation was performed using available data from monitoring networks and radiosonde data for Lisbon (Portugal) and Bordeaux (France). The simulated O_3 concentrations were compared with the hourly measurements of the European Environmental Agency's (EEA) air quality time series.

We applied the IPR tool for the location of the rural air quality monitoring station of Douro Norte in Portugal, which is located at its north-eastern part, at an altitude of 1086 m ASL. In this station O_3 exceedances are numerous, and the background values are very high. This method provides detailed process rate information for each physical process in CAMx (advection, diffusion, deposition, emissions, and chemistry). The IPR results gave us a complete description of how CAMx simulated O_3 concentrations at this rural mountainous region of Portugal.

3 Results

In 15–20 July 2016 meteorology was characterized by a blocking anticyclone located over NW France-S British Isles, forcing NE surface winds over NI and Northerlies over the coast of Portugal, favoring mesoscale circulations. These conditions, together with cloudless skies and elevated temperatures, promoted O_3 formation and its effective interregional transport. During the earlier stage of the episode, 15–17 July, WI was the most affected by exceedances in the EU O_3 target value for the protection of human health of 120 $\mu g{\cdot}m^{-3}$ as maximum daily 8 h mean, specially in the west coast of Portugal and its inland regions (Fig. 2). In contrast, NI was affected later, 17–19 July.

In order to document O_3 transport in upper layers we calculated the integrated ozone concentration up to 1600 m AGL and represented it together with wind fields at 800 m AGL, both obtained from the WRF-CAMx simulation.

Our results show that during 15 July, O_3 was transported from central IP and NI to Portugal, in both surface and upper layers, caused by NE-E winds (Fig. 2). On 16 July the model simulates O_3-enriched air masses from the Southern Atlantic coast of France being transported to NI, with higher concentrations compared to the previous day of +5 –10 $\mu g{\cdot}m^{-3}$ in surface and +5 –10 ppm·m in upper layers. Concentrations exceeding hourly concentrations of 120 $\mu g{\cdot}m^{-3}$ are simulated inland in Northern Portugal, caused by the onshore transport of pollutants after the onset of the sea-breeze in the afternoon. In the western part of NI exceedances associated to the transport with the NE winds are also simulated. During 17 July sea breeze regimes prevailed over WI and NI causing hourly O_3 concentrations higher than

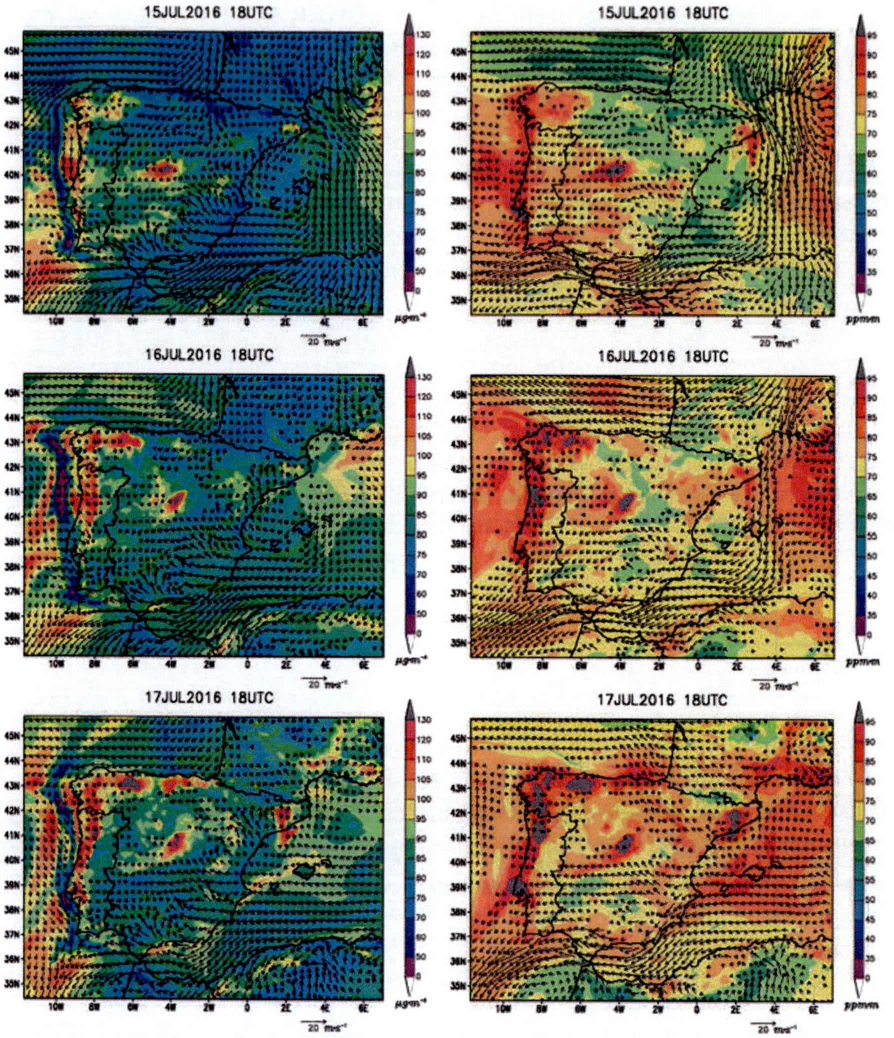

Fig. 2 Simulated O_3 concentrations (color scale) and wind fields (vectors) by WRF-CAMx in d02 at 18 UTC on 15, 16 and 17 July 2016. Left panels show the ozone and wind concentration in surface and right panels the integrated ozone concentration up to 1600 m AGL and wind at 800 m AGL. Winds less than 2 $m \cdot s^{-1}$ have been omitted

120 $\mu g \cdot m^{-3}$ in inland areas. The situation in the upper layers changed compared to the previous day, with SE weak winds during the approach of a cold front from the W.

On 18 July high O_3 concentrations were again simulated for both regions. On the following day (19 July), the cold front produced the transport of pollutants from W to E all over the IP, ending the pollution episode in WI, but not in NI, where the highest concentrations were recorded and simulated. Finally, on 20 July the pollution

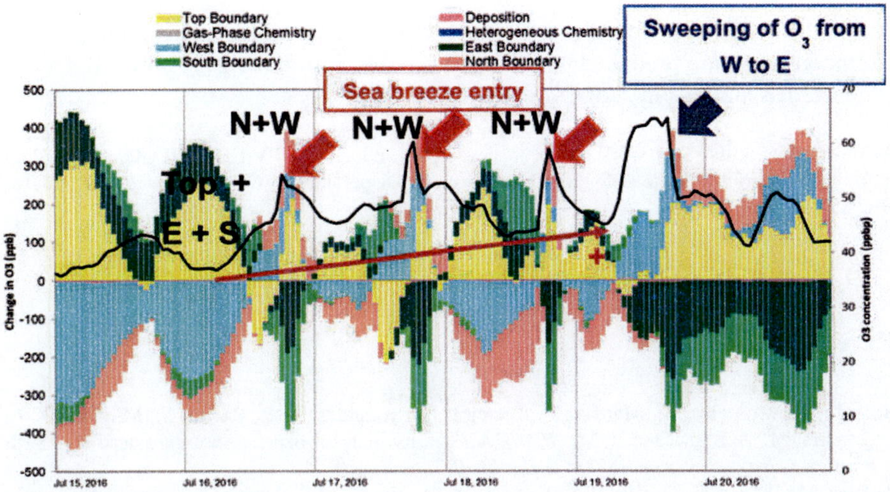

Fig. 3 Results of the contribution of different processes to the simulated O_3 concentrations in Douro Norte for the ozone pollution episode of July 2016

episode ended in both regions with the passage of colder and cleaner air masses with this front.

IPR results for the location of Douro Norte show that simulated peak concentrations in the afternoon are governed by the west and north boundary advection contribution (Fig. 3), being this net contribution of +10 ppb (+20 $\mu g{\cdot}m^{-3}$) in simulated O_3 concentrations (right axis). This fact corroborates the aforementioned influence of the sea-breeze regime in producing target value exceedances in rural or inland areas in these coastal regions. Additionally, O_3 accumulation takes place: the increasingly high daily minimum O_3 concentrations during consecutive days, i.e., from 38 ppb on 16 July to 48 ppb on 17 July, show that top, south and east boundary contribution are also relevant and they do not produce a significant drop in concentrations. This infers that O_3 transport from other regions is produced, from Southern France to the NI and from the NI and central IP to WI, as shown in Fig. 2.

4 Conclusion

O_3 episodes in NI and WI occur due to long-range and regional O_3 transport. This process occurs mainly during the first days of the episode. In WI transport takes place from the center of the IP and NI, whereas in the case of NI from the Southern Atlantic of France. This transport contributes to an increase of +5 –10 $\mu g{\cdot}m^{-3}$ of O_3 per day during the first days of the studied episode. High concentrations were registered on the coast of both regions, but maximum concentrations were recorded in inland areas due to the transport of polluted air masses within the sea breeze regimes during the afternoon. The IPR tool supports these assertions, and it confirms that afternoon peaks are produced by these see breezes, introducing O_3 polluted masses which added +20 $\mu g \cdot m^{-3}$ of O_3.

Acknowledgements This work has been developed under the UPV/EHU PIFG004-2017 PhD fellowship program: "Meteorología, Climatología y Dispersión de Contaminantes" IT1057-16, GIC 15-152.

References

Borrego, C., Monteiro, A., Martins, H., Ferreira, J., Fernandes, A. P., Rafael, S., Miranda, A. I., Guevara, M., & Baldasano, J. M. (2016). Air quality plan for ozone: An urgent need for North Portugal. *Air Quality, Atmosphere & Health, 9*(5), 447–460.

de Blas, M., Ibáñez, P., García, J. A., Gómez, M. C., Navazo, M., Alonso, L., Durana, N., Iza, J., Gangoiti, G., & de Cámara, E. S. (2019). Summertime high resolution variability of atmospheric formaldehyde and non-methane volatile organic compounds in a rural background area. *Science of the Total Environment, 647*, 862–877.

Gangoiti, G., Albizuri, A., Alonso, L., Navazo, M., Matabuena, M., Valdenebro, V., García, J. A., & Millán, M. M. (2006). Sub-continental transport mechanisms and pathways during two ozone episodes in northern Spain. *Atmospheric Chemistry and Physics, 6*, 1469–1484.

Monteiro, A., Strunk, A., Carvalho, A., Tchepel, O., Miranda, A. I., Borrego, C., Saavedra, S., Rodríguez, A., Souto, J., Casares, J., Friese, E., & Elbern, H. (2012). Investigating a high ozone episode in a rural mountain site. *Environmental Pollution, 162*, 176–189.

NCAR. (2011). Tropospheric Ultraviolet and Visible (TUV) Radiation Model|Atmospheric Chemistry Observations & Modeling (ACOM). https://www2.acom.ucar.edu/modeling/tropospheric-ultraviolet-and-visible-tuv-radiationmodel

Sáez de Cámara, E., Gangoiti, G., Alonso, L., Valdenebro V., Aksoyoglu, S., & Oikonomakis, E. (2018). Ozone source apportionment to quantify local-to-continental source contributions to episodic events in Northern Iberia. In C. Mensink & G. Kallos (Eds.), *Air Pollution Modeling and its Application XXV, Springer Proceedings in Complexity* (pp. 361–365). Springer International Publishing.

Torre-Pascual, E., Sáez de Cámara, E., Gangoiti, G., & Zuazo, I. (2021). Biogenic VOC emission modeling for Spain: Adaptation of the national forest inventory as input for MEGANv3. In C. Mensink & V. Matthias (Eds.), *Air Pollution Modeling and its Application XXVII* (pp. 45–50). Springer.

Valdenebro, V., Gangoiti, G., Albizuri, A., Alonso, L., Navazo, M., García, J., Iza, J., & Millán, M. (2011). Build-up and decay of two ozone episodes through northern Iberia and southern France—an inter-regional transport analysis. *Atmospheric Environment, 45*, 1595–1603.

Yuan, H., Dai, Y., Xiao, Z., Ji, D., & Shangguan, W. (2011). Reprocessing the MODIS leaf area index products for land surface and climate modelling. *Remote Sensing of Environment, 115*(5), 1171–1187.

Data Assimilation and Air Quality Forecasting

Recommendations and Generic Data Assimilation Tools for the Improvement of CAMS Regional Air Quality Service

Renske Timmermans, Arjo Segers, Enrico Dammers, Oriol Jorba, Dene Bowdalo, Hilde Fagerli, Alvaro Valdebenito, Augustin Colette, Gaël Descombes, Rostislav Kouznetsov, Andreas Uppstu, and Martijn Schaap

Abstract Within Copernicus Atmosphere Monitoring Service (CAMS) an ensemble of 9 chemistry transport models provides daily analyses and forecasts, as well as annual reanalyses over Europe. The CAMS_61 project was developed to provide development plans, guidelines and tools for the upgrade of the models within this regional air quality service. In this study we will present the main outcomes of the CAMS_61 project with special focus on the generic data assimilation tool developed to promote the uptake of Sentinel 4 and 5 satellite observations within the service. The tool consists of a pre- and post-processor tool for the download, selection, conversion and visualisation of the satellite data and an observation operator source code that could in principle be included in any data assimilation system for the simulation of the observations using model variables. The tool has been tested on TROPOMI data within different model/assimilation systems and on a set of simulated Sentinel 4 data.

Keywords CAMS · Air pollution · Chemistry transport model · Data assimilation

1 Introduction

Copernicus is the European Union's Earth observation programme which provides information about the state our planet and its environment for the ultimate benefit

R. Timmermans (✉) · A. Segers · E. Dammers · M. Schaap
Department on Climate, Air and Sustainability, TNO, Utrecht, The Netherlands
e-mail: renske.timmermans@tno.nl

A. Segers · H. Fagerli · A. Valdebenito
MetNorway, Oslo, Norway

O. Jorba · D. Bowdalo
Barcelona Supercomputing Center (BSC), Barcelona, Spain

A. Colette · G. Descombes
INERIS, Verneuil-en-Halatte, France

R. Kouznetsov · A. Uppstu
Finnish Meteorological Institute (FMI), Helsinki, Finland

C. Mensink and O. Jorba (eds.), *Air Pollution Modeling and its Application XXVIII*, Springer Proceedings in Complexity, https://doi.org/10.1007/978-3-031-12786-1_22

of all European citizens. The Copernicus Atmosphere Monitoring Service (CAMS) provides consistent and quality-controlled information related to air pollution and health, solar energy, greenhouse gases and climate forcing, everywhere in the world. The CAMS information services are based on modelling systems, which are enriched with satellite observation and in situ (non-satellite) data. Within CAMS an ensemble of 9 (soon to be expanded to 11) regional chemistry transport models operationally provides daily analyses and forecasts (CAMS, n.d.), as well as annual reanalyses of air quality across Europe.

The CAMS_61 project was developed to provide development plans, guidelines and tools for the upgrade of the models within this regional air quality service. In this study we will present the main outcomes of the CAMS_61 project with special focus on the generic data assimilation tool developed to promote the uptake of Sentinel 4 and 5 satellite observations within the service.

2 The CAMS_61 Project

The goal of the CAMS_61 project to provide plans, guidelines and tools for the improvement of the CAMS regional service was achieved through three main activities:

(1) In depth-assessment of the CAMS regional forecasts to identify and prioritise model developments required to improve the quality of the CAMS regional forecasts.
(2) Evaluation of the benefit and suggestion of the best practices for coupling of the regional air quality forecasts to the current analyses of (mainly) surface observations.
(3) Production of model-agnostic tools for the data assimilation of Sentinel-4 and 5p observations

In the first activity detailed evaluations of the CAMS regional modelling systems have been performed. The evaluations went beyond investigating the officially reported pollutants by addressing an extensive set of model outputs, including a wide variety of chemical species and deposition values. The use of an evaluation interface (https://aerocom-evaluation.met.no/) allowed each model team to investigate and compare their own model performance in relation to measurements and to the other ensemble members (see Fig. 1).

The systematic difference between observed and modelled black carbon and organic carbon was significantly lowered by adopting an emissions inventory including condensables (Denier Van Der Gon et al., 2015) and appropriate Elemental Carbon (EC)/Organic Carbon (OC) fractions. Further analyses allowed the identification of model specific development requirements as well as a number of general issues. The large spread in the modelled mineral dust and sea salt concentrations illustrated that the budgets of these components need further attention. The model performance for primary components (e.g. ammonia, primary PM) shows that accounting

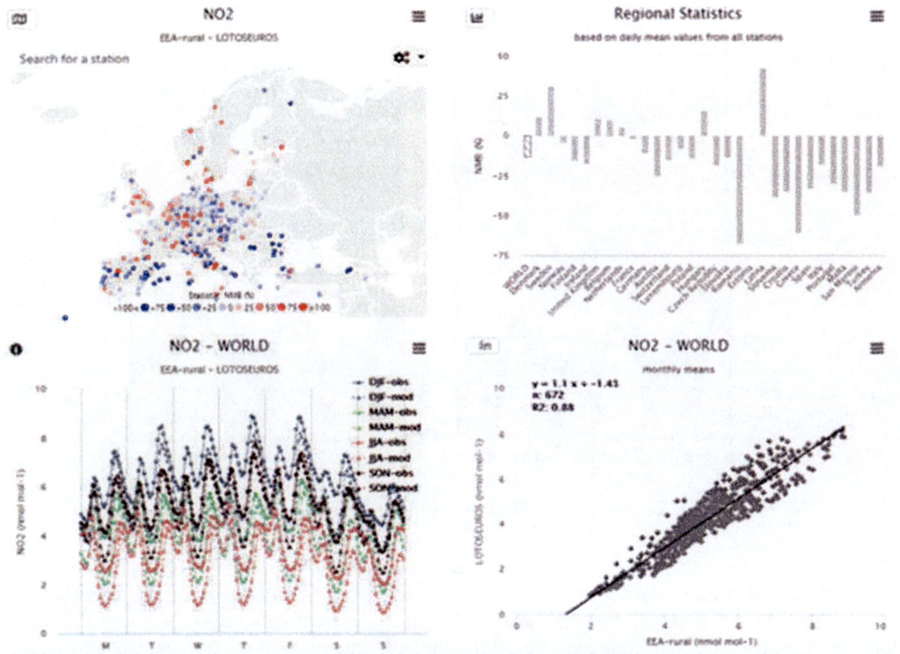

Fig. 1 Illustration of some plots from the AEROCOM evaluation interface

for the dynamic behavior of anthropogenic emissions is required to improve the regional ensemble. We also recommended to improve and detail the representation of dry and wet deposition and to develop the associated deposition products. This activity illustrated that integrating a benchmark test in the operations of the CAMS regional service, would make it transparent and traceable which development steps have caused the changes in the model performances and would provide the means to prioritize the development requirements for the participating modelling systems. Expanding the benchmarking activity towards diagnostic and dynamic evaluation is advised to substantiate the model development activities.

In the second activity the initialization of air quality forecasts obtained from analyzed distributions was addressed in a systematic way. The analyzed distributions were obtained through assimilation of (mainly) surface observations. A positive impact of initialization with analyzed distributions on e.g. RMSE disappears within a few hours (short lived species) up to 24 h for longer lived species in winter (Fig. 2).

The main improvement of the forecast scores in many cases comes from the bias correction and thus is also very dependent on the initial bias between model and observations. For initially unbiased species the effect of data assimilation on the forecasts is small. The data assimilation technique employed in two model systems in this activity also allows to estimate parameters concerning the model forcing (emissions, boundary conditions, model parameters). Propagation of these parameter estimates in some cases allowed for reaching a uniform improvement over the whole

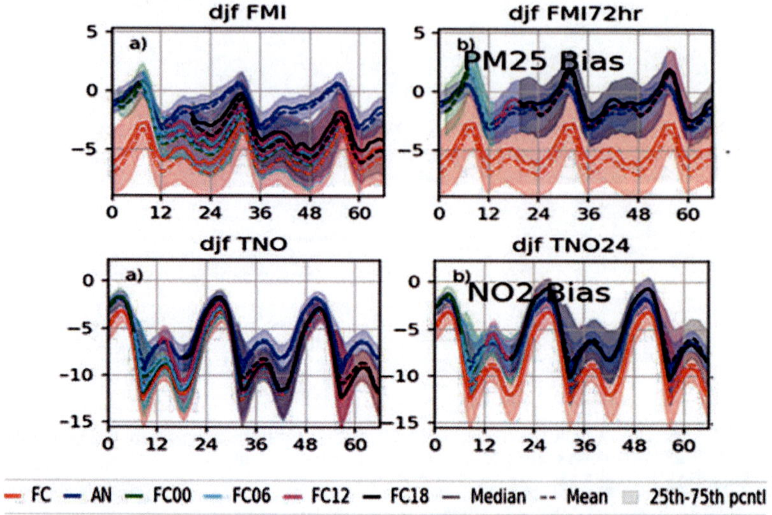

Fig. 2 Illustration of added longevity through the inheritance of adjusted emissions into the forecasts. Within SILAM system for PM2.5 (top plots) and LOTOS-EUROS system for NO_2 (bottom plots). Plots show the results from the analyse based forecasts (green, cyan, magenta and black lines, each for different start time of forecast) through initialization only (left plots) and with inheritance of emission adjustments (right plots) in comparison to free running forecasts (red lines) and analysis (blue lines). With the emission adjustments the forecasts stay close to the analysis (blue line). With only initialization the analysis based forecasts quickly return to the free-running forecast performance (red line)

forecast time (Fig. 2). It was concluded that due to the convergent nature of CTMs data assimilation strategies should prioritize parameter estimation above optimizing initial conditions.

In the third activity the CSO (CAMS Satellite Operator) tool has been developed which will be described in the next section.

3 Generic Tool for the Data Assimilation of Sentinel 5p and Sentinel 4 Data

The CSO tool is a generic tool which combines a pre-processor that can be used to download, select, and convert satellite data sets with an observation operator that simulates satellite observations (including pixel footprints, vertical column and vertical sensitivity of the instrument) from modelled concentration fields. The tool is aimed at the facilitation of the data assimilation of Sentinel 5p and future Sentinel 4 data. Version control is implemented using a GIT repository. The repository is also used for the provision of the tool to the project partners and other interested parties. Documentation on the tool including tutorial sections is available.

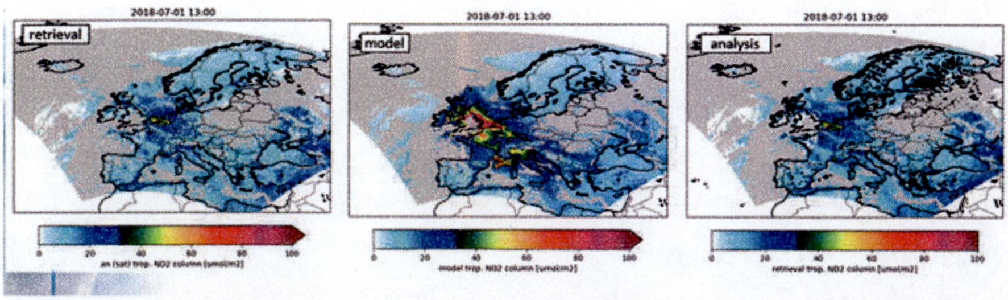

Fig. 3 Illustration of impact of assimilating the synthetic Sentinel-4 data into the EMEP model. Tropospheric NO_2 columns for one specific moment from simulated Sentinel-4 data (left plot), EMEP model (middle plot) and from EMEP model after data assimilation of simulated Sentinel-4 data (right plot)

The observation operator part of the tool was tested within five different models to produce simulated tropospheric NO_2 columns at the location of the satellite observations. For the LOTOS-EUROS and EMEP models the operator was included in the model itself, while for the other three models it has been applied as post-processing tool as is the common practice for these models.

After this testing of the CSO tool by all participating model systems, data assimilation tests with Sentinel 5p observations of NO_2, SO_2 and formaldehyde have been successfully performed within three different models (Fig. 3). To prepare for the future availability of Sentinel-4, providing geostationary data for Europe, an additional test was performed using a synthetic dataset produced using the MONARCH model for July 2018. It was possible to assimilate the data set without adaptation of the assimilation systems. The experiment with the EMEP 3D-Var system showed that the Sentinel-4 data could have impact on surface concentrations during almost all daytime hours, rather than occasionally at overpass time as is the case for Sentinel-5p. Assimilation of the synthetic Sentinel-4 data, which 4–5 times more high-quality pixels per day, was proven to be performed efficiently using domain decomposition as is currently employed in most regional models. Hence, through the development of the CSO tool the uptake of the future sentinel-4 and other data sets into the CAMS assimilation systems has been facilitated. Recommendations for research and development activities needed for assimilation-based emission estimates as part of the data assimilation process in the CAMS regional systems are provided).

4 General Conclusion

The three activities within the CAMS_61 project all led to plans or tools which have been shown to be able to improve the models within the CAMS regional service. It also clearly illustrated that advancing the CAMS regional ensemble requires a balanced and coordinated investment in emission modelling, CTM development, diagnostic evaluation and data assimilation capacities.

Acknowledgements The CAMS_61 project was funded by the European Union and implemented by ECMWF. We would like to acknowledge all CAMS regional modeling teams for their contributions to our work.

Questions and Answers

Questioner: Volker Matthias

Question: You compare models to available observations at ground level and to column concentrations from satellites. If you could wish to have additional observations to improve the model performance, what would you select?

Answer: We have seen that the lack of information on vertical profile in the satellite products can lead to issues within the data assimilation and a negative impact of the satellite products on improving the model performance at the surface. So having more information on the vertical distribution in the satellite products would be beneficial. In addition for aerosols, up till now aerosol optical depth (AOD) observations from satellites have been used for data assimilation in air quality models. The AOD is an optical measurement of all aerosols combined which complicates the assimilation within models working with aerosol mass and different species. So here on top on vertical profile information, the availability of satellite observations providing information on aerosol composition would be beneficial.

Questioner: Ronny Petrik

Question: Very interesting discussion about the assimilation of column data from satellites. You introduced the estimation of retrieval errors for the species NO_2. How do you do that for Sentinel-4 retrievals based on Sentinel-5p data.

Answer: The Sentinel-4 retrieval errors have been simulated using a random forest machine learning model that defines the retrieval error as function of simulated values from the nature run. The model was trained with existing Sentinel-5p observations using the independent variables: longitude, latitude, surface pressure, solar zenith angle. For more information I refer to the group at BSC from Oriol Jorba who performed the simulations.

References

CAMS. (n.d.). *Copernicus atmosphere monitoring service: European air quality service*. Retrieved January 28, 2022, from https://regional.atmosphere.copernicus.eu/

Denier Van Der Gon, H. A. C., Bergström, R., Fountoukis, C., Johansson, C., Pandis, S. N., Simpson, D., & Visschedijk, A. J. H. (2015). Particulate emissions from residential wood combustion in Europe—Revised estimates and an evaluation. *Atmospheric Chemistry and Physics*. https://doi.org/10.5194/acp-15-6503-2015

Status and Future Vision of the CALIOPE Air Quality Forecasting System: Support for Air Quality Policies

A. Soret, K. Serradell, M. Guevara, Carlos Pérez García-Pando, Miriam Olid, J. Mateu, J. Benavides, M. Terrado, M. T. Pay, C. Tena, Francesca Macchia, Sara Basart, Gilbert Montané, Dene Bowdalo, H. Petetin, D. Rodriguez, S. Enciso, E. M. Pérez, C. Alonso, P. de la Viesca, X. Sanyer, Ll. Alegre, X. Guinart, I. Hernandez, and Oriol Jorba

Abstract The CALIOPE forecasting system, developed and operated by the Barcelona Supercomputing Center, has been providing daily forecasts of air quality for Europe and Spain since 2008. The system currently integrates an emission processing model (HERMESv3), a meteorological model (WRF-ARW), a chemical transport model (CMAQ), and a mineral dust model (BSC-DREAM8b) that accounts for the dust contribution from North Africa. Since its initial deployment, continuous updates have been implemented to the system, with improvements in emission source estimates and adjustments in the model configuration. Since 2020, and thanks to the co-development with the Catalan Regional Administration and the Barcelona Metropolitan Transport Authority, new approaches and methodologies have been implemented, including a new nested domain at higher spatial resolution (1×1 km^2) over Catalonia, to further support decision-making. Additionally, beyond the mesoscale domains, the CALIOPE-Urban model has been incorporated into the modelling chain to simulate air quality at the urban street-scale in the city of Barcelona. In parallel to those improvements, the emission estimates have been improved through updates in emission activity data from the industrial and on-road traffic sectors. Lastly, a data-fusion method has been implemented to correct urban

A. Soret (✉) · K. Serradell · M. Guevara · C. P. García-Pando · M. Olid · J. Mateu · J. Benavides · M. Terrado · C. Tena · F. Macchia · S. Basart · G. Montané · D. Bowdalo · H. Petetin · D. Rodriguez · S. Enciso · O. Jorba
Barcelona Supercomputing Center (BSC), 08034 Barcelona, Spain
e-mail: albert.soret@bsc.es

C. P. García-Pando
Catalan Institution for Research and Advanced Studies, ICREA, Barcelona, Spain

M. T. Pay
Universitat de Barcelona, Barcelona, Spain

E. M. Pérez · C. Alonso · P. de la Viesca · X. Guinart · I. Hernandez
Generalitat de Catalunya, Barcelona, Spain

X. Sanyer · Ll. Alegre
Autoritat del Transport Metropolità, Barcelona, Spain

C. Mensink and O. Jorba (eds.), *Air Pollution Modeling and its Application XXVIII*, Springer Proceedings in Complexity, https://doi.org/10.1007/978-3-031-12786-1_23

air quality fields and reduce biases affecting the forecasting system. In this contribution, we will present the most recent developments applied to the CALIOPE forecasting system and its application as an air quality management tool. We will also discuss future user-driven developments that take both scientific and technical aspects into consideration, including the migration of the mesoscale system to the on-line MONARCH model, a dissemination strategy aiming to promote the wider use of CALIOPE predictions through a new portal designed to improve the user experience, and the further integration of the system into decision-making processes.

Keywords Atmospheric services · Co-development · Forecast system · Integrated knowledge

1 Introduction

As stated by the World Health Organization (WHO) and the European Environmental Agency (EEA), air pollution is the biggest environmental risk to health in the European Union (EU). Thus, there is a need to design adequate policies to improve air quality and reduce the negative impacts of pollution on health. From the legislative point of view, the air quality directive (2008/50/EC) requires Member States to regularly assess air quality and implement plans in case of exceedances values. Therefore, attention is increasingly focusing on the use of air quality models combined with monitoring stations. In this context, modelling becomes an essential tool for air quality assessment, management and planning (e.g. through the design of plans to guarantee the fulfilment of air quality objectives).

2 Starting Point: The CALIOPE Air Quality Forecasting System

The CALIOPE forecasting system (www.bsc.es/caliope), developed and operated by BSC, provides daily forecasts of air quality for Europe (12×12 km^2) and Spain (4×4 km^2) since 2008. The system currently integrates an emission processing model (HERMESv3), a meteorological model (WRF-ARW v3.6), a chemical transport model (CMAQ v5.0.2), and a mineral dust model (BSC-DREAM8bv2). Since its initial deployment, continuous updates have been implemented with improvements in emission source estimates and adjustments in the model configuration. During all this time, the CALIOPE system has proven to be a suitable air quality management tool, allowing to: (1) analyse past events and get a better understanding of the emissions and the dynamic conditions in a particular region, their implications in terms of air quality levels, and their impact on different societal challenges such as health; (2) produce actionable operational forecasts, e.g. useful to alert vulnerable population groups when air quality is deficient, and (3) analyse the impacts of different strategies, e.g. measures to abate anthropogenic emissions in order to improve air quality.

3 Co-development of a Service

In recent years, the amount of available information and methods about air quality and emission modelling and assessment has rapidly increased. To ensure that this information is effectively and appropriately developed and used, a strong collaboration between researchers and air quality managers is fundamental. Since 2020, and thanks to the co-development with the Directorate of Environmental Quality and Climate Change of the Catalan Government and the Metropolitan Transport Authority of Barcelona, new approaches and methodologies have been implemented to translate knowledge into action.

The objectives of this collaboration are:

- To establish a permanent forum for air quality management, including the definition of policymakers' needs and the discussion of scientific evidence
- To set up a modelling framework (the CALIOPE system) operating at different spatial scales
- To support air quality management in developing and implementing mitigation plans to reduce outdoor air pollution
- To promote capacity building activities aiming at providing guidance and ensuring optimal use of the CALIOPE system
- To discuss and propose recommendations on future priorities to enhance the impact of research activities.

Within the context of this science-policy collaboration, new methods have been developed and analysed (see Sects. 3.1–3.5 below) taking advantage of the different expertise in our transdisciplinary working group, which not only pursues the production of scientifically-relevant research results but also actionable information that can be exploited by a wide range of users, including other scientific disciplines, public authorities, businesses and the general public.

3.1 Air Quality at the Urban Street Scale

As the majority of the world's population lives in cities, urban areas are crucial spots to tackle pollution episodes. Air quality in urban areas is affected by processes happening at different spatial and temporal scales. Thus, to simulate these dynamics, we developed the CALIOPE-Urban model (Benavides et al., 2019). CALIOPE-Urban couples the mesoscale air quality forecasting system CALIOPE with the urban roadway dispersion model, R-LINE. First, we defined a nested domain at 1×1 km^2 resolution in CALIOPE. Then, for each regional 1×1 km^2 grid cell, meteorological data from WRF and background concentrations from CMAQ are used as input combined with traffic emissions from the HERMES emission model at road link level. R-LINE has been adapted to estimate pollutant dispersion processes within street canyons. Local meteorology and background concentrations are estimated based on the urban geometry and atmospheric stability for each street. Key progresses that have been recently developed and tested include the set-up of the urban model in forecast mode; the air quality assessment for different emission scenarios (Benavides et al., 2021); and the coupling of the urban model with a traffic simulator (Rodriguez-Rey et al., 2021) in order to evaluate different traffic restriction measures, such as the Barcelona city Hall's policies to improve air quality in the city.

3.2 Implementation of Data-Fusion

Data-fusion techniques aim at improving model predictions by combining observations with the model outputs. Unlike data assimilation methods, which typically restrict initial conditions in numerical forecast models, data-fusion methods merge raw forecasts with one or more additional datasets to improve the forecasting skills in a post-processing step. In this framework, we have recently developed a data-fusion method consisting of a two-step bias correction.

The first step corrects the raw model forecasts at the monitoring stations using historical observations (Petetin et al., 2020). Such corrections are performed at each station using either popular bias-correction models, such as the Kalman filter, or innovative machine learning algorithms based on multiple ancillary variables. In the second bias correction step, a geostatistical data-fusion method extends the corrected forecasts to areas where no observations are available based on the spatial patterns provided by the model output (Armengol et al., 2021).

3.3 Improvement of Emission Estimates

The representability of industrial and road transport emissions, the two major sources of NO_x and PM emissions in Catalonia, was improved in HERMESv3 (Guevara et al., 2020) through an exchange of knowledge, methods and data within the working group. A better definition of emission sources is key to define appropriate air quality management plans and support policymaking.

Emissions from industrial sources were updated based on the digital catalogue of Catalan industrial point sources, which contains detailed information for small, medium and large industrial point sources, technologies and fuels used, geographical location and measured-based emission rates, among others. This digital catalogue allowed to improve the characterisation of point source emissions in HERMESv3 in two key aspects. First, the number of sources considered for the estimation of emissions, which increased from 143 to 3,268. Although large point sources were

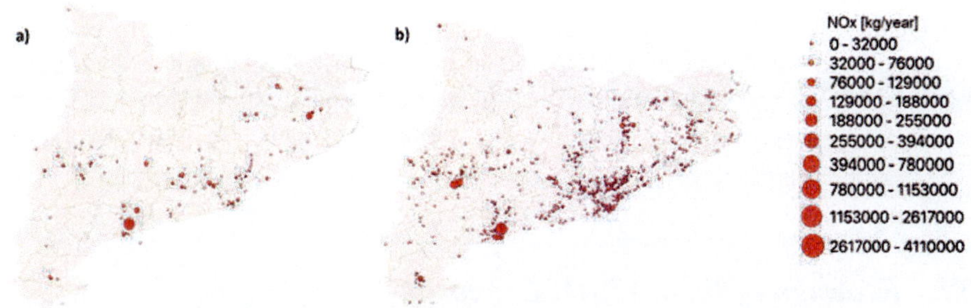

Fig. 1 NO_x emissions from industrial point sources [kg/year] before (**a**) and after (**b**) the update

already considered in HERMESv3 (e.g. cement plants, power plants or refineries), the updating allowed to include small and medium combustion plants that were previously disregarded due to the lack of information. Second, the detailed information associated to each point source in terms of stack properties and working hours allowed to improve the vertical and temporal distribution of the emissions. Figure 1a, b represent the distribution of NO_x point source emissions in Catalonia before and after the update, respectively. When updating the point source database in HERMESv3, associated NO_x emissions increased 54%.

Emissions from the road transport sector were improved by integrating updated information on annual average daily traffic and average speed values per road link for the province of Barcelona. This information was derived from a traffic simulator maintained by the ATM and calibrated considering measured data from traffic count stations. The inclusion of this detailed network in HERMESv3 allowed improving the representation of traffic emissions in urban areas, not only in Barcelona but also in adjacent municipalities such as Badalona, L'Hospitalet or Santa Coloma, which were not considered before due to a lack of information. When updating the transport network, the travelled-km in Catalonia increased around 10% in both urban and interurban areas.

3.4 Migration of the Mesoscale System to the On-Line MONARCH Model

The MONARCH is a fully online multiscale chemical weather prediction system for regional and global-scale applications with telescoping nest capabilities developed by BSC since 2008 (Badia et al., 2017). It uses the NMMB model developed by the NOAA's National Centers for Environmental Prediction (NCEP) as a meteorological driver and couples online state-of-the-art chemistry modules to solve the tropospheric chemistry of the atmosphere. MONARCH is the operational model of the Barcelona Dust Forecast Center of the World Meteorological Organization and will become a new member of the multi-model ensemble system of the Copernicus Regional

Air Quality production service in June 2022. A key milestone in BSC forecasting activities is the deployment of MONARCH in the CALIOPE system replacing the WRF-CMAQ models. This effort will allow maintaining a unique system and incorporate recent advancements regarding new pollutant species (i.e., bioaerosols) or data assimilation products.

3.5 Redesign of the CALIOPE Website

When scientists are able to effectively communicate information beyond the scientific audiences, it builds support for science, and encourages information uptake and more informed decision-making at all levels, from the government to local communities to individuals. Within this context, the redesign of the CALIOPE website started by gaining a better understanding of the needs of the users, making the website easier to navigate and displaying information in a clearer way; thus, creating a friendlier experience for the users. The methodology applied in the redesign was based on the Double Diamond process created by the British Design Council, commonly used in the User Experience (UX) design process to co-explore different options before refining and narrowing them down to select the best one.

During the research process, different users of the CALIOPE system were interviewed and a competitor analysis was performed to assess the services offered by other similar tools (Fig. 2). The information gathered was analysed using an affinity map that helped to track the most important insights and define two UX personas representing the main types of users of the website. The first persona represents the non-expert users interested in exploring the forecasts and identifying in an easy way how the air quality is in the area they are interested in. The second persona represents the experienced users that are interested in more detailed information, such as the evaluation of the forecast, the related meteorological information or the resolution of the models. To separate both types of users and provide a more personalised experience, registration will be asked to access a more detailed view of the dashboard. At the same time, this will allow to gather data on the users interested in air quality issues and to have more information on experienced users of CALIOPE.

Fig. 2 Low and high fidelity prototype used for the rounds of tests with users

References

Armengol, J., Rodriguez-Rey, D., Benavides, J., Jorba, O., Guevara, M., Pérez García-Pando, C., & Soret Miravet, A. (2021). Advances on urban air quality modeling: bias correction approach for estimated annual NO_2 levels and macroscopic traffic simulators for scenario planning, EGU General Assembly 2021, EGU21-14178. https://doi.org/10.5194/egusphere-egu21-14178

Badia, A., Jorba, O., Voulgarakis, A., Dabdub, D., Pérez García-Pando, C., Hilboll, A., Gonçalves, M., & Janjic, Z. (2017). Description and evaluation of the Multiscale Online Nonhydrostatic AtmospheRe CHemistry model (NMMB-MONARCH) version 1.0: Gas-phase chemistry at global scale. *Geoscientific Model Development, 10*, 609–638. https://doi.org/10.5194/gmd-10-609-2017

Benavides, J., Snyder, M., Guevara, M., Soret, A., Pérez García-Pando, C., Amato, F., Querol, X., & Jorba, O. (2019). CALIOPE-Urban v1.0: Coupling R-LINE with a mesoscale air quality modelling system for urban air quality forecasts over Barcelona city (Spain). *Geoscience Model Development, 12*, 2811–2835. https://doi.org/10.5194/gmd-12-2811-2019

Benavides, J., Guevara, M., Snyder, M., Rodríguez-Rey, D., Soret, A., Perez Garcia-Pando, C., & Jorba, O. (2021). On the impact of excess diesel NO_x emissions upon NO_2 pollution in a compact city. *Environmental Research Letters, 16*, 024024. https://doi.org/10.1088/1748-9326/abd5dd

Guevara, M., Tena, C., Porquet, M., Jorba, O., & Pérez García-Pando, C. (2020). HERMESv3, a stand-alone multi-scale atmospheric emission modelling framework—Part 2: The bottom–up module. *Geoscientific Model Development, 13*, 873–903. https://doi.org/10.5194/gmd-13-873-2020

Petetin, H., Bowdalo, D., Soret, A., Guevara, M., Jorba, O., Serradell, K., & Pérez García-Pando, C. (2020). Meteorology-normalized impact of the COVID-19 lockdown upon NO_2 pollution in Spain. *Atmospheric Chemistry and Physics, 20*, 11119–11141. https://doi.org/10.5194/acp-20-11119-2020

Rodriguez-Rey, D., Guevara, M., Linares, M. P., Casanovas, J., Salmerón, J., Soret, A., Jorba, O., Tena, C., & Pérez García-Pando, C. (2021). A coupled macroscopic traffic and pollutant emission modelling system for Barcelona. *Transportation Research Part D, 92*, 102725. https://doi.org/10.1016/j.trd.2021.102725

Question and Answer

QUESTIONER	Lasse Johansson (Finnish Meteorological Institute)
QUESTION	About CALIOPE-Urban, how do you nest a local scale Gaussian model in a regional scale one, so that you avoid e.g. double-counting of emissions?
ANSWER	Within CALIOPE-Urban, R-LINE and CALIOPE are coupled offline. One of this approach's main challenges is obtaining background concentrations from the mesoscale model without double counting traffic emissions in regional- and street-scale models. To this aim, we use the upwind urban background concentration from the regional model to avoid the double-counting of traffic emissions when coupling the mesoscale with the street-scale model. This method is comprehensively explained in Benavides et al. (2019).

Data Fusion for the Improvement of Low-Cost Air Quality Sensors

Theodosios Kassandros, Evangelos Bagkis, and Kostas Karatzas

Abstract Aim of this study is to develop a calibration procedure through Machine Learning to upgrade the low-cost air quality sensor performance and investigate the generalization of this function over a specific area towards air quality data fusion.

Keywords Air quality · Low-cost sensors · Machine learning · Data fusion

1 Introduction

Bad air quality (AQ) has a negative impact on peoples' quality of life. The small number of monitoring stations used for the official AQ monitoring and the operationally available air pollution modelling tools still leave open space for improving local AQ knowledge. The KASTOM project (www.air4me.eu) is developing a versatile and flexible air quality monitoring and forecasting system by deploying an IoT-oriented network of low-cost AQ sensor nodes (LCAQSN), while in parallel developing a state-of-the-art emission modeling module combined with state-of-the-art three-dimensional AQ models. LCAQSN can cover larger areas due to their low cost but are lacking the necessary accuracy.

2 Materials

The Greater Thessaloniki Area (GTA) is the second largest urban agglomeration in Greece hosting more than 1 million inhabitants. The KASTOM project has installed 33 LCAQSN in the GTA including: (a) Particle sensors (PM10–PM2.5: PMS5003, Beijing Plantower Co., Ltd.), (b) sensors for gaseous pollutants (NO_2, O_3 and CO: Alphasense Ltd., U.K.) and meteorological sensors (Air Temperature, Relative Humidity and Air Pressure, BME280 Bosch Sensortec, Germany).

T. Kassandros (✉) · E. Bagkis · K. Karatzas
Environmental Informatics Research Group, School of Mechanical Engineering, Aristotle University of Thessaloniki, Thessaloniki, Greece
e-mail: tkassand@physics.auth.gr

C. Mensink and O. Jorba (eds.), *Air Pollution Modeling and its Application XXVIII*, Springer Proceedings in Complexity, https://doi.org/10.1007/978-3-031-12786-1_24

Fig. 1 LCAQSN network in the GTA (Dots). ▢: KORD ✪: AGSOF

Table 1 Dataset description

NodeSet (Nset)	FusionSet (Fset)	Target variables
PM10, PM2.5, PM1, CO, O_3, NO_2, relative humidity, temperature, pressure	NodeSet + wind speed, wind direction, precipitation, friction velocity, saturation, saturation ratio, traffic	PM10, O_3, NO_2

In this study, we have collocated six nodes with two reference stations (Fig. 1) in Agias Sofias (AGSOF) and Kordelio (KORD) areas, classified by the European Environment Agency as an urban traffic and urban industrial station respectively.

The initial dataset (NodeSet) consists of six nodes measurements (Node1–3 located in AGSOF and Node4–6 located in KORD) for the period of 21/12/2019–10/03/2020 and the reference stations measurements for PM10, O_3 and NO_2 (NO_2 measurements in KORD omitted due to missing value problems). The additional dataset (FSet) included meteorological modeling (WRF) and free traffic flow data (Salanova et al., 2018). All variables are presented in Table 1.

3 Methods

The first step of the computational procedure aimed at generating a set of features, capturing the maximum amount of information. We therefore applied time lags (from 1 to 12 h) and rolling—aggregation statistics (6 and 12 h) to all the variables, leading to

161 features for the Nset and 401 features for the Fset. To reduce noise introduced by features, a feature reduction procedure was followed employing the Random Forest Feature Importance (RFFI) method. We then employed a Machine Learning (ML)-oriented modeling approach, making use of the reference station measurements as target parameters (PM10, O_3 and NO_2) to calibrate and upgrade the KASTOM nodes performance. Models were trained in the two subsets, for each sensor and location. A Gradient Boosting algorithm was used (Friedman, 2001), combining the outputs sequentially from individual regression trees, where each new tree helps to correct errors made by a previously trained tree.

To evaluate the initial performance of the LCAQSN, the Pearson Correlation Coefficient (r) and Coefficient of Divergence (CoD) were calculated. The ML models were evaluated using a fivefold time forward cross validation on a rolling basis, using the Coefficient of Determination (R^2) and the Relative Expanded Uncertainty (REU), following the methodology described in the Guide to the Demonstration of Equivalence of Ambient Air Monitoring Methods (EUD, 2008). According to the European Air Quality Directive, uncertainties for "class 1 sensor" or indicative measurements are 50, 25, 30% and for "class 2 sensor" or objective measurements are 100, 75, 75% for PM10, NO_2 and O_3 respectively.

4 Results

Field calibration of an LCAQSN network requires the individual nodes to perform identical to each other, this being the first condition to apply the same calibration function. This was checked with the aid of the CoD versus Pearson (Fig. 2). All PM10 sensors scored very high Pearson and very low CoD thus behaving identical, but the gas sensors, and especially O_3 sensors in the AGSOF, displaying a more diverse behavior therefore suggesting that in this case, the generalization of the calibration functions could be more challenging.

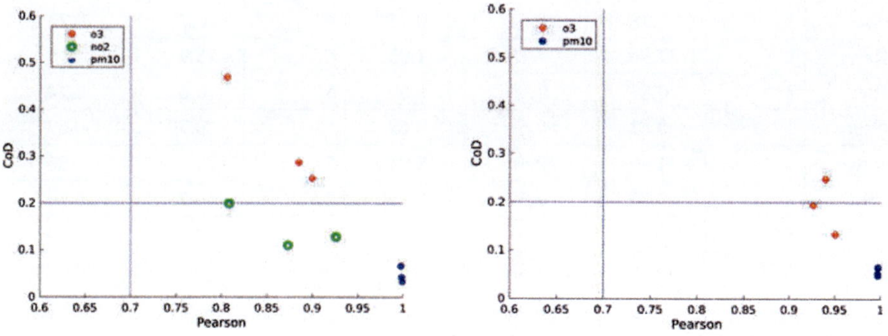

Fig. 2 Pearson against CoD. Left: AGSOF, right: KORD

The RFFI selected the most relevant features, mostly the ones deriving from the KASTOM nodes' measurements, but also meteorological factors deriving from modeling (Fig. 3). On the other hand, traffic related features are only chosen in the AGSOF location (an urban traffic station). Also, traffic features seem to influence more NO_2 and PM10 than O_3.

.

While raw measurements display extremely poor scores against reference measurements, the computational procedure and the XGBoost shows promising results (Table 2). In most of the cases the use of the Fset leads to better output than the use of the Nset, though by a small margin.

In terms of REU, the calibrated PM10 can be considered as "class 1 sensor" in both locations, while the calibrated O3 are above the desired threshold but have still improved their performance and be considered as "class 2 sensor" (Fig. 4).

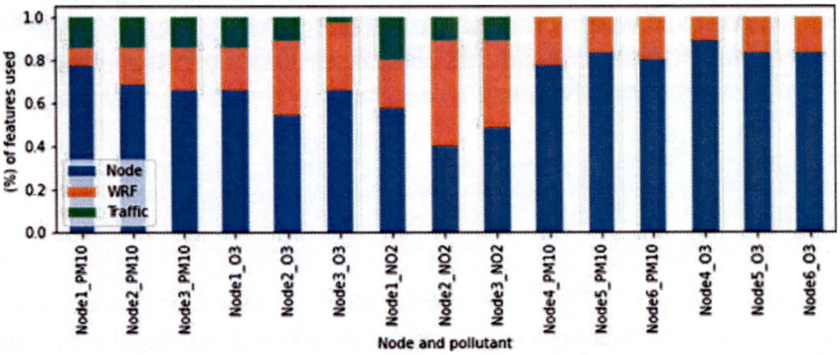

Fig. 3 Feature selected from the RFFI by category for each node

Table 2 R^2 score for XGBoost and raw measurements. Bold: best performance per sensor

Set	Node	PM10	NO_2	O_3	Node	PM10	O_3
Fset	1	**0.82**	**0.69**	**0.76**	4	**0.8**	**0.72**
Nset	1	0.78	0.61	0.75	4	0.78	**0.72**
Raw	1	−27.96	−4.28	−1.95	4	−25.93	−12.18
Fset	2	**0.81**	**0.65**	**0.69**	5	**0.8**	**0.74**
Nset	2	**0.81**	0.6	**0.69**	5	0.79	0.69
Raw	2	−24.43	−10.46	−13.95	5	−23.95	−1.36
Fset	3	**0.82**	**0.67**	**0.74**	6	**0.77**	**0.74**
Nset	3	0.78	0.6	0.73	6	**0.77**	**0.74**
Raw	3	−33.35	−1.35	−0.3	6	−20.54	0.67

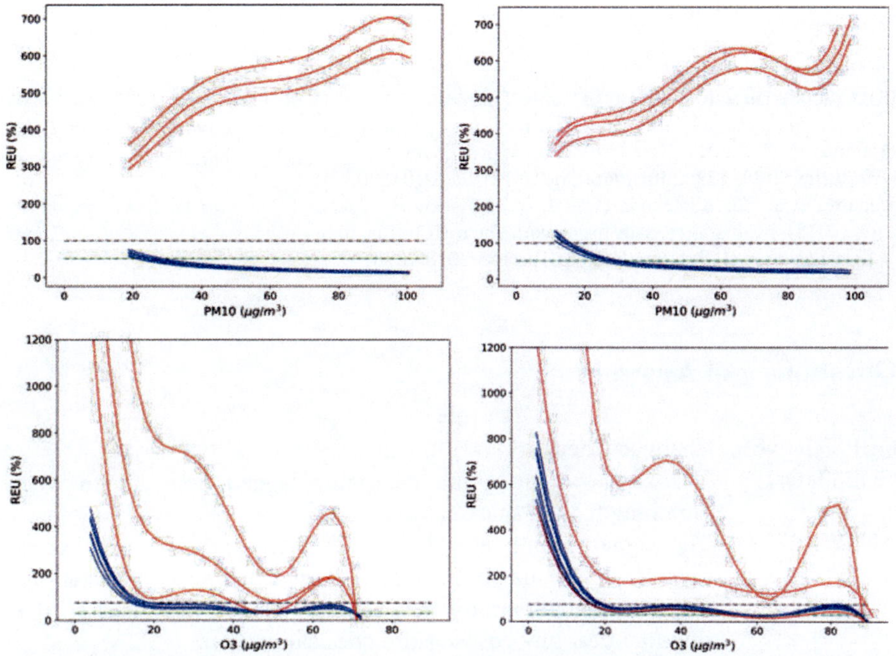

Fig. 4 REU for PM10 (up) and O_3 (down) in AGSOF (left) and KORD (right). Red lines: raw measurements, blue lines: calibrated measurements, black line: "class 2 sensor" threshold, green line: "class 1 sensor" threshold

5 Conclusions

The intercomparison of LCAQSN for a small time period, proves that PM10 sensors are behaving similar in the same locations and the proposed computational calibration procedure can upgrade their performance as indicative measurements for regulatory purposes, while it may be possible to apply the same approach to the rest of the network. For NO_2 and O_3, while the calibration functions can improve the sensors' response, the desired REU levels couldn't be reached. In every case data fusion is improving results and therefore more data sources and additional effort towards better fusion should be considered.

Acknowledgements This research has been co-financed by the European Union and Greek national funds through the Operational Program Competitiveness, Entrepreneurship and Innovation, under the call RESEARCH—CREATE—INNOVATE. Project code T1EDK-01697; project name Innovative system for air quality monitoring and forecasting (KASTOM, www.air4me.eu).

References

EUD. (2008). Directive 2008/50/EC of the European Parliament and of the Council of 21 May 2008 on ambient air quality and cleaner air for Europe. *Official Journal of the European Union L152*

Friedman, J. H. (2001). Greedy function approximation: A gradient boosting machine. *Annals of Statistics*, 1189–1232.https://doi.org/10.1214/aos/1013203451

Salanova et al., 2018Salanova Grau J. M., Mitsakis E., Tzenos P., Stamos I., & Aifadopoulou, G. (2018). Multisource data framework for road traffic state estimation. *Journal of Advanced Transportation*, 1–9. https://doi.org/10.1155/2018/9078547

Questions and Answers

QUESTIONER Zhaoyue Chen

QUESTION Thanks, how could you determine lagged hour length when enriching feature space?

ANSWER The lagged length was determined after trial-and-error experiments, while it has been observed from previous computational exercises by our group that no more than 24 hours lagged is important for low-cost sensor nodes calibration.

QUESTIONER Bas Mijling

QUESTION Low-cost sensors are calibrated at two different sites. What would happen if the sensor location snapped? Does the calibration obtained at site 1 is applicable at site 2?

ANSWER This is a very interesting question and can be answered thoroughly only if further research is applied. From our knowledge of the field and ongoing experiments, applying a calibrated function from Agias Sofias to Kordelio and vice versa is yielding good results in terms of uncertainty and R^2 for PM2.5 and PM10, and acceptable metrics for O_3. Although the question about the spatial generalizability of the calibration function cannot be answered with only two reference stations collocated with the low-cost sensors. Data from a third collocated reference station, not included in this study, show more ambiguous behavior and thus applying functions by proximity or by type of station (urban, suburban, traffic, background, etc.) or applying one generalized calibration function trained in all available locations, would be considered for calibrating the whole network of 33 devices.

An Evaluation of Data-Driven Models

Fabian Lenartz, Pascal Joassin, Marie Dury, and Pierre Crespin

Abstract The increased availability of atmospheric measurements and machine learning techniques allows any data scientist to make his own air quality model and the most common application in this domain is the short-term prediction of pollution levels at monitoring sites, by means of regression techniques, including neural networks. If a sufficiently large number of live measurements are provided, e.g. via the portals of monitoring stations or low-cost sensors systems such as the one from the European Environmental Agency, the Sensor. Community or some commercial air quality station manufacturers, a geostatistical interpolation could be performed, yielding to air pollution forecast maps similar to the ones from much more complex modelling systems.

Keywords Machine learning · Short-term forecast · PM10 · PM2.5 · Low-cost sensor systems

1 Introduction

Machine learning (ML) techniques have become more and more popular in the last decade. They are now part of our everyday life; they can take the form of a chat bot, recognize and track a variety of objects or persons on CCTV cameras, defeat the best human players of checkers, chess and go, and, obviously, they can also predict air quality.

Zhang et al. (2012) indicate that statistical models trained or fitted to historical air quality and meteorological data were already developed in the 1970s to forecast air pollution in urban areas. Despite some common drawbacks shared by these earlier and the more recent methods, such as their inability to predict unknown situations and to provide a better understanding of chemical and physical processes, regression techniques and artificial neural networks are part of the major approaches used in real-time air quality forecasting. Despite these limitations, Fierens (2020) have shown that

F. Lenartz (✉) · P. Joassin · M. Dury · P. Crespin
ISSeP, Rue du Chéra 200, 4000 Liège, Belgium
e-mail: f.lenartz@issep.be

C. Mensink and O. Jorba (eds.), *Air Pollution Modeling and its Application XXVIII*, Springer Proceedings in Complexity, https://doi.org/10.1007/978-3-031-12786-1_25

it was possible to get an insight of the COVID-19 lockdown impact on air quality in Belgium by comparing actual measurements during the restriction period and results of a random forest model trained 2 years before, hence representative of a regular situation in terms of emissions.

In their review of 47 articles on air pollution epidemiology using data mining (DM) or ML, Bellinger et al. (2017) highlight that (i) these studies mostly focus on outdoor, rather than indoor, air pollutants, (ii) their usual objective is to forecast/predict epidemiological values/outcomes, rather than to generate hypotheses or apportion sources, and (iii) their application is most widely based on classification and/or regression techniques, rather than clustering or association mining algorithms. The most common methods used in forecast studies reported in this survey are artificial neural networks (ANN), decision trees (DT), k-means (kM), k-nearest neighbor (kNN), random forest (RF) and support vector machine (SVM).

In his master thesis, which focuses on air quality prediction—not necessarily related to epidemiological research—Lepperød (2019) also mentions ANN and RF among the most popular techniques but also adds recurrent neural networks (RNN) and gradient boosting (GB) to the list, as well as the combination of some of them.

In the present study we test DT, SVM, bagged and boosted trees (which are close to RF and GB), as well as different NN for the forecast of outdoor PM10 concentrations, and compare their performance. In the discussion, we deliver some thoughts about the geographical extension of the results and the use of information about the error on the ML predictions.

2 Material and Method

Since it is difficult to get all detailed information about a particular ML method implementation and its parameterization, we chose not to reproduce the exact works that can be found in the literature and to use the MathWorks® algorithms of both the Statistics and Machine Learning and the Deep Learning toolboxes. Furthermore, the Regression Learner app interface (Fig. 1) makes it very easy to test a variety of algorithms with no a priori knowledge, and then to generate the code of the corresponding model function to be tuned and used for the validation process.

Although shared data set may be available, we chose to apply these ML techniques on measurements from the Walloon monitoring network. More precisely, we have collected the validated PM2.5 and PM10 measurements at the monitoring stations of Angleur, Herstal, Vielsalm and Jemeppe (Belgium) for the years 2014 to 2016. Instead of using meteorological forecast from another source, we have smoothed out actual measurements of wind speed and direction at 30 m height and temperature at 3 m, through a 6-h moving average, and we have shifted the time series one day in the past.

The initial training period started on January 1st, 2014 at 00:00 and ended on December 31st, 2015 at 23:30, whereas the testing period was a rolling 5-day period covering the whole years of 2016. The response parameter was the PM10 values at

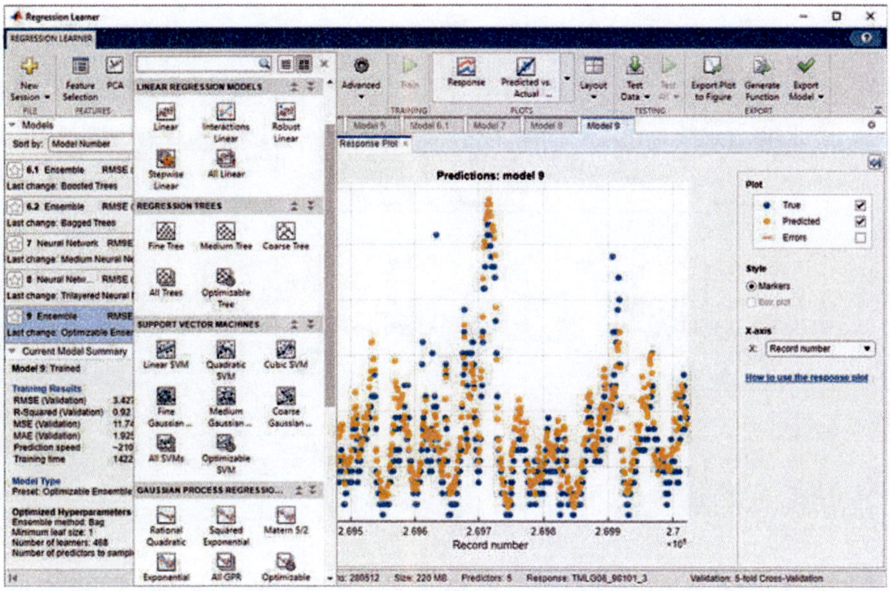

Fig. 1 Matlab® regression app interface

one station, while the predictors were the PM2.5 concentrations at all stations and the surrogate meteorological forecast described above.

The considered algorithms were the decision tree, the medium Gaussian SVM, the boosted trees, the bagged trees, an optimizable ensemble of trees, a medium NN and a trilayered NN.

3 Results

Performance of the various algorithms was consistent between training and testing periods, although the error was larger during the generalization phase, as can be seen in Fig. 2 and Table 1.

With appropriate geostatic tools, like DIVAnd presented by Barth et al. (2014), it is possible to take into consideration the forecast error at each location and weigh its relative importance by specifying an adequate signal-to-noise ratio, as can be seen in the example of Fig. 3, where the same data points are interpolated using different relative error variances.

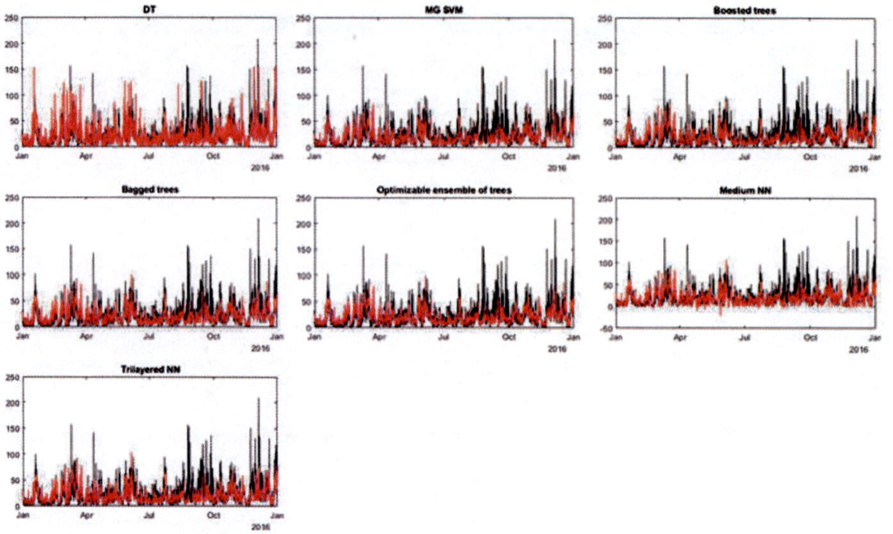

Fig. 2 Time series measured (black) and predicted (red) in the test period using different ML algorithms

Table 1 RMSE of the different algorithms

	DT	MG SVM	Boosted trees	Bagged trees	Opt. Ens	Medium NN	Trilayered NN
Training	10.3	9.0	10.5	9.6	9.8	9.1	8.9
Testing	12.6	9.9	11.1	10.3	10.5	12.3	11.3

4 Conclusion

Availability of ML libraries in most common high level programming languages makes it possible for coders to rapidly develop their own data-driven air quality forecast model. In this study various algorithms were trained, tested and compared with each other for the prediction of PM10 concentrations based on PM2.5 measurements. The typical RMSE value obtained was about 10 $\mu g/m^3$ and shown to be useful information for the mapping of these predictions.

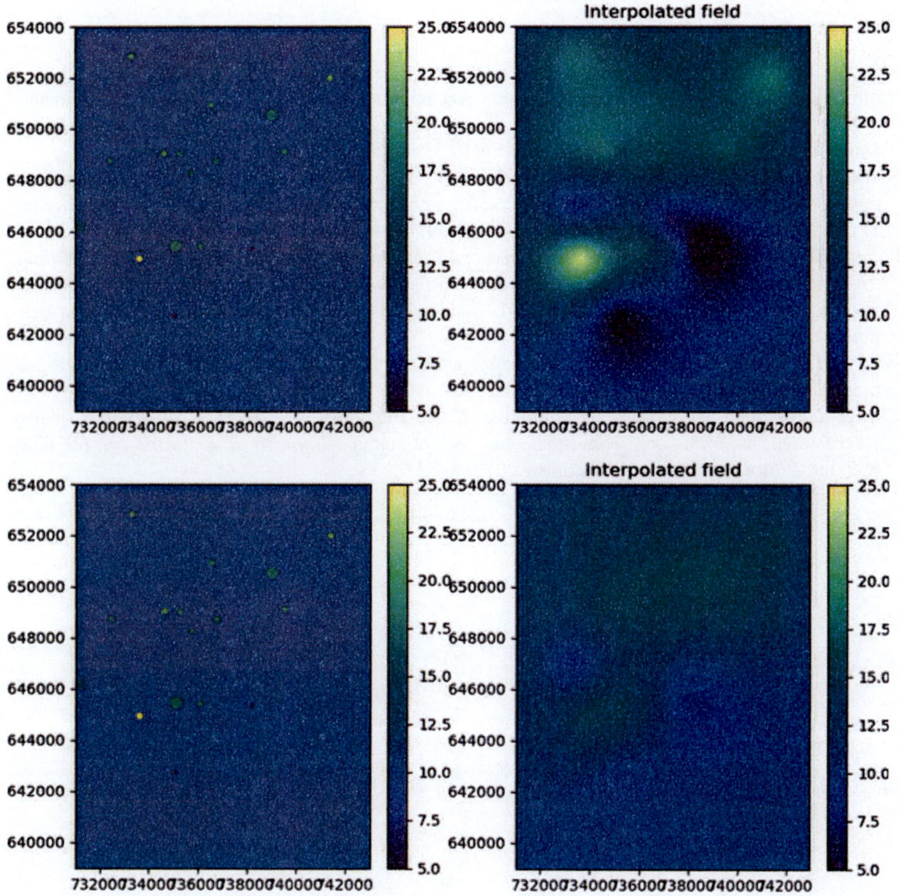

Fig. 3 Left, data point locations and values, larger circles denote reference analyser data, smaller ones LCSS data. Top right, interpolated field using a relative error variance of 0.01 for all points. Bottom right, interpolated field using a relative error variance of 0.01 for reference instruments and of 1 for LCSS

Acknowledgements This work has been carried out as a complementary task of the Microcapteurs project, within the framework of the plan Environnement-Santé funded by the Walloon Ministry for the Environment, Nature, Forestry, Rural Affairs and Animal Welfare. We would like to thank the SPW ARNE and the AwAC for their collaboration in the Microcapteurs project.

References

Barth, A., Beckers, J. M., Troupin, C., Alvera-Azcárate, A., & Vandenbulcke, L. (2014). Divand-1.0: N-dimensional variational data analysis for ocean observations. *Geoscientific Model Development, 7*(1), 225–241. https://doi.org/10.5194/gmd-7-225-2014

Bellinger, C., Mohomed Jabbar, M. S., Zaïane, O., & Osornio-Vargas, A. (2017). A systematic review of data mining and machine learning for air pollution epidemiology. *BMC Public Health, 17*(1), 907.https://doi.org/10.1186/s12889-017-4914-3

Fierens, F. (2020). *Effects of COVID-19 induced lockdown measures on air quality in Belgium: An assessment using a Random Forest model.* https://www.irceline.be/en/documentation/publications/scientific-reports/effects-of-covid-19-induced-lockdown-measures-on-air-quality-in-belgium/view

Lepperød, A. J. (2019). *Air quality prediction with machine learning.* NTNU. https://ntnuopen.ntnu.no/ntnu-xmlui/handle/11250/2626165

Zhang, Y., Bocquet, M., Mallet, V., Seigneur, C., & Baklanov, A. (2012). Real-time air quality forecasting, part I: History, techniques, and current status. *Atmospheric Environment, 60*, 632–655. https://doi.org/10.1016/j.atmosenv.2012.06.031

Assimilation of Surface Ozone Measurements to WRF-Chem—Impact on the Model Capability to Predict Peak Concentrations

Małgorzata Werner, Jakub Guzikowski, and Maciej Kryza

Abstract Ozone in the troposphere has negative effects on vegetation and human health, which causes that forecasts of ozone concentrations are developed and used by policy makers and society to assess air quality. In this study we used the WRF-Chem model and GSI assimilation tool to analyse the impact of surface 3D-Var data assimilation on modelled O_3 concentrations over Central Europe, with a focus on Poland. We run two simulations for summer 2015: (1) BASE—no data assimilation, and (2) SURF—the WRF-Chem output from the previous day was modified by the GSI assimilation system. The results show that data assimilation has not a significant positive impact on mean error measures such as factor of two, mean bias, and mean gross error. However, the results also show that the SURF simulation better reflects peaks of high O_3 concentrations, which is especially important in terms of air pollution forecasting and prevention of the harmful effects of pollutants.

Keywords WRF-Chem · Ozone · Data assimilation · Surface observations

1 Introduction

Tropospheric ozone (O_3) is an important secondary air pollutant produced by photochemical oxidation of volatile organic compounds (VOC) and carbon monoxide (CO) in the presence of nitrogen oxides (NO_x). It plays a key role in the chemical processes occurring in the atmosphere, and in climate change (Griffiths et al., 2020). It also behaves as a pollutant with negative effects on vegetation and human health (Cohen et al., 2017; UNEP, 2015). The regulatory O_3 levels are still exceeded in a number of European cities and are especially a concern in growing urban areas. This is why forecasts of ozone concentrations are developed to be used by policy makers and society to assess air quality for the forthcoming hours/days and to make decisions on

M. Werner (✉) · J. Guzikowski · M. Kryza
Faculty of Earth Sciences and Environmental Management, University of Wroclaw, Ul. Kosiby 8, 51-621 Wroclaw, Poland
e-mail: malgorzata.werner@uwr.edu.pl

C. Mensink and O. Jorba (eds.), *Air Pollution Modeling and its Application XXVIII*, Springer Proceedings in Complexity, https://doi.org/10.1007/978-3-031-12786-1_26

future behavior. It is especially important to correctly predict high pollution concentrations, as it allows prevention of the harmful effects of pollutants on human health (Marécal et al., 2015a, 2015b).

One of the methods to improve air pollution forecasts is data assimilation (DA). DA consists of using hybridization methods of measurement data and modelling results to constraint the model prediction as close as possible to the observed data (Menut & Bessagnet, 2019).

In this study we aim to verify whether assimilation of surface ozone concentrations to the WRF-Chem transport model can improve model capability to predict peak ozone concentrations over Central Europe (https://prognozy.uni.wroc.pl/), where critical levels of ozone concentrations are frequently exceeded during late spring and summer.

2 Data and Methods

2.1 The Modelling System and the Experiment Design

We used the Weather Research and Forecasting model with Chemistry (WRF-Chem, version 3.9) and Gridpoint Statistical Interpolation (GSI). WRF-Chem was run with two nested domains—an outer domain that covers Europe on a 12 km × 12 km grid and an inner domain that focuses on Poland at 4 km × 4 km resolution. We used the Regional Acid Deposition Model, version 2 (RADM2) for the gas phase chemistry and the Goddard Chemistry Aerosol Radiation and Transport scheme (GOCART) for aerosols. Full description of the model configuration is given in (Werner et al., 2019). The simulations were driven by the Global Forecasting System (GFS) meteorological data, available every 3 h, at a 0.5° × 0.5° spatial resolution. Emissions data include TNO MACC III for Europe and national high resolution (1 km × 1 km) database for Poland, provided by the Chief Inspectorate of Environmental Protection (CIEP). We used the 3D-Var component of the GSI system, developed at the national Centers for Environmental Prediction (Pagowski et al., 2010, 2014) to integrate surface observations into modelled initial conditions.

The simulations were run for August 2015, 1st–31st with chemical cycling included. Forecasts started at 12:00 UTC and lasted forecasts. We run two simulations:

(1) BASE—no data assimilation, the WRF-Chem model output from the last hour of the previous simulation was used to initialize the next simulation,
(2) SURF—with data assimilation, the WRF-Chem output from the previous day was modified by the GSI assimilation system that integrated surface measurements to the initial conditions.

Observations of ground O_3 concentrations from 91 stations (64 urban, 8 suburban and 19 rural) were provided by the Chief Inspectorate of Environmental Protection.

2.2 Model Verification

We used the standard error measures to compare modelled O_3 concentrations with observations: factor of two (FAC2, the fraction of points greater than 0.5 times and less than 2 times the measured value), mean bias (MB), mean gross error (MGE), normalised mean bias (NMB), normalised mean gross error (NMGE), and correlation coefficient (R). The mean statistics for all stations and the entire simulation periods were calculated based on the first 24 h of each forecast. The statistics were calculated for all types of stations together and individually for stations located in urban, suburban and rural area.

3 Results

Mean error statistics show that the WRF-Chem model underestimates measured O_3 concentrations for the analysed period (Table 1). MB is below zero for both runs (BASE and SURF) and for all types of stations (urban, suburban and rural). Underestimation is at a similar level at different types of stations—NMB ranges from −0.24 for urban to −0.31 for suburban stations for the BASE run. Correlation coefficient for BASE varies from 0.60 for urban to 0.68 for rural stations. Application of surface data assimilation has not a significant positive impact on the mean statistic errors. FAC2, MB and MGE are poorer for the SURF compared to the BASE simulation for all types of stations. A small improvement of correlation coefficient (from 0.60 to 0.62) is for urban stations.

Table 1 Mean error statistics for O_3 concentrations for August 2015; based on hourly data for all stations and according to type of area

	ALL					
SIM	FAC2	MB	MGE	NMB	NMGE	R
BASE	0.79	−18.44	29.54	−0.24	0.38	0.62
SURF	0.54	−23.83	37.21	−0.30	0.47	0.62
	URBAN					
BASE	0.77	−17.01	29.08	−0.22	0.38	0.60
SURF	0.54	−22.83	36.47	−0.30	0.48	0.62
	SUBURBAN					
BASE	0.76	−26.11	32.66	−0.31	0.39	0.65
SURF	0.49	−31.12	41.07	−0.37	0.49	0.63
	RURAL					
BASE	0.84	−19.78	29.65	−0.24	0.36	0.68
SURF	0.57	−23.92	37.96	−0.29	0.46	0.61

However, mean time series plot (Fig. 1) indicates that the SURF simulation better represents peaks of high O_3 concentrations than BASE. Simultaneously, the SURF simulates minimum values too low, which was indicated above by e.g. FAC2 statistic (poorer for SURF than for BASE). Time series of O_3 concentrations for individual stations (Fig. 2) confirm that the SURF run performs better for the high concentrations at different types of stations and gives too low values for the periods with lower O_3 concentrations.

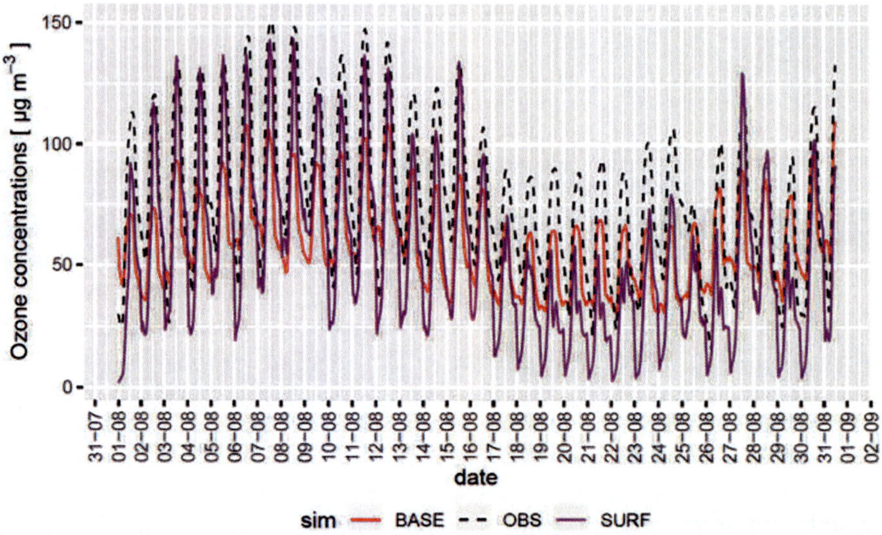

Fig. 1 Time series (mean for all stations) of modelled and measured hourly O_3 concentrations

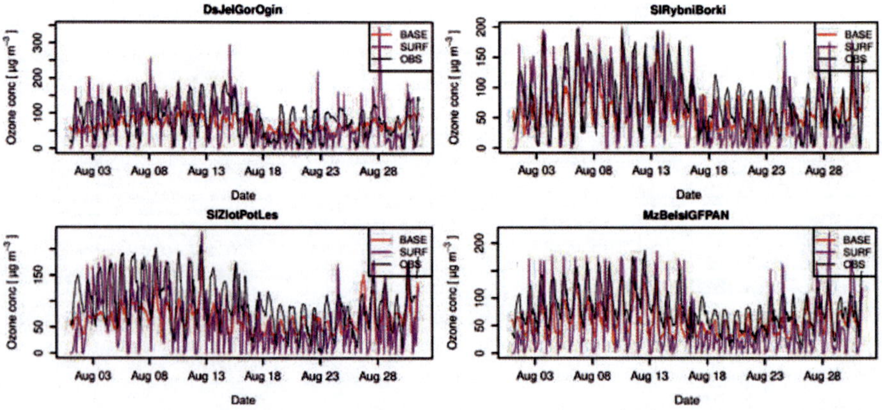

Fig. 2 Time series of modelled and measured O_3 concentrations for selected stations in different type of area—urban (upper) and rural (lower)

4 Summary and Conclusions

In this study we used the WRF-Chem model and GSI assimilation tool to analyse the impact of surface 3D-Var data assimilation on modelled O_3 concentrations. Two simulations were run for the summer period of 2015 over Central Europe and focused on the area of Poland. First of them did not include DA (BASE run) and for the second run (SURF) the WRF-Chem output from the previous day was modified by the GSI assimilation system. Mean error statistics show that the WRF-Chem model with the RADM2 gas chemistry mechanisms underestimates measured O_3 concentrations for both runs and all types of stations, which is similar to WRF-Chem results over Europe reported by Mar et al. (2016) or Visser et al. (2019). DA has not a significant positive impact on mean error measures such as FAC2, MB and MGE. However, the results also show that the SURF simulation better reflects peaks of high O_3 concentrations, which is especially important in terms of air pollution forecasting and prevention of the harmful effects of pollutants.

Acknowledgements This research was funded by the Polish National Science Centre project No. UMO-2016/23/B/ST10/01797. The study was supported by the European Union's Horizon 2020 research and innovation program under grant agreement No. 856599.

TNO MACC III data base was provided free of charge by TNO Netherlands (https://www.tno.nl/en/about-tno/contact/, direct contact to TNO). Emissions data for Poland were provided by the Chief Inspectorate of Environmental Protection (http://www.gios.gov.pl/en/, direct contact to the institution). These data were prepared within the project: "Supporting the air quality assessment system with application of modelling of PM10, PM2.5, SO_2, NO_2, B(a)P for the years 2015, 2016 and 2017".

References

Cohen, A. J., Brauer, M., Burnett, R., Anderson, H. R., Frostad, J., Estep, K., Balakrishnan, K., Brunekreef, B., Dandona, L., Dandona, R., Feigin, V., Freedman, G., Hubbell, B., Jobling, A., Kan, H., Knibbs, L., Liu, Y., Martin, R., Morawska, L., ... Forouzanfar, M. H. (2017). Estimates and 25-year trends of the global burden of disease attributable to ambient air pollution: An analysis of data from the Global Burden of Diseases Study 2015. *Lancet, 389*, 1907–1918. https://doi.org/10.1016/S0140-6736(17)30505-6

Griffiths, P. T., Keeble, J., Shin, Y. M., Abraham, N. L., Archibald, A. T., & Pyle, J. A. (2020). On the changing role of the stratosphere on the tropospheric ozone budget: 1979–2010. *Geophysical Research Letters, 47*, e2019GL086901. https://doi.org/10.1029/2019GL086901

Mar, K. A., Ojha, N., Pozzer, A., & Butler, T. M. (2016). Ozone air quality simulations with WRF-Chem (v3.5.1) over Europe: Model evaluation and chemical mechanism comparison. *Geoscientific Model Development, 9*, 3699–3728. https://doi.org/10.5194/gmd-9-3699-2016

Marécal, V., Peuch, V.-H., Andersson, C., Andersson, S., Arteta, J., Beekmann, M., Benedictow, A., Bergström, R., Bessagnet, B., Cansado, A., Chéroux, F., Colette, A., Coman, A., Curier, R. L., Denier van der Gon, H. A. C., Drouin, A., Elbern, H., Emili, E., Engelen, R. J., ... Ung, A. (2015a). A regional air quality forecasting system over Europe: The MACC-II daily ensemble production. *Geosci. Model Dev., 8*, 2777–2813. https://doi.org/10.5194/gmd-8-2777-2015

Marécal, V., et al. (2015b). A regional air quality forecasting system over Europe: The MACC-II daily ensemble production. *Geoscientific Model Development, 8*, 2777–2813. https://doi.org/10.5194/gmd-8-2777-2015

Menut, L., & Bessagnet, B. (2019). What can we expect from data assimilation for air quality forecast? Part I: Quantification with academic test cases. *Journal of Atmospheric and Oceanic Technology, 36*, 269–279. https://doi.org/10.1175/JTECH-D-18-0002.1

Pagowski, M., Grell, G. A., McKeen, S. A., Peckham, S. E., & Devenyi, D. (2010). Three-dimensional variational data assimilation of ozone and fine particulate matter observations: Some results using the weather research and forecasting-chemistry model and grid-point statistical interpolation. *Quarterly Journal Royal Meteorological Society, 136*, 2013–2024. https://doi.org/10.1002/qj.700

Pagowski, M., Liu, Z., Grell, G. A., Hu, M., Lin, H.-C., & Schwartz, C. S. (2014). Implementation of aerosol assimilation in Gridpoint Statistical Interpolation v. 3.2 and WRF-Chem v. 4.3.1. *Geoscientific Model Development Discuss, 7*, 2483–2500. https://doi.org/10.5194/gmdd-7-2483-2014

UNEP. (2015). UNEP Annual Report 2015, GE.

Visser, A. J., Folkert Boersma, K., Ganzeveld, L. N., & Krol, M. C. (2019). European NOx emissions in WRF-Chem derived from OMI: Impacts on summertime surface ozone. *Atmospheric Chemistry and Physics, 19*, 11821–11841. https://doi.org/10.5194/ACP-19-11821-2019

Werner, M., Kryza, M., & Guzikowski, J. (2019). Can data assimilation of surface PM2.5 and Satellite AOD improve WRF-Chem Forecasting? A case study for two scenarios of particulate air pollution episodes in Poland. *Remote Sensing, 11*. https://doi.org/10.3390/rs11202364

Forecasting PM2.5 Concentrations with uEMEP and EMEP4PL for Poland

Maciej Kryza, Malgorzata Werner, Bruce Rolstad Denby, Qing Mu, Tymoteusz Sawiński, and Arkadiusz Remut

Abstract In this work we use the EMEP4PL model in combination with the high-resolution uEMEP model. EMEP4PL is an application of the EMEP MSC-W chemistry transport model for the area of Poland. The model uses two nested domains. The parent domain covers Europe with 12 km × 12 km. The nested domain is centered over Poland with a 4 km × 4 km grid. uEMEP is based on Gaussian modelling principles and allows downscaling of the EMEP4PL model results to very high spatial resolution using the local fraction approach. In this work, we have applied the uEMEP model to forecast hourly PM2.5 concentrations for the area of SW Poland with a spatial mesh of 250 m × 250 m. Both the EMEP4PL and uEMEP results are compared with two measuring networks. First, we compare the results with data gathered at six reference stations operated by the Chief Inspectorate of Environmental Protection. Second, we use data from the newly established LIFE/MappingAir network. This network uses low-cost sensors and provides data from 25 sites located in SW Poland. This network provides more data from residential areas, which are often the hot-spots for PM2.5 concentrations. We have shown that the uEMEP forecasts have smaller bias and higher index of agreement if compared to EMEP for the test period from April to end of June 2021.

M. Kryza (✉) · M. Werner · T. Sawiński · A. Remut
Faculty of Earth Sciences and Environmental Management, University of Wroclaw, Ul Kosiby 8, 51-621 Wroclaw, Poland
e-mail: maciej.kryza@uwr.edu.pl

M. Werner
e-mail: Malgorzata.werner@uwr.edu.pl

T. Sawiński
e-mail: tymoteusz.sawinski@uwr.edu.pl

A. Remut
e-mail: arkadiusz.remut@uwr.edu.pl

B. R. Denby · Q. Mu
The Norwegian Meteorological Institute, Henrik 5 Mohns Plass 1, 0313 Oslo, Norway
e-mail: brucerd@met.no

Q. Mu
e-mail: qingm@met.no

C. Mensink and O. Jorba (eds.), *Air Pollution Modeling and its Application XXVIII*, Springer Proceedings in Complexity, https://doi.org/10.1007/978-3-031-12786-1_27

Keywords uEMEP model · PM2.5 forecasting · Air quality modelling

1 Introduction

Poland belongs to the group of the European countries with the most severe air quality problems. The main issue is related to particulate matter with aerodynamic diameter of 2.5 (PM2.5) and 10 (PM10) μm (Krynicka & Drzeniecka-Osiadacz, 2013; Reizer & Juda-Rezler, 2016). This is often linked to a large share of emission from residential combustion, where various types of fuels of different quality are used (Bebkiewicz et al., 2020; Kryza et al., 2020).

In this work we have used the EMEP4PL linked with the uEMEP model to forecasts hourly PM2.5 concentrations for SW Poland. The models were run for the test period from 1st of April 2021 to end of June 2021. The results were compared with the PM2.5 measurements gathered at two different measuring networks: national air quality network operated by the Chief Inspectorate of Environmental Protection (CIEP) and the network operated by the University of Wroclaw (UWr), which uses low-cost sensors.

2 Data and Methods

2.1 *EMEP4PL and uEMEP Model Configuration*

Application and configuration of the EMEP4PL model for Poland was presented by Werner et al. (2018) We have used the same model setting, but changed the emission inventory. For Poland, we have used KOBIZE national emission inventory. For other countries within the model domain we have applied EMEP 0.1° × 0.1° emission inventory. Both KOBIZE and EMEP emission inventories used were developed for year 2018. The model was run using meteorological forecasts provided by the Weather Research and Forecasting model (WRF). Both EMEP4PL and WRF use the same model configuration. The forecasts were run for the 72 h lead time, starting each day at 00 UTC using Global Forecasting System 0.25° × 0.25° data as initial and boundary conditions. For EMEP4PL, chemistry cycling was applied, where forecast from the previous day was used as initial data for a given day.

The urban EMEP (uEMEP) model description was presented by Rolstad Denby et al. (2020) and Wind et al. (2020). In this work, the uEMEP model is used for downscaling the EMEP modelled PM2.5 concentrations into 250 m × 250 m for SW Poland. We have used national high-resolution SNAP sector 2 emission inventory as proxy data for uEMEP downscaling, together with road network map provided by OpenStreet map, which was used as proxy for traffic.

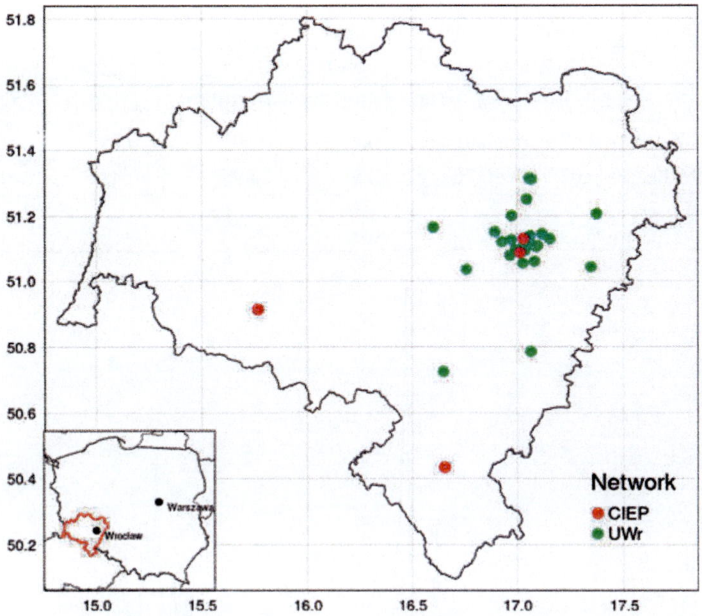

Fig. 1 Location of the PM2.5 measuring sites used for model's evaluation

2.2 *Model Evaluation*

Measurements were gathered at two distinct air quality networks. First, data from national air quality network operated by CIEP was used. For SW Poland, there are 4 stations (1 traffic, 3 urban background) with hourly PM2.5 measurements. Second network was established within the LIFE/MAPPINGAIR project and operated by UWr. There are 25 sites in SW Poland, which use the low-cost sensors and provide PM2.5 measurements every one minute. The sensors are regularly calibrated by comparison with the reference data. This network is denser (25 sites), and stations are often located in small cities, where residential combustion is significant. Location of the stations is presented in Fig. 1.

Model errors are calculated for the first 24 h of the forecasts. The errors are summarized using mean error [ME], index of agreement [IOA] and Pearson correlation coefficient (COR).

3 Results

The uEMEP and EMEP4PL model's performance is summarized in Table 1. The uEMEP model bias is reduced if compared to the EMEP4PL model. This is observed for both CIEP and UWr networks. The bias is smaller for UWr network if compared

Table 1 uEMEP and EMEP4PL model performance

	n	ME [$\mu g\ m^{-3}$]		IOA [–]		COR [–]	
		uEMEP	EMEP4PL	uEMEP	EMEP4PL	uEMEP	EMEP4PL
CIEP	7786	−4.43	−6.01	0.65	0.59	0.55	0.59
UWr	49,920	−0.02	−0.76	0.62	0.59	0.48	0.46

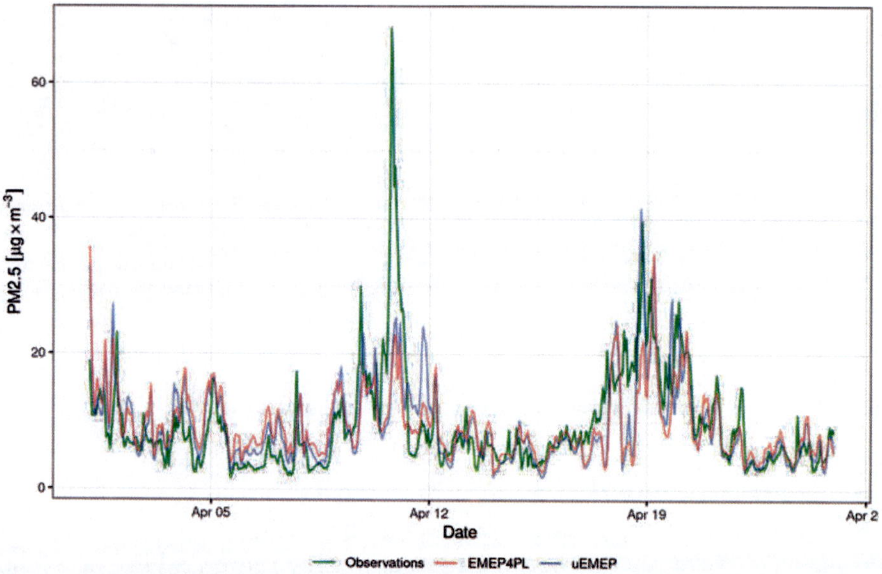

Fig. 2 uEMEP and EMEP4PL model performance for April for Trzebnica station

with CIEP for both models. The IOA and COR are higher for the uEMEP. The only exception is for correlation coefficient and CIEP network, which is smaller for uEMEP if compared to EMEP4PL.

The uEMEP and EMEP4PL is compared with observed PM2.5 concentrations for Trzebnica station for April. April was selected as there were some peak PM2.5 concentrations observed related with air temperature drops and increased residential combustion (e.g. on April 11 or 18). Both model's failed to resolve the first peak on April 11. Second peak, on 18th April, was well resolved by the uEMEP, while is missing for the EMEP4PL model (Fig. 2).

4 Summary and Conclusions

In this work we have shown the performance of the uEMEP and EMEP4PL forecasts for PM2.5 for SW Poland. The uEMEP forecasts have lower bias if compared with EMEP4PL, which is a strong advantage. Bias reduction is observed on both measuring networks used for model evaluation in this study.

Next steps will include application of the uEMEP and EMEP4PL for PM2.5 forecasting during the winter season. This is the season with the highest PM2.5 concentrations in this area.

Acknowledgements This research was funded by the Polish National Science Centre project no UMO-2017/25/B/ST10/01041. The study was supported by the European Union's Horizon 2020 research and innovation programme under grant agreement No. 856599. Emission data for Poland was provided by The National Centre for Emissions Management (KOBiZE). Calculations were performed at Wroclaw Centre for Supercomputing and Networking, Grant No. 170.

References

Bebkiewicz, K., Chłopek, Z., Doberska, A., Kanafa, M., Kargulewicz, I., Olecka, A., Rutkowski, J., Skośkiewicz, J., Waśniewska, S., Zasina, D., Zimakowska-Laskowska, M., & Żaczek, M. (2020). Poland's Informative Inventory Report 2020.

Krynicka, J., & Drzeniecka-Osiadacz, A. (2013). Analysis of variability in PM 10 concentration in the Wrocław agglomeration. *Polish Journal of Environmental Studies, 22*, 1091–1099.

Kryza, M., Werner, M., Dudek, J., & Dore, A. J. (2020). The effect of emission inventory on modelling of seasonal exposure metrics of particulate matter and ozone with the WRF-Chem model for Poland. *12*, 1–16. https://doi.org/10.3390/su12135414

Reizer, M., & Juda-Rezler, K. (2016). Explaining the high PM10 concentrations observed in Polish urban areas. *Air Quality, Atmosphere and Health, 9*, 517–531. https://doi.org/10.1007/s11869-015-0358-z

Rolstad Denby, B., Gauss, M., Wind, P., Mu, Q., Grøtting Wærsted, E., Fagerli, H., Valdebenito, A., & Klein, H. (2020). Description of the uEMEP_v5 downscaling approach for the EMEP MSC-W chemistry transport model. *Geoscientific Model Development, 13*, 6303–6323. https://doi.org/10.5194/gmd-13-6303-2020

Werner, M., Kryza, M., & Wind, P. (2018). High resolution application of the EMEP MSC-W model over Eastern Europe—Analysis of the EMEP4PL results. *Atmospheric Research, 212*. https://doi.org/10.1016/j.atmosres.2018.04.025

Wind, P., Rolstad Denby, B., & Gauss, M. (2020). Local fractions—A method for the calculation of local source contributions to air pollution, illustrated by examples using the EMEP MSC-W model (rv4_33). *Geoscientific Model Development, 13*, 1623–1634. https://doi.org/10.5194/gmd-13-1623-2020

Influence of Meteorology on Fine Particles Concentration in an Urban Center in Southeast of Brazil

A. K. C. Ribeiro, A. L. Emrich, B. L. V. Maia, D. D. Marques, M. C. A. F. Ramos, V. A. Torres, E. S. Galvão, and T. T. A. Albuquerque

Abstract The negative impacts of particulate matter (PM), mainly the fractions of particles with a diameter <10 μm (PM10), and particles with a diameter <2.5 μm (PM2.5) on human health are well documented on technical literature. As PM concentrations can build up in periods of limited dispersion, understanding how meteorological conditions influence air pollution levels could help cities plan better air quality control and health policies. This study aimed to evaluate the influence of weather conditions on the concentration (daily mean) of fine particulate matter in Belo Horizonte, Betim, and Contagem, located in Brazil's Southeast region, using data from six meteorological stations between January 2015 and December 2017. Non-parametric methods were used to check statistically significant differences between the dry season (April to September) and wet season (October to March) and to correlate particulate concentration with meteorological parameters (precipitation, temperature, humidity, and wind speed). The results showed that PM concentration in the dry season was significantly higher than in the wet season, with p-values <0.05 for all data groups. Significant correlations were found between PM concentration and most of the assessed parameters. The data also shows a non-adequacy to linear regression, which indicates a possible non-linear or codependent relationship between PM concentration and meteorological parameters. Further studies with multiple variable analyses could help develop adequate models for PM concentration prediction.

Keywords Air quality · Influence of meteorology · Particulate matter · Statistics analysis

A. K. C. Ribeiro (✉) · A. L. Emrich · B. L. V. Maia · D. D. Marques · M. C. A. F. Ramos · V. A. Torres
Graduate Program in Sanitation, Environment and Water Resources, Federal University of Minas Gerais, Av. Pres. Antônio Carlos, Belo Horizonte 6627, 31270-901, Brazil
e-mail: amandakcr@ufmg.br

E. S. Galvão
Department of Environmental Engineering, Federal University of Espírito Santo, Av. Fernando Ferrari, Vitória, Espírito Santo 514, Brazil

T. T. A. Albuquerque
Department of Sanitary and Environmental Engineering, Federal University of Minas Gerais, Av. Pres. Antônio Carlos, Belo Horizonte 6627, 31270-901, Brazil

C. Mensink and O. Jorba (eds.), *Air Pollution Modeling and its Application XXVIII*, Springer Proceedings in Complexity, https://doi.org/10.1007/978-3-031-12786-1_28

1 Introduction

The Metropolitan Region of Belo Horizonte (MRBH) is an urban region in Southeast of Brazil, and it is the third most populous metropolitan region in Brazil with approximately 6 million inhabitants (IBGE—Instituto Brasileiro de Geografia e Estatística, 2020). It has become a strong commercial, industrial, cultural and tourist reference in the state of Minas Gerais, with a high potential for atmospheric pollutants emissions. Vehicular emissions are great contributors to air pollution (Albuquerque et al., 2019; Pacheco et al., 2017), and the MRBH is marked by the absence of effective public policies to improve air quality. In addition, there is no any encouragement for the use of alternative and collective transport instead private use. The interaction between emissions and the atmosphere determines the level of air quality, which can change over the years due to changes in both emissions and meteorological conditions. These levels of air quality are directly related to public health. In Brazil, the number of deaths between 2006 and 2016 associated with the air pollution increased by 14% (MS—Ministério da Saúde, 2019), while the population grew approximately 9% in the same period (IBGE—Instituto Brasileiro de Geografia e Estatística, 2019). Recent studies (Andreão & Albuquerque, 2021; Andreão et al., 2020) also highlight the urgent need to improve air quality policies in Brazil, which revealed that the concentrations of particulates are above the guidelines of the World Health Organization (WHO) (WHO—World Health Organization, 2006), with the estimate that a total of 48,700 ($\pm$7,570) deaths related to air pollution could have been avoided between 2014 and 2018 if the concentration of PM2.5 had not violated the WHO recommendations.

In this way, understanding how meteorological conditions influence air pollution levels could help MRBH plan better air quality control and health policies. Therefore, this study aims to evaluate the influence of meteorological conditions (dry and wet seasons) on the concentration of fine particulate matter between 2015 and 2017, mainly the fractions of particles with a diameter <10 μm (PM10), and particles with a diameter <2.5 μm (PM2.5), in the cities of Belo Horizonte, Betim, and Contagem, all located in the MRBH, southeastern Brazil.

2 Metodology

Hourly concentrations of PM10, PM2,5 and meteorological parameters (temperature, wind speed, precipitation and relative humidity) were obtained from the State Environmental Foundation (In Portuguese Fundação Estadual do Meio Ambiente—FEAM) (FEAM—Fundação Estadual do Meio Ambiente, 2020) air quality monitoring network. For this study, we considered only 6 of 14 MRBH's air quality monitoring stations. Daily average data were calculated considering the periods between 00:00 and 23:59 h of each day, and 24 h-accumulated precipitation data were considered as daily precipitation.

For the statistical analysis, the following methods were applied: (i) Anderson test to assess normality in the data; (ii) As the behavior from data as non-parametric variables, the Mann–Whitney U test was used to verify the difference of the concentrations between the dry (April to September) and wet (October to March) seasons; (iii) Spearman's correlation test was applied to analysis of the correlation between pollutants concentrations and meteorological data; and (iv) simple linear regression was used to investigate the nature of the correlations.

3 Results and Discussion

The daily averages of PM_{10} and $PM_{2.5}$ between 2015 and 2017 are shown in Fig. 1. The results depict a similar behavior over the years, with higher concentrations in the dry seasons for both PM_{10} and $PM_{2.5}$. Considering a significance level of 5% ($\alpha = 0.05$), the Mann–Whitney's analysis confirmed that the concentrations in the dry period are significantly higher than in the wet season (p-values < 0.05).

It is important to highlight that $PM_{2.5}$ data had a percentage of missing data greater than 25% for all stations. For PM_{10}, precipitation and relative humidity, these parameters also had a percentage with no data greater than 25% in S2 and S5 stations. It demonstrates failures in the monitoring network and results in non-representative data.

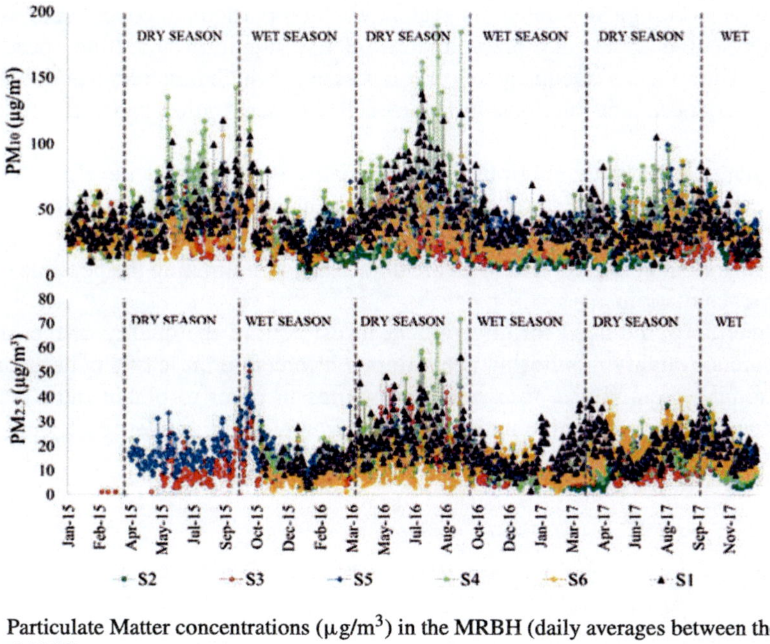

Fig. 1 Particulate Matter concentrations (μg/m³) in the MRBH (daily averages between the years 2015 and 2017)

The analysis of Spearman's correlation test for a significance level of 5% (Table 1) shows a significant correlation (p-values < 0.05) for most cases, with negative association (rho < 0) for precipitation, relative humidity and wind speed, which suggests that the higher the levels of these meteorological parameters, the lower the fine particle concentrations. These results confirm that the increase of the air humidity and precipitation are associated to wet deposition processes of particulate matter.

Although the absence of data from the two most relevant months of the rainy season (December and January) in many of the stations have compromised the analysis of the correlation coefficient for the precipitation variable, nonetheless the differences between the seasons were significant. However, the present analysis seeks to correlate the impacts in a linear and individual way but in the atmosphere all processes happen in a dynamic way. With the elaboration of scatter charts, the low adequacy of a linear regression model for the correlations was verified, indicating overall unsatisfactory fit of the linear model even for all parameters.

4 Conclusion

The results showed significantly higher concentrations of particles for all stations in the dry period, considering the years 2015, 2016 and 2017. In addition, significant correlations were observed between the concentration of particles in the atmosphere and the meteorological parameters evaluated. Such particulate concentrations tend to decrease with increasing precipitation, relative humidity and wind speed. The data also shows a non-adequacy to linear regression, which indicates a possible non-linear or codependent relationship between PM concentration and meteorological parameters.

Further studies with multiple variable analyses could help develop adequate models for PM concentration prediction. It is also suggested that future studies should be carried out considering moving averages over time and that consider a longer historical period of data, with the objective of minimizing the possible effects of atypical situations.

Furthermore, the need for government investment in the quality of the MRBH automatic air quality monitoring network is reinforced, or at least a better management, aiming to minimize measurement failures in order to obtain more realistic results for studies that aim to understand the dynamics and influences of air quality for the state of Minas Gerais.

Table 1 Spearman correlation test results for particulate matter concentrations and meteorological parameters ($\alpha = 0.05$)

Station	Correlation parameters	Meteorological parameters							
		Rainfall		Temperature		Relative humidity		Wind speed	
		PM10	PM2.5	PM10	PM2.5	PM10	PM2.5	PM10	PM2.5
S1	p-value	NA	NA	>0.05	<0.05	<0.05	<0.05	<0.05	<0.05
	rho	NA	NA	−0.0472	−0.2956	−0.3463	−0.2756	−0.1912	−0.2670
S2	p-value	<0.05	<0.05	>0.05	>0.05	<0.05	<0.05	<0.05	<0.05
	rho	−0.1670	−0.2538	0.0253	−0.1419	−0.4550	−0.3446	−0.1494	−0.1985
S3	p-value	NA	NA	>0.05	<0.05	<0.05	<0.05	<0.05	<0.05
	rho	NA	NA	0.0222	0.1050	−0.1386	−0.2590	−0.4159	−0.2858
S4	p-value	<0.05	<0.05	<0.05	>0.05	<0.05	<0.05	<0.05	<0.05
	rho	−0.3354	−0.2971	−0.1060	−0.1482	−0.3792	−0.3446	−0.3068	−0.3310
S5	p−value	<0.05	<0.05	<0.05	<0.05	<0.05	<0.05	<0.05	<0.05
	rho	−0.1874	−0.2142	0.1135	0.1467	−0.4930	−0.4120	−0.2218	−0.3318
S6	p-value	<0.05	<0.05	>0.05	<0.05	<0.05	<0.05	<0.05	<0.05
	rho	−0.2762	−0.2091	0.0449	−0.2754	−0.3046	−0.2150	−0.2557	−0.1828

References

Albuquerque, T. T. A., West, J., Andrade, M. F., et al. (2019). Analysis of PM2.5 concentrations under pollutant emission control strategies in the metropolitan area of São Paulo, Brazil. *Environmental Science and Pollution Research, 26*, 33216–33227.

Andreão, W., & Albuquerque, T. T. A. (2021). Avoidable mortality by implementing more restrictive fine particles standards in Brazil: An estimation using satellite surface data. *Environmental Research, 192*, 110288.

Andreão, W. L., Pinto, J. A., Pedruzzi, R., et al. (2020). Quantifying the impact of particle matter on mortality and hospitalizations in four Brazilian metropolitan areas. *Journal of Environmental Management, 270*, 110840.

FEAM—Fundação Estadual do Meio Ambiente. (2020). Qualidade do ar. Retrieved October 2020, from http://www.feam.br/qualidade-do-ar/dados

IBGE—Instituto Brasileiro de Geografia e Estatística. Estimativas da população. Retrieved Sept 2020, from https://www.ibge.gov.br/estatisticas/sociais/populacao/9103-estimativas-de-populacao.html?=&t=o-que-e

IBGE—Instituto Brasileiro de Geografia e Estatística. (2019). IBGE divulga as estimativas da população dos municípios para 2019. Retrieved July 2020, from https://agenciadenoticias.ibge.gov.br/agencia-sala-de-imprensa/2013-agencia-denoticias/releases/25278-ibge-divulga-as-estimativas-da-populacao-dos-municipios-para2019

MS—Ministério da Saúde. (2019). Saúde Brasil 2018: uma análise da situação de saúde e das doenças e agravos crônicos: desafios e perspectivas. Retrieved July 2020, from https://bvsms.saude.gov.br/bvs/publicacoes/saude_brasil_2018_analise_situacao_saude_doencas_agravos_cronicos_desafios_perspectivas.pdf

Pacheco, M. T., Parmigiani, M. M. M., Andrade, M. F., et al. (2017). A review of emissions and concentrations of particulate matter in the three major metropolitan areas of Brazil. *Journal of Transport and Health, 4*, 53–72.

WHO—World Health Organization. (2006). Air quality guidelines: Global update 2005. Retrieved July 2020, from https://www.euro.who.int/__data/assets/pdf_file/0005/78638/E90038.pdf

Modelling Air Pollution in a Changing Climate

Climate Change Impact on Source Contributions to the Air Quality in Aveiro Region

Sílvia Coelho, Joana Ferreira, David Carvalho, Ana Isabel Miranda, and Myriam Lopes

Abstract The full understanding of the links between air pollution sources and concentrations and climate change will be a challenge of future research. In this scope, the WRF-CAMx modelling system was applied to Aveiro Region (coastal central region of Portugal), recognized as one of the regions affected by some poor air quality events in Portugal and vulnerable to climate change effects. A summer period for the recent-past and the medium-term future climate, considering the new CMIP6 SSP5-8.5 scenario, was simulated, using the CAMx source apportionment tool to quantify the contribution of several source activities to ambient pollution, and their influence area type. Results show a decrease in the PM10 and NO_2 concentrations in the future. The source-oriented analysis reveals a strong contribution of the industrial and transport sectors for NO_2, while the background areas are those with the greatest influence for both pollutants.

S. Coelho (✉) · J. Ferreira · A. I. Miranda · M. Lopes
CESAM and Department of Environment and Planning, University of Aveiro, 3810-193 Aveiro, Portugal
e-mail: silviacatarina@ua.pt

J. Ferreira
e-mail: jferreira@ua.pt

A. I. Miranda
e-mail: miranda@ua.pt

M. Lopes
e-mail: myr@ua.pt

D. Carvalho
CESAM and Department of Physics, University of Aveiro, 3810-193 Aveiro, Portugal
e-mail: david.carvalho@ua.pt

C. Mensink and O. Jorba (eds.), *Air Pollution Modeling and its Application XXVIII*, Springer Proceedings in Complexity, https://doi.org/10.1007/978-3-031-12786-1_29

1 Introduction

Air pollution and climate change (CC) are serious environmental and health problems nowadays and it is expected that this issue will continue for the next few decades. As air quality is strongly dependent on meteorological conditions, it is therefore sensitive to CC and, according to the latest IPCC report (IPCC, 2021), CC will have complex effects on chemistry, transport and deposition of local air pollutants. This highlights the need for further studies that assess the future CC effects on air quality, with a high-resolution level, to support the identification of early climate and air pollution adaptation strategies. Notwithstanding, before the adaptation strategies definition, a comprehensive assessment of current air pollution and their main source contributions must be made. Source apportionment tools can provide valuable information by quantifying the contribution of different source categories (e.g.: transport, domestic heating, industrial activities) and source regions (e.g.: local, urban, metropolitan areas, counties) to several pollutant concentrations.

This work presents a diagnosis of the current and future air quality, and investigates the main source contributions, for a summer period in the Aveiro Region, in central Portugal.

2 Modelling Application

A modelling system which includes the Weather Research and Forecasting Model (WRF) (Skamarock et al., 2019) and the Comprehensive Air Quality Model with Extensions (CAMx) (ENVIRON, 2020), was applied to an urban domain over the Aveiro Region (central region of Portugal), recognized as one of the regions affected by some poor air quality events in Portugal. Both models were applied using three online-nested domains, with increasing resolution from a coarser domain of 30 km horizontal resolution, to the innermost domain of 1.2 km horizontal resolution focusing on the Aveiro Region (Fig. 1a).

WRF, forced by the Max Planck Institute for Meteorology Earth System Model version 1.2 (MPI-ESM1.2-h) (Mauritsen et al., 2019), was applied for two summer periods, one statistically representative of the recent-past (2014) and the other statistically representative of the medium-term future climate (2055). For the future climate, the new SSP5-8.5 greenhouse-gases scenario (Kriegler et al., 2017) was applied, which considers a stringent air pollution control but without CC mitigation. The considered emission reduction values were: 9% for CO; 16% for NH_3; 15% for NOx; 6% for NMVOC; 30% for PM10; 54% for PM2.5; and 36% for SO_2. WRF outputs were used as inputs to the CAMx, with its source apportionment tool (PSAT) (Yarwood et al., 2007), to quantify the contribution of several source activities to ambient pollution and their influence area type (Fig. 1b).

CAMx initial and boundary conditions were provided by the global chemical model CAM-Chem (Emmons et al., 2020). Anthropogenic emissions were taken

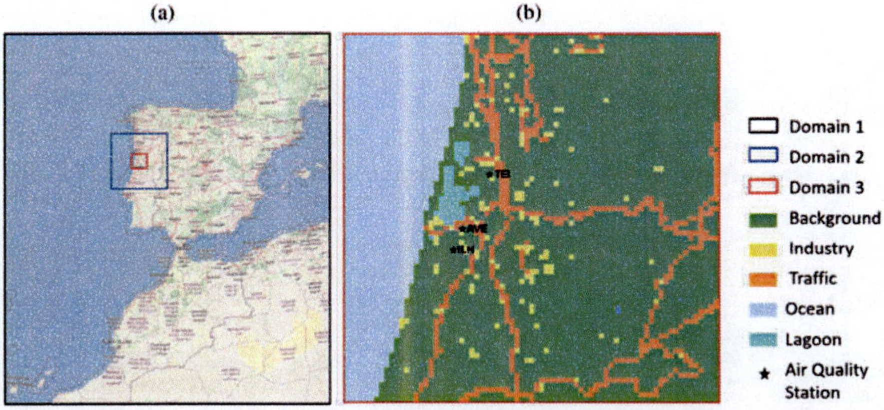

Fig. 1 **a** CAMx domains; **b** innermost domain, including the location of the air quality stations and the influence area type (traffic, industrial, background, ocean, and lagoon)

from the EMEP European emission inventory (EMEP, 2017), and spatially disaggregated following the Ferreira et al. (2020) approach and speciated into the CB6 chemical mechanism species considered in the CAMx simulation (Yarwood et al., 2010). Future emissions were estimated according to Sá et al. (2015) approach. For the PSAT application, as a first approach, the analysis focused on a source contribution. Based on the national emission inventories (Pereira et al., 2018), the more relevant emission sectors (SNAP categories) in Aveiro Region were defined, namely: (i) residential and commercial combustion (S2); (ii) industrial combustion and processes (S3&4), (iii) road transport (S7); and (iv) other mobile sources (S8). All the remaining sectors were grouped and treated as one (OTH). The PSAT receptor points were defined as the location of three air quality monitoring stations (see Fig. 1b), namely: (i) Aveiro (AVE—urban traffic); (ii) Ílhavo (ILH—suburban background); and (iii) Teixugueira (TEI—suburban industrial). As a second and complementary approach, a PSAT analysis focusing on the contributions by region was performed. Regions, i.e., influencing areas, were defined by the land use characterization of each cell of the inner domain, as shown in Fig. 1b. The different types of areas considered were: (i) traffic; (ii) industrial; (iii) background; (iv) ocean; and (v) lagoon.

3 Air Quality Characterization

The air quality characterization, for both PM10 and NO_2, was based on the spatial distribution concentration maps and on the source contribution analysis results.

The NO_2 and PM10 concentration fields (figures not shown) show similar spatial patterns for the different periods. For both PM10 and NO_2, higher concentrations were estimated over the most industrialized areas, in particular over the largest

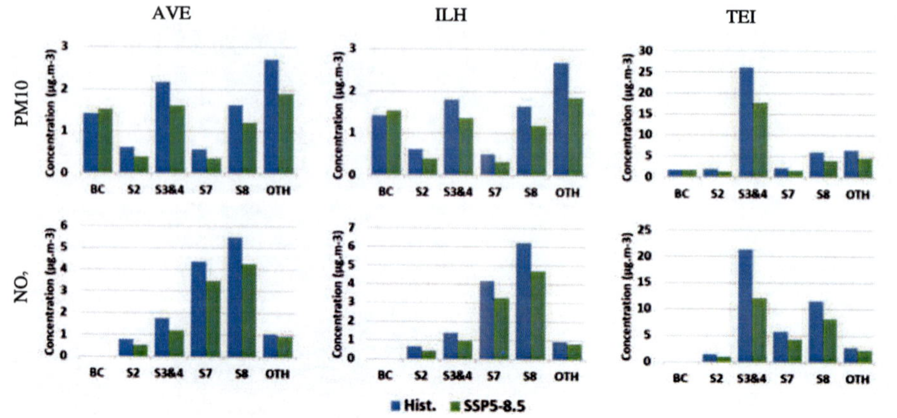

Fig. 2 Summer average contributions of each source group for NO_2 and PM10 concentrations, for the recent-past (Hist.) and future climate (SSP5-8.5)

one located next to the Teixugueira air quality station (Fig. 1b). For the recent-past climate, the maximum of the summer mean concentrations of PM10 and NO_2 was around 85 $\mu g.m^{-3}$ and 80 $\mu g.m^{-3}$, respectively. Concerning the future climate, these maximum concentrations decrease around 25% for both pollutants, following the emissions reductions considered in the SSP5-8.5 scenario. The air pollutant dispersion patterns follow the predominant wind directions from north and northwest.

The contribution of each source activity group for PM10 and NO_2 concentrations, in the three PSAT receptors, for both recent-past and mid-term future climate, is presented in Fig. 2.

When comparing the results for the recent-past and the future climate, all emission sectors contribute to the concentration decrease obtained for the future climate. Only boundary conditions (BC) tend to increase the PM10 concentrations in the future.

For both pollutants, Aveiro (urban traffic) and Ílhavo (suburban background) receptors show similar results, unlike Teixugueira (suburban industrial) receptor. Concerning PM10, for Aveiro and Ílhavo, none of the analyzed sectors presents a predominant contribution. On the other hand, for NO_2, there is a clear predominance of the transportation sectors (S7 and S8). Teixugueira shows a great contribution, for both PM10 and NO_2, of the industrial combustion and processes (S3&4).

To complement this source-oriented approach and get an overview on how the characteristics of the region influence its air quality, the analysis by area type is presented in Fig. 3, that shows the contribution of each influence area for PM10 and NO_2 concentrations, in the three PSAT receptors.

When analyzing contributions based on the influence area type, although the three receptors are linked with air quality stations with different influences, all receptors have a predominance of contribution from areas with background influence. This indicates that the air quality in the region is more influenced by the background levels of air pollution than the local emission sources as transports and industry. It

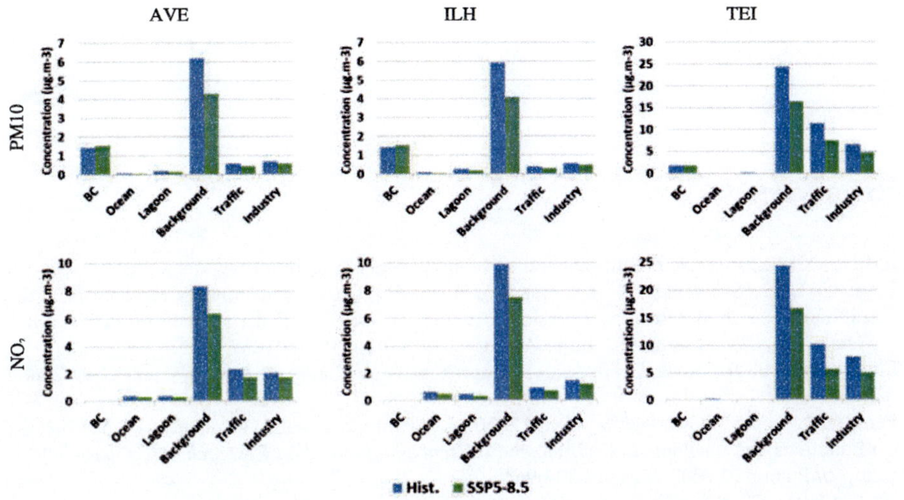

Fig. 3 Summer average contributions of each influence area type for NO_2 and PM10 concentrations, for the recent-past (Hist.) and future climate (SSP5-8.5)

is to be noted that the Porto metropolitan area, the second largest urban area in the country, is located North of the studied region at 60 km distance and the dominant winds are from North-Northwest. Teixugueira is the location where the air quality is also influenced by the transport and industrial areas, for both the present and the future. This industrial place is located close (about 2 km) to two main highways.

4 Conclusion

This work presents the WRF-CAMx modelling system as a tool to perform a diagnosis of current and future air quality and investigates the main source contributions. For both pollutants, NO_2 and PM10, results show a decrease in concentration in the future. Although the results do not show a predominance of any emission sector contribution for PM10 concentrations, for NO_2, the strong contribution of the industrial and transport sectors is noticeable, both in recent-past and future climate. The source-oriented analysis by area, for all cases, show a greater contribution from background areas. These results are relevant for the definition of emission reduction scenarios, that could be simulated using the same modeling system. Future work will also consider longer periods and additional greenhouse gases scenarios.

Acknowledgements Thanks are due for the financial support to the contract grants of J. Ferreira [2020.00622.CEECIND] and D. Carvalho [CEECIND/01726/2017], and the PhD grant of S. Coelho [SFRH/BD/137999/2018] and CESAM [UIDB/50017/2020 + UIDP/50017/2020] to FCT/MCTES through national funds, and the co funding by the FEDER, within the PT2020 Partnership Agreement and Compete 2020. Thanks are also due to HERA project, funded by the EU H2020 Research and Innovation Programme under the grant agreement no. 825417.

References

EMEP. (2017). European monitoring and evaluation programme. http://www.emep.int/

Emmons, L. K., Schwantes, R. H., Orlando, J. J., et al. (2020). The chemistry mechanism in the community earth system model version 2 (CESM2). *Journal of Advances in Modeling Earth Systems, 12*, e2019MS001882. https://doi.org/10.1029/2019MS001882

ENVIRON. (2020). CAMx. User's guide. Comprehensive air quality model with extensions. Version 7.10.

Ferreira, J., Lopes, D., Rafael, S., et al. (2020). Modelling air quality levels of regulated metals: Limitations and challenges. *Environmental Science and Pollution Research, 27*, 33916–33928. https://doi.org/10.1007/s11356-020-09645-9

IPCC. (2021). Climate change 2021: The physical science basis. Contribution of Working Group I to the Sixth Assessment Report of the Intergovernmental Panel on Climate Change. Cambridge University Press (in press).

Kriegler, E., Bauer, N., Popp, A., et al. (2017). Fossil-fueled development (SSP5): An energy and resource intensive scenario for the 21st century. *Global Environmental Change, 42*, 297–315. https://doi.org/10.1016/J.GLOENVCHA.2016.05.015

Mauritsen, T., Bader, J., Becker, T., et al. (2019). Developments in the MPI-M earth system model version 1.2 (MPI-ESM1.2) and its response to increasing CO2. *Journal of Advances in Modeling Earth Systems, 11*, 998–1038. https://doi.org/10.1029/2018MS001400

Pereira, T. C., Seabra, T., Pina, A,, et al. (2018). Portuguese Informative Inventory Report 1990–2016. 616

Sá, E., Ferreira, J., Carvalho, A., & Borrego, C. (2015). Development of current and future pollutant emissions for Portugal. *AtmosPheric Pollution Research, 6*, 849–857. https://doi.org/10.5094/APR.2015.094

Skamarock, W. C., Klemp, J. B., Dudhia, J., et al. (2019). A description of the advanced research WRF version 4. NCAR Tech. Note NCAR/TN-556+STR. Boulder, Colorado, USA.

Yarwood, G., Jung, J., Whitten, G. Z., et al. (2010). Updates to the carbon bond mechanism for version 6 (CB6). In *9th Annual CMAS Conference*. Chapel Hill, NC.

Yarwood, G., Morris, R. E., & Wilson, G. M. (2007). Particulate matter source apportionment technology (PSAT) in the CAMx photochemical grid model. In C. Borrego & A.-L. Norman (Eds.), *Air Pollution Modeling and Its Application XVII* (pp. 478–492). Springer.

Model Assessment and Verification

Intercomparison and Sensitivity Analysis of Gas-Phase Dry Deposition Schemes

Franco López, María Gonçalves-Ageitos, Dene Bowdalo, and Oriol Jorba

Abstract The dry deposition of gaseous species, such as ozone, sulphur dioxide and nitrogen oxides, is a major pathway in its removal from the atmosphere. Chemical Transport Models (CTM) commonly use a resistance approach to determine the deposition velocity of these gases. The objective of this study is twofold, first to compare different dry deposition schemes used in state-of-the-art CTMs, and second to evaluate the sensitivity of such schemes to key input parameters (meteorological conditions, soil type, leaf area index, reference values for non-stomatal resistances). The canopy resistance accounts for most of the resistance in deposition velocity, therefore its formulation has been used for the intercomparison in the present study. The selected schemes are based on the formulations of Wesely (Atmos Environ 23:1293–1304, 1989), Emberson et al. (Towards a model of ozone deposition and stomatal uptake over Europe, Oslo: Det Norske Meteorologiske Institutt., 2000) or Zhang et al. (Atmos Chem Phys 3:2067–2082, 2003) with subsequent modifications accounting for the effect of phenology, photoactive radiation, vapor pressure deficit and soil–water potential. The stomatal resistance is the component of the canopy resistance related to the uptake of gases by the leaves. As it is commonly calculated for ozone and then extrapolated to other gases, the parameterization for ozone has been studied. The sensitivity analyses have been carried out by modifying the input parameters shared by most of the schemes. They are related mainly to the physiology of the vegetation and the process of gas exchange at the stomata. Meteorological inputs derived from ERA5 are used in the intercomparison. The simulated ozone deposition velocities are compared with available observational data. The dry deposition schemes show strong differences in the seasonal cycle, both in the vegetated and non-vegetated land use classes. The identification of critical parameters helps constraining dry deposition schemes incorporated in atmospheric models.

Keywords Dry deposition velocity · Parametrization · Ozone

F. López (✉) · M. Gonçalves-Ageitos · D. Bowdalo · O. Jorba
Barcelona Supercomputing Center (BSC), Barcelona, Spain
e-mail: franco.lopez@bsc.es

F. López · M. Gonçalves-Ageitos
Department of Project and Construction Engineering, Universitat Politècnica de Catalunya –BarcelonaTech (UPC), Barcelona, Spain

C. Mensink and O. Jorba (eds.), *Air Pollution Modeling and its Application XXVIII*, Springer Proceedings in Complexity, https://doi.org/10.1007/978-3-031-12786-1_30

1 Introduction

Air quality is a major concern nowadays, as air pollution is responsible for negative impacts on human health and ecosystems. Aerosol and gas phase pollutants are routinely monitored in Europe through the observations of air quality networks. Besides, the diagnosis and prediction of air pollutant levels are commonly supported by numerical models, which allow defining and assessing air quality management strategies.

Chemical Transport Models (CTMs) represent the life cycle of air pollutants in the atmosphere, including emission, transport, chemical transformation, and wet and dry deposition processes. The dry deposition of gaseous species, such as ozone, sulphur dioxide or nitrogen oxides, is a major pathway in its removal from the atmosphere, but the mechanism that guides this process is not fully understood yet.

CTMs commonly use a resistance approach to determine the deposition velocity of trace gases, defining various in-series or in-parallel deposition pathways. These pathways consider an aerodynamic resistance, a common or surface-dependent quasi-laminar resistance and a canopy resistance.

The objective of this study is twofold, first to compare different dry deposition schemes used in state-of-the-art CTMs, and second to evaluate the sensitivity of such schemes to key input parameters.

2 Methodology

The intercomparison in the present study focuses on the formulation of the canopy resistance among models, as it is the most influencing resistance. We selected seven schemes (A to G) used in eulerian air quality models developed at Europe and North America, which ultimately rely on three main formulations (Table 1). Schemes A to C follow the Wesely (1989) dry deposition formulation, which defines the stomatal, cuticular, lower canopy and ground pathways. The scheme A represents the annual cycle defining two seasons, and the scheme B considers four seasons. The scheme C corrects a reference value for the stomatal resistance through an effective Leaf Area Index (LAI), which depends on LAI, zenith angle and cloud fraction. The other schemes (D to G) exclude the lower canopy pathway, but comprise modifications accounting for the effect of phenology, photoactive radiation, vapor pressure deficit and soil–water potential. The schemes D to F are based on Emberson et al. (2000), which considers as input parameters the LAI and the surface area index (SAI), while the scheme G, based on Zhang et al. (2003), only includes the former. As many of the models calculate the canopy resistance for ozone and then extrapolate the results to other gases, the parameterization for ozone is studied here.

The seven schemes for ozone dry deposition velocity were implemented as box models. The sensitivity analyses cover the year 2018 and meteorological inputs, derived from ERA5 (Copernicus Climate Change Service (C3S), 2021), were

Table 1 Dry deposition schemes' characteristics. T: temperature, SR: solar radiation, eLAI: effective leaf area index, PAR: photoactive radiation, VPD: vapor pressure deficit, PHE: phenology, SWP: soil–water potential

Dry deposition scheme	Dry deposition formulation	Deposition pathways	Stomatal resistance's correction factors	Tabulated resistance values
A	Wesely	4	T, SR	By land use and season (2)
B	Wesely	4	T, SR	By land use and season (4)
C	Wesely	4	T, eLAI	By land use
D	Emberson	3	T, PAR, VPD, PHE	By land use
E	Emberson	3	T, PAR, VPD, PHE, SWP	By land use
F	Emberson	3	T, PAR, VPD	By land use
G	Zhang	3	T, PAR, VPD, SWP	By land use

obtained for a European location not directly affected by coastal phenomena (47° N, 4° E). All the schemes use tabulated resistances that depend on the land use classification, therefore several executions were made to consider a diversity of land uses. The input parameters considered to test the sensitivity of the deposition velocity were those shared by most of the schemes, namely: temperature at 2 m above ground, relative humidity at 2 m above ground, solar radiation, LAI, and the cuticular and ground surface resistances. These are related mainly to the physiology of the vegetation and the gas exchange process at the stomata.

Additionally, simulations of ozone deposition velocities were conducted and compared with available observational data over Europe and North America, on a monthly basis, considering the local meteorology and the land use reported at these sites (Hole et al., 2004; Vuolo et al., 2017; Wu et al., 2016; Zona et al., 2014). The observation periods were 11 months in 2011 at Belgium (51°06′44″ N, 3°51′02″ E), 8 months between 2012 and 2013 at France (48°51′ N, 1°58′ E), 33 months between 2000 and 2003 at Norway (60°22′ N, 11°4′ E), and 60 months between 2008 and 2013 at Canada (44°19′ N, 79°56′ W).

3 Results and Discussion

The baseline simulations of the dry deposition velocity show a high variability for the annual cycle among the schemes used, with peaks occurring at different times (Fig. 1). However, the schemes also share some common traits, for instance on vegetated surfaces, the deposition velocity is higher in spring–summer, while on non-vegetated surfaces it follows a constant annual profile (*data not shown*).

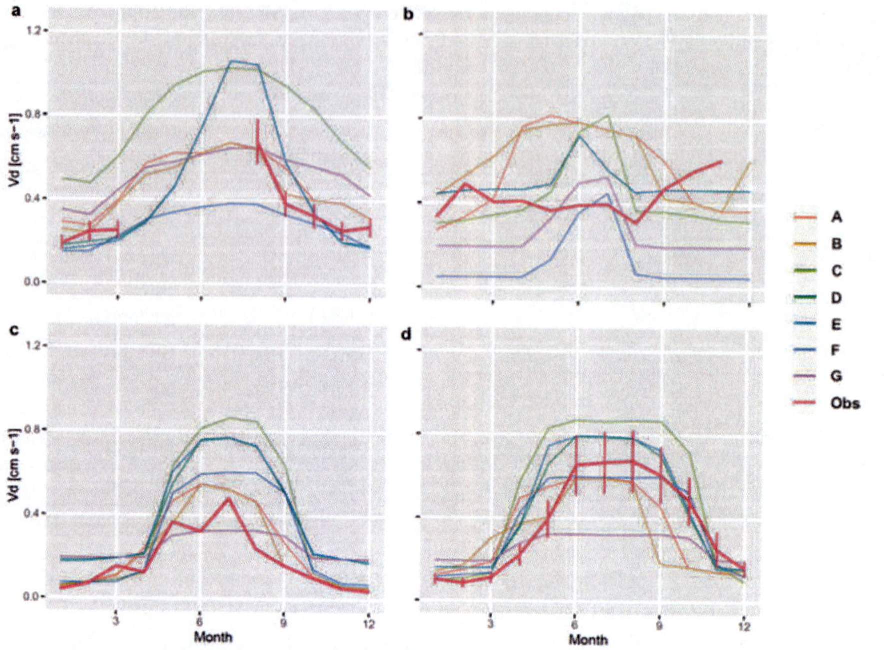

Fig. 1 Monthly mean ozone dry deposition velocity (Vd, cm s^{-1}) simulated in the box models (A–G) compared to observational data (Obs, thick pink line), over **a** grassland in France, **b** crop in Belgium, **c** coniferous forest in Norway and **d** coniferous forest in Canada. If available, mean ± sd are shown in the observational data with error bars

The evaluation of dry deposition velocities against observations is a challenging task, due to the scarcity of measurements. For this work, we use as a reference the retrieved ozone dry deposition velocities in four sites in Europe and North America. The evaluation with data from France, Norway and Canada suggests that, even if all schemes appropriately follow the deposition velocity seasonality, they show both under- and over- estimations of the retrieved values, and this happens both in winter and summer. The closer values between observational and simulated data are obtained with the schemes based on either Wesley (1989) or Emberson et al. (2000) formulations. The simulated values did not match the observations at the site in Belgium, which falls in a land use class characterized as temperate crop (Fig. 1). In fact, almost none of the schemes captured the observed seasonal cycle, showing negative correlation coefficients.

Then, the sensitivity analyses allowed us to determine the key parameters that modulate the deposition velocity, selected as those that increased at least two times this variable (Table 2). On any land use, temperature, and on vegetated surfaces the LAI, if used in the scheme, are critical parameters in the modulation of the dry deposition velocity of ozone. The reference values for cuticular and ground surface resistances are key parameters as well, on vegetated and non-vegetated surfaces, respectively. Therefore, an accurate definition of these key parameters would allow increasing the robustness of the models. Some of them are inherent to the models' formulation, i.e., the cuticular and ground surface resistances are usually tabulated values. Others come either from tabulated values or external data sources, e.g., the LAI. Given its relevance in defining the resistance for deposition, observation-constrained data (e.g., from remote sensing) would be recommended.

4 Conclusions

State of the art Chemical Transport Models commonly represent the dry deposition of gas-phase species through resistance approaches and define deposition pathways to the vegetation and to the ground surface. These dry deposition schemes rely on three different main formulations, with subsequent modifications. Regardless of their common approach, the deposition velocities derived from these schemes show high variability in magnitude and seasonality over vegetated surfaces. For ozone deposition velocity, and by extension for other gases, temperature and LAI constitute the key input parameters.

The evaluation of the dry deposition schemes performance is challenging due to the scarcity of observational data. The comparison conducted in this work, with the available information, shows that the schemes are better at reproducing the seasonality of the ozone deposition velocity than its magnitude, with under- or over-estimations depending on the scheme and period.

In addition to the refinement of the key input parameters that have proven relevant to define the dry deposition velocity, there is a need for more observational data to better constrain the process representation in air quality models.

To complement the analysis presented here, additional sensitivity tests will be performed with a 3D eulerian air quality model to assess the impact of the deposition parameters on the modelled air quality levels.

Table 2 Sensitivity analysis of meteorological and vegetation related parameters. The change in dry deposition velocity equal to or higher than 2 times is indicated for the corresponding land use categories (V: vegetated surfaces; NV: non-vegetated surfaces)

Dry deposition scheme	Temperature (15/−15 °C)	Relative humidity (90/10%)	Solar radiation (700/100 W m^{-2})	Leaf area index (5/0 m^2 m^{-2})	Cuticular resistance (200/9000 s m^{-1})	Ground surface resistance (100/3500 s m^{-1})
A	V, NV	–	V	–	V	V, NV
B	V, NV	–	V	–	V	V, NV
C	V, NV	–	–	V	V	V, NV
D	V, NV	–	–	V	V	V, NV
E	V, NV	V	–	V	V	V, NV
F	V	–	–	V	V	V, NV
G	V, NV	–	–	V	V	V, NV

Acknowledgements The authors acknowledge the support from the Ministerio de Ciencia, Innovación y Universidades (MICINN) as part of the BROWNING project RTI2018-099894-B-I00, and the computer resources at MareNostrum and the technical support provided by Barcelona Supercomputing Center (AECT-2021-1-0027).

References

Copernicus Climate Change Service. (C3S). (2021). *ERA5: Fifth generation of ECMWF atmospheric reanalyses of the global climate.* https://cds.climate.copernicus.eu/

Emberson, L., Simpson, D., Tuovinen, J.-P., Ashmore, M., Cambridge, H., Research Note No. 42. (2000). Towards a model of ozone deposition and stomatal uptake over Europe. Det Norske Meteorologiske Institutt.

Hole, L., Semb, A., & Tørseth, K. (2004). Ozone deposition to a temperate coniferous forest in Norway; gradient method measurements and comparison with the EMEP deposition module. *Atmospheric Environment, 38*(15), 2217–2223.

Vuolo, R., Loubet, B., Mascher, N., Gueudet, J.-C., Durand, B., Laville, P., Zurfluh, O., Ciuraru, R., Stella, P., & Trebs, I. (2017). Nitrogen oxides and ozone fluxes from an oilseed-rape management cycle: The influence of cattle slurry application. *Biogeosciences, 14*, 2225–2244.

Wesely, M. (1989). Parameterization of surface resistances to gaseous dry deposition in regional-scale numerical models. *Atmospheric Environment, 23*(6), 1293–1304.

Wu, Z., Staebler, R., Vet, R., & Zhang, L. (2016). Dry deposition of O_3 and SO_2 estimated from gradient measurements above a temperate mixed forest. *Environmental Pollution, 210*, 202–210.

Zhang, L., Brook, R., & Vet, R. (2003). A revised parameterization for gaseous dry deposition in air-quality models. *Atmospheric Chemistry and Physics, 3*, 2067–2082.

Zona, D., Gioli, B., Fares, S., De Groote, T., Pilegaard, K., Ibrom, A., & Ceulemans, R. (2014). Environmental controls on ozone fluxes in a poplar plantation in Western Europe. *Environmental Pollution, 184*, 201–210.

The WMO Barcelona Dust Regional Center: Linking Research with the Development of Dust User-Oriented Services

Sara Basart, Ernest Werner, Emilio Cuevas, and Carlos Pérez García-Pando

Abstract Sand and Dust Storms (SDS) play a significant role in different aspects of the Earth system and can represent a severe hazard for life, health, property, environment and economy. SDS can severely disrupt communications, energy production and transportation. The Barcelona Dust Regional Center manages and coordinates the activities related to sand and dust storms of the World Meteorological Organization (WMO) for research and operations. The Center provides access to available dust products and coordinates a network of collaborators (researchers, data providers and users' communities) in Northern Africa, the Middle East and Europe. The activities of the Center focus on facilitating access to the available dust information. The network around the Center promotes scientific collaborations that aim to deepen our understanding of the dust cycle and its variability, along with its impacts on key socio-economic sectors. Additionally, one of the core activities of the Center is to build the capacity of end-users with the aim of promoting the use of dust products to address the risks associated with airborne dust. Here, we introduce two examples of the use of SDS information accessible through the Barcelona Dust Regional Center website. First, we revise the benefits of using multi-model products for the early prediction of extreme events as the ones occurred in February 2020 in the Canary Islands and February 2021 in Europe. On the other hand, we introduce the results

S. Basart (✉) · C. P. García-Pando
Barcelona Supercomputing Center (BSC), 08034 Barcelona, Spain
e-mail: sara.basart@bsc.es

C. P. García-Pando
e-mail: carlos.perez@bsc.es

E. Werner
State Meteorological Agency of Spain, Barcelona, Spain
e-mail: ewernerh@aemet.es

E. Cuevas
Izaña Atmospheric Research Center-State Meteorological Agency, Santa Cruz de Tenerife, Spain
e-mail: ecuevasa@aemet.es

C. P. García-Pando
Catalan Institution for Research and Advanced Studies (ICREA), Barcelona, Spain

C. Mensink and O. Jorba (eds.), *Air Pollution Modeling and its Application XXVIII*, Springer Proceedings in Complexity, https://doi.org/10.1007/978-3-031-12786-1_31

for 2019–2020 of the SDS-WAS Warning Advisory System for Burkina Faso which is a tailored daily product based on colour-coded maps that started in October 2018.

1 Introduction

Atmospheric dust consists of tiny mineral particles, originating from soils in the arid and semi-arid parts of the Earth. When lifted by the wind, dust can be transported over distances of many thousands of kilometers to be finally deposited on the land, vegetation and glaciers, or into the ocean. The Intergovernmental Panel on Climate Change (IPCC, 2021) recognizes dust as a major component of atmospheric aerosols, one of the essential climate variables that characterize the Earth's climate. Dust particles are increasingly considered by atmospheric researchers to have important effects on weather, as they are involved in feedback processes related with atmospheric dynamics, clouds and precipitation formation.

Dust is gaining visibility because of its direct impacts on society and the economy. Airborne dust can carry irritating spores, bacteria, viruses and persistent organic pollutants. In areas located in or downwind of arid regions, sand and dust storms present serious risks to human health, the environment and infrastructure. Impacts on health include respiratory and cardiovascular problems, eye infections, and increased risk of infectious diseases in some regions, such as meningitis in Sahel and valley fever in Arizona. Dust can also transport nutrients to other parts of the world like the oceans, where it can affect marine biomass production. Other impacts include negative effects on ground transport, aviation, agriculture and the generation of solar energy.

Recognising the need for international coordination of the diverse community that deals with the societal impacts of sand and dust storms, in 2007 the World Meteorological Organization (WMO) took the lead and worked with international partners to develop and implement a Sand and Dust Storm Warning Advisory and Assessment System (SDS-WAS). The mission of SDS-WAS is to enhance the ability of countries to deliver timely and high- quality sand and dust storm forecasts, observations, information and knowledge to users through an international partnership of research and operational communities (Nickovic et al., 2014; Terradellas et al., 2015). Here, it is presented the WMO Barcelona Dust Regional Center and two examples of the use of SDS information accessible through the Barcelona Dust Regional Center website.

2 The WMO Barcelona Dust Regional Center

The WMO Barcelona Dust Regional Center manages and coordinates the research activities and operations of the WMO related to sand and dust storms by hosting the SDS-WAS Regional Center for Northern Africa, Middle East and Europe and the

Regional Specialized Meteorological Center with activity specialisation on Atmospheric Sand and Dust Forecast (Basart et al., 2019). The WMO Barcelona Dust Regional Center was created in 2010 thanks to the formal agreement of two Spanish institutions: the Meteorological State Agency of Spain (AEMET) (https://www.aemet.es/) and the Barcelona Supercomputing Center (BSC) (www.bsc.es). The Center provides access to available dust products and coordinates a network of collaborators (researchers, data providers and users' communities) in Northern Africa, the Middle East and Europe. The activities of the center focus on facilitating access to the available dust information. The network around the Center promotes scientific collaborations that aim to deepen our understanding of the dust cycle and its variability, along with its impacts on key socio-economic sectors. Additionally, one of the core activities of the Center is to build capacity of end-users with the aim of promoting the use of dust products to address the risks associated with airborne dust.

The WMO Barcelona Dust Regional Center provides access to high-quality dust information for the benefit of society. This information is useful to predict the occurrence of Sand and Dust Storms (SDS), as well as to manage their effects and impacts. In this context, the Center offers a wide range of dust products, both models and observations that serve the need for detailed dust information on a regional scale. One of the most important activities of the Regional Center is the multi-model intercomparison (Fig. 1) and forecast evaluation, which is deemed an indispensable service to the users and an invaluable tool to assess model skills. Currently, the Regional Center collects daily dust forecasts from up to 15 models run (at regional and global domains) by international partners. This unique multi-model ensemble is used to provide added-value products to the users (Fig. 2). In the next sections, it is presented some examples of the use of dust information for specific purposes: as a tool of prediction of extreme events (see Sect. 2.1) and warning advisory product (Sect. 2.2).

2.1 *Extreme Sand and Dust Storms*

In this section, we will present the forecast results of the multi-model dust ensemble for two extreme sand and dust events that occurred in February 2020 in Canary Islands and in February 2021 in Europe. Both events had a large impact and they were reported extensively in local, national and international social media and news.

The 22–24 February 2020 dust outbreak (see Fig. 1) was the most intense dust episode so far recorded in the Canary Islands in the last 30 years (Cuevas et al., 2021) achieving surface concentrations over 3000 $\mu g/m^3$ and reducing visibility to less than 10 m. The event was widely reported especially because of the closure of airports and the enormous problems caused to tens of thousands of tourists who suffered the cancellation of their flights. On 5–8 February 2021, African dust was transported from northeast Algeria caused reddish skies in large parts of Europe (see Fig. 2). This event coincided with snow and rain giving the Pyrenees and the Alps an exceptional brownish aspect.

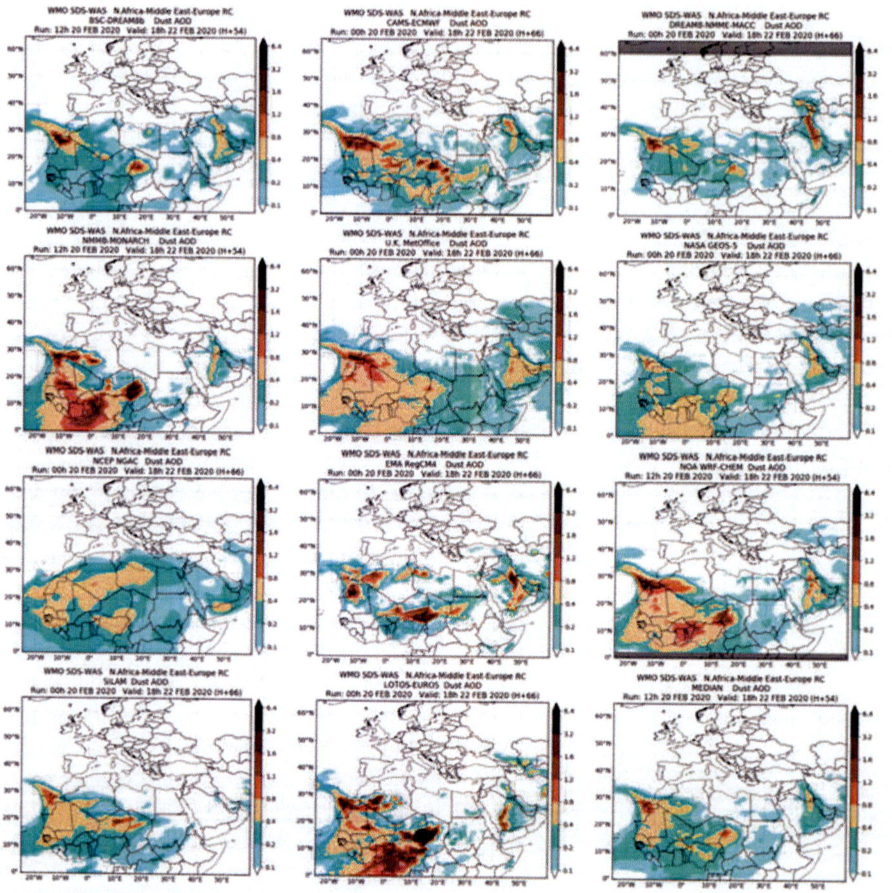

Fig. 1 Dust optical depth at 550 nm (DOD) forecast for 22nd February 2020 at 18UTC. The image displays the eleven dust forecast models participating in the multi-model dust intercomparison. The last panel corresponds to the multi-model product that corresponds to the median of the model ensemble. Extracted from https://dust.aemet.es/

During both events, dust forecasts were predicting the arrival and the extension of the African dust plume although the maximum peaks were underestimated as it is indicated in daily comparison of the total-column dust optical depth (DOD) using remote-sensing retrievals from sun-photometric (AERONET) and satellite (MODIS) measurements (not shown here). During extreme events is when it is observed a large spread between the forecast results (see standard deviation and range in Fig. 2).

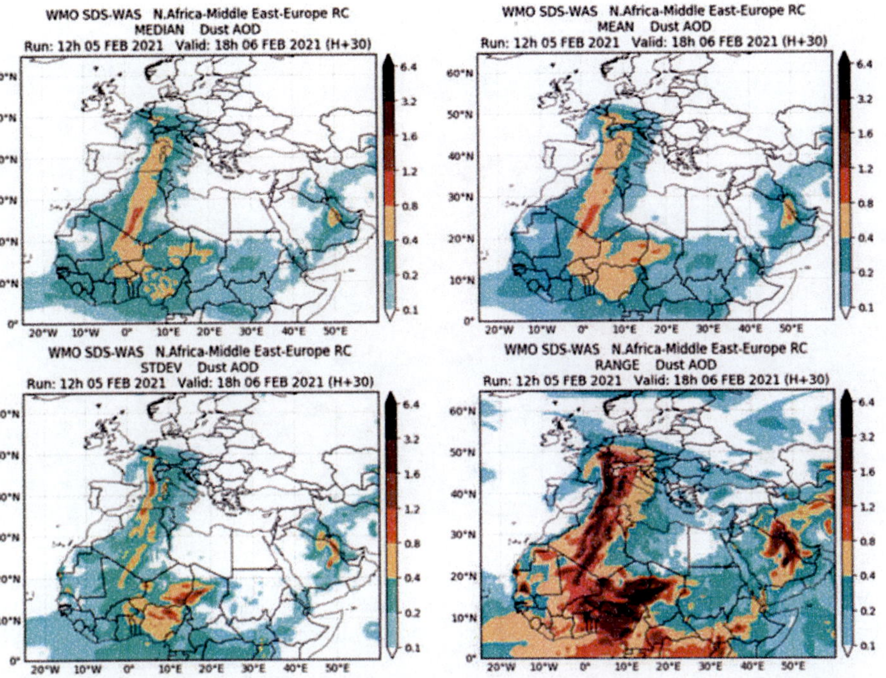

Fig. 2 Dust optical depth at 550 nm (DOD) forecast for 6th February 2021 at 18UTC. The image displays the results of the multi-model calculations based on the mean, median, standard deviation and rang result over the ensemble. Extracted from https://dust.aemet.es

2.2 *Warning Advisory System*

Early Warning Systems are well recognized as a critical life-saving tool for meteorological related hazards as floods, droughts, storms or bushfires but also sand and dust storms. In October 2018, the WMO Barcelona Dust Regional Center launched the Warning Advisory System (WAS) for preventing about the occurrence of extreme sand and dust storms in Burkina Faso in the Sahel (Terradellas et al., 2018) within the framework of the CREWS initiative. The designed WAS is a universally understood system based on colour-coded maps (see Fig. 3) that uses the results of the dust surface concentration multi-model median forecast. The warning level for each region is set according to the percentiles obtained from 5-years dust surface concentration multi-model ensemble time series. This clear, concise information is expected to help planning any activity vulnerable to airborne dust or activate services and procedures aimed at the mitigation of damages caused in agriculture, public health, or any other vulnerable sector. This WAS is expected to be expanded to other African countries in the Sahel.

The verification of the WAS over Burkina Faso is based in the comparison with visibility observations (Camino et al., 2015) because there is no routinely air quality

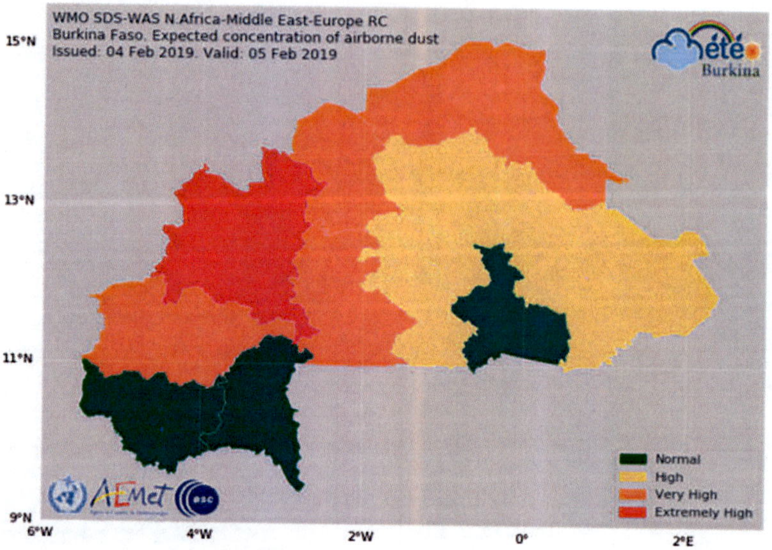

Fig. 3 Warning Advisory map for Burkina Faso valid for 5th February 2019. The image displays the result for the different provinces considering four categories of sand and dust storms considering its intensity (i.e. normal, high, very high and extremely high). Extracted from https://dust.aemet.es

measurement in the country. The preliminary results of the comparison over a visibility site in northeastern Burkina Faso during the dry season (from November 2019 to March 2020, when dust activity is maximum in this region), showed that the designed WAS can accurately predict the intensity of the events over 75% of the days (67% corresponds to normal days). Otherwise, around 20% of the days the WAS tend to over predict the maximum dust concentration peak and 5% of the days is under predicting their intensity.

3 Next Steps

Over the past decade, there is an increasing need for SDS accurate information and predictions to support early warning systems, and preparedness and mitigation plans. To facilitate the access to accurate and adapted information to specific user's needs, the WMO Barcelona Dust Regional Center website is being completely redesigned. The website's design includes all the efforts done in the last years to gather a large number of dust datasets and resources; as well as to understand the user's needs and building a more user-friendly interface for accessing to the dust products.

Acknowledgements Special thanks to all researchers, data providers and collaborators of the WMO SDS-WAS NAMEE Regional Node. The Copernicus, CREWS, and the COST Action inDust (COST Action CA16202) projects are also acknowledged. S. Basart and C. Pérez García-Pando also acknowledge the AXA Research Fund for funding aerosol research at the Barcelona Supercomputing Center through the AXA Chair on Sand and Dust Storms.

References

Basart, S., Nickovic, S., Terradellas, E., Cuevas, E., García-Pando, C. P., García-Castrillo, G., ... & Benincasa, F. (2019). The WMO SDS-WAS Regional Center for Northern Africa, Middle East and Europe. In *E3S Web of Conferences* (Vol. 99, p. 04008). EDP Sciences.

Camino, C., Cuevas, E., Basart, S., Alonso-Pérez, S., Baldasano, J. M., Terradellas, E., ... & Berjón, A. (2015). An empirical equation to estimate mineral dust concentrations from visibility observations in Northern Africa. *Aeolian Research, 16*, 55–68.

Cuevas, E., Milford, C., Barreto, A., Bustos, J. J., García, R. D., Marrero, C. L., Prats, N., Bayo, C., Ramos, R., Terradellas, E., Suárez, D., Rodríguez, S., de la Rosa, J., Vilches, J., Basart, S., Werner, E., López-Villarrubia, E., Rodríguez-Mireles, S., Pita Toledo, M. L., González, O., Belmonte, J., Puigdemunt, R., Lorenzo, J.A., Oromí, P., & del Campo-Hernández, R. (2021). Desert dust outbreak in the Canary Islands (February 2020): Assessment and impacts. In E. Cuevas, C. Milford, & S. Basart. (Eds.), State Meteorological Agency (AEMET), Madrid, Spain and World Meteorological Organization, Geneva, Switzerland, WMO Global Atmosphere Watch (GAW), Report No. 259, WWRP 2021–1, 2021.

IPCC. (2021). Climate change 2021: The physical science basis. Contribution of Working Group I to the Sixth Assessment Report of the Intergovernmental Panel on Climate Change [Masson-Delmotte, V., Zhai, P., Pirani, A., Connors, S. L., Péan, C., Berger, S., Caud, N., Chen, Y., Goldfarb, L., Gomis, M. I., Huang, M., Leitzell, K., Lonnoy, E., Matthews, J. B. R., Maycock, T. K., Waterfield, T., Yelekçi, O., Yu, R., & Zhou, B. (Eds.)]. Cambridge University Press. In Press.

Nickovic, S., et al. (2014). *SDS-WAS science and implementation plan: 2011–2015*. World Meteorological Organization.

Terradellas, E., et al. (2015). Airborne dust: A hazard to human health, environment and society. *World Meteorological Bulletin, 64*(2), 42–46.

Terradellas, E., Basart, S., Benincasa, F., & Werner, E. (2018). Warning advisory system for sand and dust storm in Burkina Faso, WMO Barcelona Dust Regional Center. Technical report, No. SDS-WAS-2018–001.

Performance of SILAM Model in Respect to Peak Concentrations of Trace Gases Attributable to Regional Point Sources

Terje Tammekivi, Marko Kaasik, Raido Kiss, Heikki Junninen, Urmas Hõrrak, and Steffen Noe

Abstract The SILAM model was tested against the measurements of ground-based monitoring stations, when the peaks of SO_2 and NO_2 concentrations from tall industrial stacks occurred in cold anticyclonal weather with frequent ground-based thermal inversions. It was found that along with adequacy in terms of meso-scale transport, the ground-based inversions, hindering the vertical transport, may appear challenging for SILAM to reproduce the near-surface concentrations of air pollutants, as the depth and strength of inversions are hard to forecast. Further investigations are needed for proper improvements in setup or even parametrizations of the model.

Keywords Downward transport · SILAM · Surface inversion · Sulphur dioxide · Nitrogen oxides

T. Tammekivi (✉) · M. Kaasik · R. Kiss · H. Junninen · U. Hõrrak
Institute of Physics, University of Tartu, Tartu, Estonia
e-mail: terje.tammekivi@ut.ee

M. Kaasik
e-mail: marko.kaasik@ut.ee

R. Kiss
e-mail: raido.kiss@ut.ee

H. Junninen
e-mail: heikki.junninen@ut.ee

U. Hõrrak
e-mail: urmas.horrak@ut.ee

M. Kaasik · R. Kiss
Estonian Environmental Research Centre, Tallinn, Estonia

S. Noe
Institute of Forestry and Engineering, Estonian University of Life Sciences, Tartu, Estonia
e-mail: steffen.noe@emu.ee

C. Mensink and O. Jorba (eds.), *Air Pollution Modeling and its Application XXVIII*, Springer Proceedings in Complexity, https://doi.org/10.1007/978-3-031-12786-1_32

1 Introduction

Thermal inversions hinder the vertical dispersion of air pollutants. Besides the well-known stagnation conditions, which can tremendously worsen the air quality in urban areas, the surface inversion can, in contrast, prevent penetration of pollutants to the surface layer from above. Although less prevalent, such a situation often appears, when the gaseous admixtures or fine particles are injected to the upper boundary layer by buoyant plumes from tall stacks. In this study we investigate the ability of the SILAM model (Sofiev et al., 2006, 2015) to reproduce the selected episodes of wintertime thermal inversions, when low near-surface concentrations were observed despite the polluted plume propagating higher above.

The SILAM model was used in forward and inverse (adjoint) dispersion modes:

1. in adjoint mode to understand the origin of isolated peaks in concentrations of air pollutants measured in rural background stations;
2. in forward mode to understand how much the thermal inversion hindered the downward dispersion of the plume.

This paper presents two case analyses, when pollution plumes of SO_2 and NO_x from Estonian oil-shale-fired thermal power plants and industries were detected as isolated peaks in monitoring stations (Fig. 1). Despite large reduction in recent decades, the Estonian oil-shale-based power plants and industries still emit amounts of SO_2 and NO_x traceable for hundreds of kilometres downwind. The computation schemes of SILAM in forward and inverse modes are described in detail in papers by Sofiev et al. (2006, 2015).

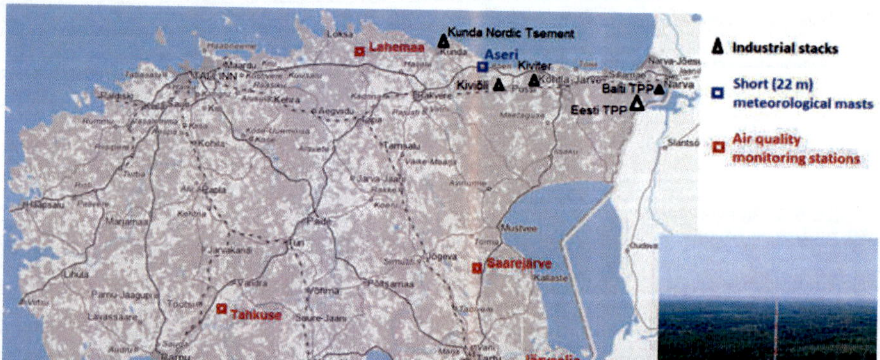

Fig. 1 Map of continental part of Estonia with main industrial point sources (black triangles), rural air pollution monitoring stations (red squares) and meteorological masts (blue squares)

2 Model and Methods

SILAM version 5.7 (System for Integrated modeLling of Atmospheric composition) is a global to meso-scale dispersion model developed to assess atmospheric composition, and air quality, apply emergency decision support applications, and provide inverse dispersion problem solution. The dynamic core of the model for the 3D transport of air pollution constituents includes both Lagrangian and Eulerian advection–diffusion formulations (Verstraeten et al., 2021).

In this study the SILAM model was used to calculate the dispersion of SO_2 and NO_x from regional point sources over the territory of Estonia. From the ground surface to the free troposphere, the layer thicknesses 25, 50, 100, 200, 400, 750, 1200 and 2000 m were applied in model runs. The Eulerian dispersion core with Galperin advection algorithm is applied, which is stable, efficient computationally and preserves the sub-grid mass distribution information (Sofiev et al., 2015). SILAM model runs were performed in the Estonian domain with nearly 5 km resolution and with 70–250 m high stack sources in North-East Estonia. In this study the SILAM model was driven by HIRLAM (High Resolution Local Area Modelling for numerical weather prediction) meteorological data provided by the Estonian Weather Service. No pollution emission sources from outside of Estonia were included and no boundary fields of meteorological conditions were used.

A series of air pollution events (pronounced peaks lasting a few hours each) in early spring of 2018 detected in Tahkuse, Lahemaa, Saarejärve and Järvselja air quality monitoring stations are studied. The stratification of temperature and trace gases was measured in a tall mast at Järvselja (58.2772° N, 27.3097° E) at heights 30, 50, 70, 90 and 110 m above the surface (Noe et al., 2015, 2016) and in addition the temperature in the surface layer (2–22 m) in masts at Uulu (58.2482° N, 24.5889° E) and Aseri (59.4443° N, 26.7635° E).

First, we computed the SO_2 and NO_x concentration "footprints" (sensitivity distributions) for Tahkuse rural background station in a south-western Estonia (58.5295° N, 24.9314° E), Lahemaa station (59.5153° N, 25.9292° E) in northern and Saarejärve station (58.7014° N, 26.7546° E) in the eastern part of the country. Then we modelled the dispersion from the strongest point sources in North-East of Estonia, which contribution was indicated in adjoint runs of model.

3 Results and Discussion

In Fig. 2 there are presented measured and modelled rural background concentrations of sulphur dioxide during an episode of peaks in late winter of 2018, which was characterised by steady anticyclone, snow-covered underlying surface and subzero temperatures. SILAM predicted several high peaks, which weren't recorded in monitoring stations. In some cases, a rather similar shape of concentration peak was measured, but at much lower concentration (Saarejärve, March 4; Tahkuse, March 5; Lahemaa, March 7).

Temperature stratification measurements in Uulu and Aseri sites (Fig. 3) show that strong surface inversions were present at nights under consideration and nearly neutral boundary layer appeared only for a few hours in daytime. To understand whether the overestimation of the vertical turbulent transfer rate is the reason for the modelled "false peaks" in the surface layer, the measurement results from a tall mast were analysed. During the near midnight on 28 march 2018 in Järvselja station, high levels of SO_2 and NO_2 were observed at heights 70–110 m. Model runs show impact of stack sources from NE Estonia, factory of Kunda Nordic Tsement in particular (Fig. 4).

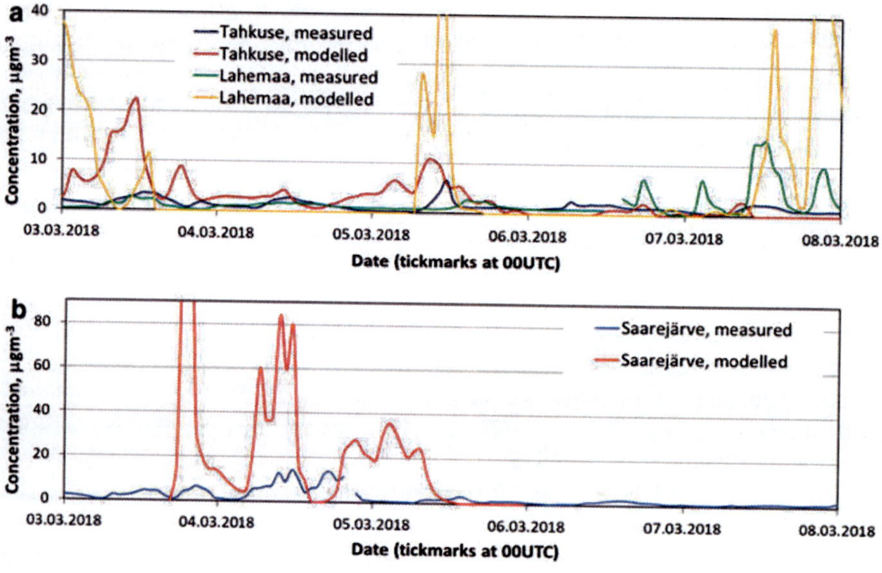

Fig. 2 Modelled and measured concentrations of SO_2 in Tahkuse and Lahemaa (**a**) and Saarejärve (**b**) monitoring stations. Locations see Fig. 1

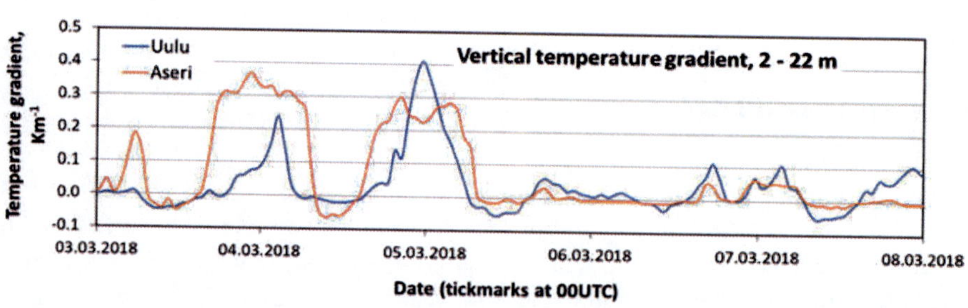

Fig. 3 Temperature distribution in surface layer, measured in low masts at Uulu and Aseri. Locations see Fig. 1

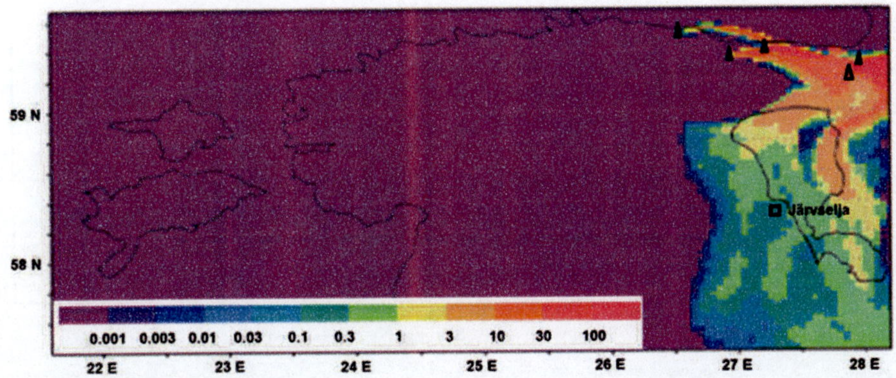

Fig. 4 Modelled SO_2 concentrations ($\mu g m^{-3}$) on 28.03.2018, 00:00 UTC at the lowest model level

Temperature profile during the event of March 27/28 (Fig. 5) changed from nearly-neutral in daytime to definitely stable at night, whereas the model predicted a much thinner inversion layer than it was measured. Accordingly, the night-time concentration profiles of SO_2 and NO_2 show rather low surface-layer concentrations, whereas much higher concentrations from the plume occurred higher above. In accordance with a thicker inversion layer, the measured night-time SO_2 profile is sharper, although model results show the same tendency, too. Huge night-time variations in measured concentrations most probably indicate narrow meandering polluted plumes in clean air. However, the model at given resolution is not able to reproduce such a feature. The concentrations of NO_x vary in smaller range and its measured profiles are smoother, suggesting at least partially different, less elevated sources. Night-time overestimation of NO_x may occur due to overestimated source term, as robust, annual-average-based stack emission data were used in modelling.

4 Conclusions

Occasions of meso-scale transport of air pollutants from tall stacks with temperature inversion below, hindering the downward transport, and thus making the plume "invisible" by ground-based monitoring stations, were identified in this study.

The SILAM model is a useful, but not yet perfect tool to detect the industrial sources of air pollutants contributing to the short-term peak concentrations at remote sites, when the surface-based temperature inversions occur. This phenomenon is rather challenging for SILAM. Further investigations, including more case studies, are needed to better understand the shortages in predicting the vertical mixing in surface-based inversion layer and tune the model setup or model parameterizations, if needed, to achieve better performance in such a case.

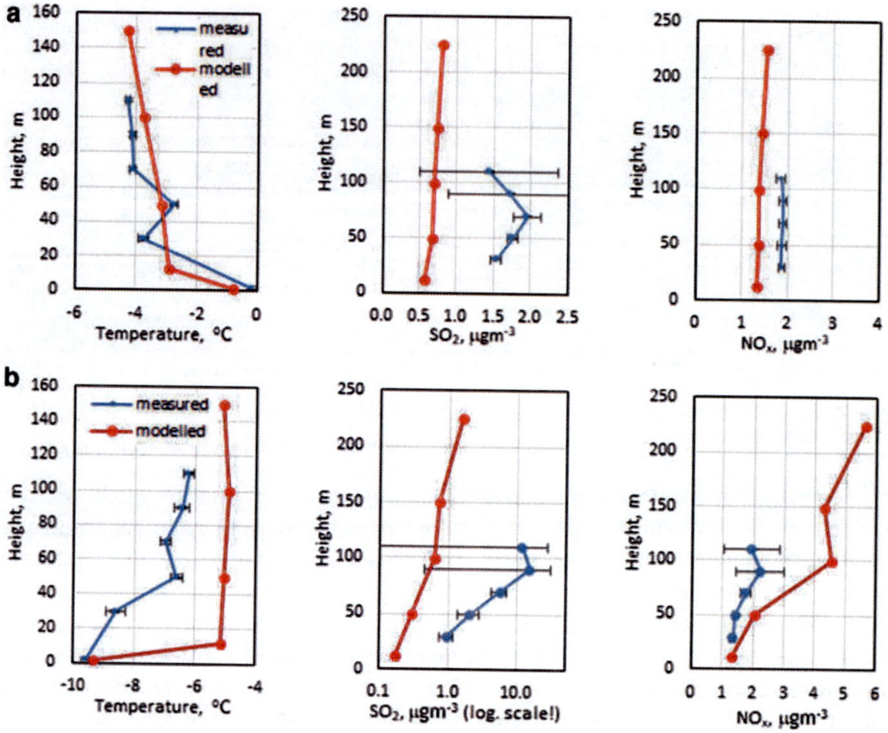

Fig. 5 Three-hourly average temperature, SO_2 and NO_x profiles in Järvselja mast on the afternoon of 27.03.2018 (**a**) and near midnight of 28.03.2018 (**b**). Standard deviation bars of fast-response sensor measurements are shown also

Acknowledgements This research is funded by the Estonian Ministry of Education and Science, research grant PRG714 and co-funded by Estonian Environmental Research Centre, contributing to implementation of the SILAM model on the Airviro monitoring platform. The measurements at Tahkuse station are supported by the Estonian Environment Agency contract LLTFY18213 contributing to the fulfilment of the National Environmental Monitoring Program. SMEAR Estonia measurements and data provision are supported by the Estonian Ministry of Sciences projects (grant nos. P180021, P180274, P200196) and by the Estonian Research Infrastructures Roadmap Project "Estonian Environmental Observatory" (3.2.0304.11-0395).

References

Noe, S. M., Niinemets, Ü., Krasnova, A., Krasnov, D., Motallebi, A., Kängsepp, V., Jõgiste, K., Hõrrak, U., Komsaare, K., Mirme, S., Vana, M., Tammet, H., Bäck, J., Vesala, T., Kulmala, M., Petäjä, T., & Kangur, A. (2015). SMEAR Estonia: Perspectives of a large-scale forest ecosystem—Atmosphere research infrastructure. *Forestry Studies, Metsanduslikud Uurimused, 63*, 56–84.

Noe, S. M., Krasnov, D., Krasnova, A., Cordey, H. P. E., & Niinemets, Ü. (2016). Seasonal variation and characterisation of reactive trace gas mixing ratios over a hemi-boreal mixed forest site in Estonia. *Boreal Environment Research, 21*, 332–344.

Sofiev, M., Siljamo, P., Valkama, I., Ilvonen, M., & Kukkonen, J. (2006). A dispersion modelling system SILAM and its evaluation against ETEX data. *Atmospheric Environment, 40*, 674–685. https://doi.org/10.1016/j.atmosenv.2005.09.069

Sofiev, M., Vira, J., Kouznetsov, R., Prank, M., Soares, J., & Genikhovich, E. (2015). Construction of the SILAM Eulerian atmospheric dispersion model based on the advection algorithm of Michael Galperin. *Geoscientific Model Development, 8*, 3497–3522. https://doi.org/10.5194/gmd-8-3497-2015

Verstraeten, W. W., Kouznetsov, R., Hoebeke, L., Bruffaerts, N., Sofiev, M., & Delcloo a, A. W. (2021). Modelling grass pollen levels in Belgium. *Science of the Total Environment, 753*, 141903. https://doi.org/10.1016/j.scitotenv.2020.141903

Aerosols in the Atmosphere

MONARCH Regional Reanalysis of Desert Dust Aerosols: An Initial Assessment

Enza Di Tomaso, Jerónimo Escribano, Sara Basart, Paul Ginoux, Francesca Macchia, Francesca Barnaba, Francesco Benincasa, Pierre-Antoine Bretonnière, Arnau Buñuel, Miguel Castrillo, Emilio Cuevas, Paola Formenti, María Gonçalves-Ageitos, Oriol Jorba, Martina Klose, Lucia Mona, Gilbert Montané, Michail Mytilinaios, Vincenzo Obiso, Miriam Olid, Nick Schutgens, Athanasios Votsis, Ernest Werner, and Carlos Pérez García-Pando

Abstract Aerosol reanalyses are a well-established tool for monitoring aerosol trends, for validation and calibration of weather chemical models, as well as for the enhancement of strategies for environmental monitoring and hazard mitigation. By providing a consistent and complete data set over a sufficiently long period, they address the shortcomings of aerosol observational records in terms of temporal and spatial coverage and aerosol speciation. These shortcomings are particularly severe for dust aerosols. A 10-year dust aerosol regional reanalysis has been recently produced on the Barcelona Supercomputing Center HPC facilities at the high spatial resolution of 0.1°. Here we present a brief description and an initial assessment of this data set. An innovative dust optical depth data set, derived from the MODIS Deep Blue products, has been ingested in the dust module of the MONARCH model by means of a LETKF with a four-dimensional extension. MONARCH ensemble

E. Di Tomaso (✉) · J. Escribano · S. Basart · F. Macchia · F. Benincasa · P.-A. Bretonnière · A. Buñuel · M. Castrillo · M. Gonçalves-Ageitos · O. Jorba · M. Klose · G. Montané · V. Obiso · M. Olid · C. Pérez García-Pando
Barcelona Supercomputing Center (BSC), Barcelona, Spain
e-mail: enza.ditomaso@bsc.es

P. Ginoux
Geophysical Fluid Dynamics Laboratory (GFDL), Princeton, NJ, USA

F. Barnaba
Consiglio Nazionale delle Ricerche-Istituto di Scienze dell'Atmosfera e del Clima (CNR-ISAC), Bologna, Italy

E. Cuevas
Izaña Atmospheric Research Center, AEMET, Santa Cruz de Tenerife, Spain

P. Formenti
LISA, UMR CNRS 7583, Université Paris-Est-Créteil, Université de Paris, Institut Pierre-Simon Laplace (IPSL), Créteil, France

M. Gonçalves-Ageitos
Universitat Politècnica de Catalunya - BarcelonaTech (UPC), Department of Project and Construction Engineering, Terrassa, Spain

C. Mensink and O. Jorba (eds.), *Air Pollution Modeling and its Application XXVIII*, Springer Proceedings in Complexity, https://doi.org/10.1007/978-3-031-12786-1_33

has been generated by applying combined meteorology and emission perturbations. This has been achieved using for each ensemble member different meteorological fields as initial and boundary conditions, and different emission schemes, in addition to stochastic perturbations of emission parameters, which we show is beneficial for dust data assimilation. We prove the consistency of the assimilation procedure by analyzing the departures of the assimilated observations from the model simulations for a two-month period. Furthermore, we show a comparison with AERONET coarse optical depth retrievals during a period of 2012, which indicates that the reanalysis data set is highly accurate. While further analysis and validation of the whole data set are ongoing, here we provide a first evidence for the reanalysis to be a useful record of dust concentration and deposition extending the existing observational-based information intended for mineral dust monitoring.

Keywords Dust · Aerosol regional reanalysis · Aerosol data assimilation · Modis deep blue · Aerosol speciation

1 Introduction

Desert dust plays a key role in the Earth's system: it influences the atmospheric radiation balance by scattering and absorbing solar and terrestrial radiation (Miller et al., 2014), it affects cloud formation and chemistry (Cziczo et al., 2013), it acts as a fertilizer when deposited over the ocean and land (Jickels et al., 2005; Yu et al., 2015). This has implications both in the short-term (in particular upon air quality and visibility, but also upon weather), and in the long-term by producing a radiative

M. Klose
Karlsruhe Institute of Technology (KIT), Institute of Meteorology and Climate Research (IMK-TRO), Department Troposphere Research, Karlsruhe, Germany

L. Mona · M. Mytilinaios
Consiglio Nazionale delle Ricerche-Istituto di Metodologie per l'Analisi Ambientale (CNR-IMAA), Rome, Italy

N. Schutgens
Department of Earth Science,Vrije Universiteit Amsterdam, 1081 HVAmsterdam, Netherlands

A. Votsis
Department of Governance and Technology for Sustainability (BMS-CSTM),University of Twente, Enschede, Netherlands

Finnish Meteorological Institute (FMI), Weather and Climate Change Impact Research, Helsinki, Finland

E. Werner
State Meteorological Agency (AEMET), Madrid, Spain

C. Pérez García-Pando
Catalan Institution for Research and Advanced Studies, (ICREA), Barcelona, Spain

forcing on the climate system. Atmospheric dust concentrations vary considerably in space and time. Different types of measurements monitor its cycle and are done either in-situ or remotely by satellite, air-borne and ground-based sensors. Dust observations are best exploited by being combined with model simulations either to provide optimal initial conditions (analyses) to model forecast or to monitor current and past states of the atmosphere through the production of reanalyses.

We present here an initial assessment of a regional reanalysis of desert dust aerosol which has been produced by the Barcelona Supercomputing Center. The reanalysis spans a 10-year period, from 2007 to 2016, has a horizontal resolution of 0.1° latitude × 0.1° longitude, and a 3-hourly temporal resolution. It provides a regional reconstruction of past dust conditions for Northern Africa, the Middle East and Europe, obtained by combining satellite remote sensing observations of dust with a dynamical model.

2 Methodology

The dust reanalysis is based on three pillars: (i) the Multiscale Online Non-hydrostatic AtmospheRe CHemistry model (MONARCH; Klose et al., 2021; Pérez García-Pando et al., 2011); (ii) the Local Ensemble Transform Kalman Filter (LETKF) data assimilation scheme (Di Tomaso et al., 2017; Hunt et al., 2007; Schutgens et al., 2010); (iii) coarse-mode dust optical depth (DOD) at 550 nm retrieved from the Moderate Resolution Imaging Spectroradiometer (MODIS) Deep Blue Level 2 products (Collection 6; Ginoux et al., 2012; Pu & Ginoux, 2016). MONARCH ensemble has been generated for data assimilation purposes with the use of different meteorological initial and boundary conditions (MERRA-2 with ERA5 soil and the ECMWF ERA-Interim reanalysis) and emission schemes, additionally to perturbations of model emission parameters. Previous studies showed that combined meteorology and aerosol source ensembles are necessary to produce sufficient spread in outflow regions. Similarly, our test experiments showed the benefit of using different meteorology and emission schemes for dust data assimilation (see Fig. 1).

3 Results

We describe here an initial validation of the reanalysis coarse DOD at 550 nm for a period of 2012 in terms of data assimilation inner diagnostics (statistics of departures from assimilated observations) and in terms of a comparison against independent observations that have not been assimilated. A more detail analysis can be found in (Di Tomaso et al., 2021) and (Mytilinaios et al. 2022, in prep.). Figure 2 shows an example for July 9 2012 of assimilated retrievals of coarse DOD, model ensemble mean first-guess and analysis DOD at 550 nm, while Fig. 3 reports the BIAS, RMSE, CORR, FRGE of the DOD at 550 nm of a free-run experiment (with no data assimilation),

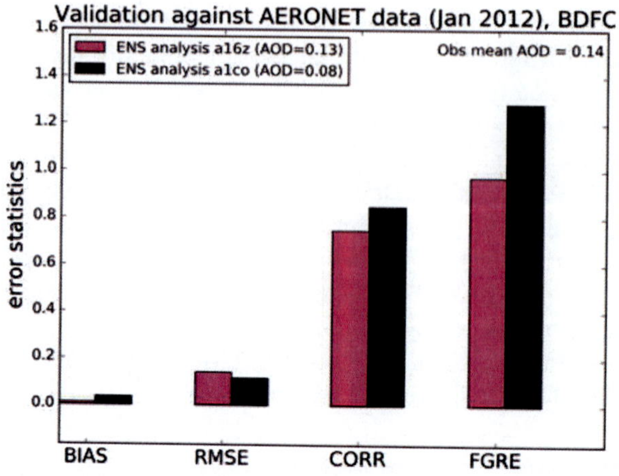

Fig. 1 Mean bias (BIAS), root mean square error (RMSE), Pearson correlation coefficient (CORR), and fractional gross error (FRGE) calculated against AERONET retrievals for the DOD at 550 nm of an assimilation experiment with emission parameter perturbation (magenta) and with the use of different emission schemes and meteorology initial and boundary conditions, additionally to the perturbation of emission parameters (black)

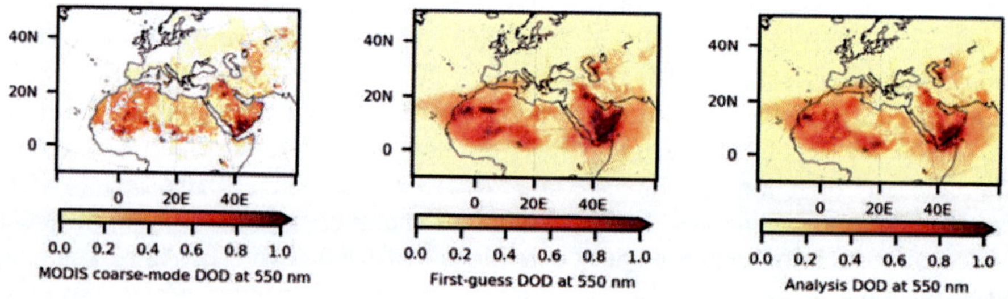

Fig. 2 Assimilated retrievals of coarse DOD at 550 nm from the Aqua MODIS Deep Blue Collection 6 Level 2 product (left), DOD at 550 for the ensemble mean first-guess (centre) and ensemble mean analysis (right) for July 9 2012

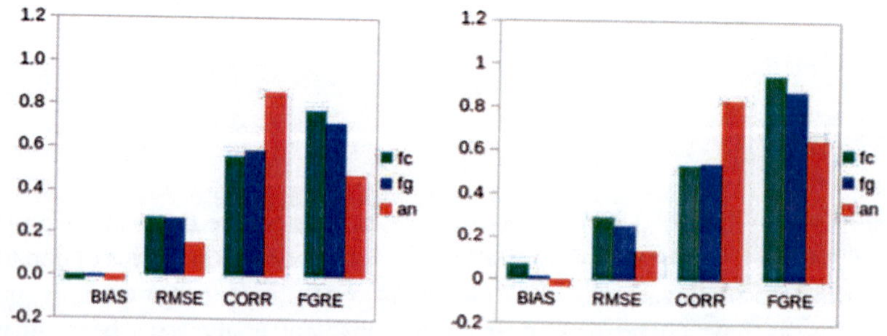

Fig. 3 Bias, RMSE, CORR, FRGE calculated against MODIS assimilated retrievals for the coarse DOD at 550 nm of a free-run experiment (fc; green), the first-guess (fg; blue) and the analysis (an; red) for a summer month (August; left) and a winter month (February; right) of 2012. Model fields are ensemble mean

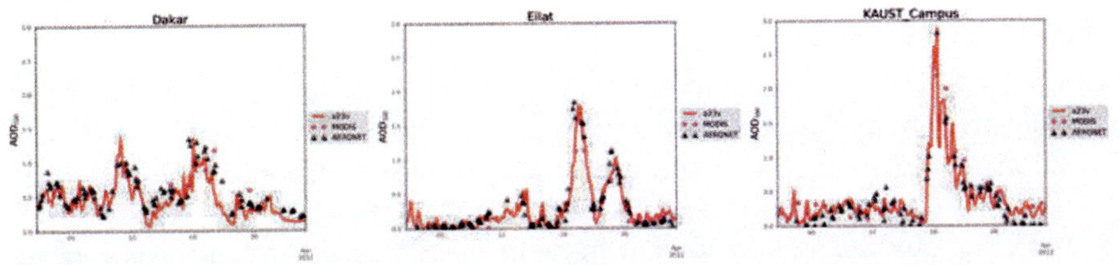

Fig. 4 Time series of coarse DOD values for a period of 2012 at three AERONET stations for the mean analysis (red), for AERONET SDA AOD (black triangles) and for MODIS coarse DOD retrievals

of the first-guess and of the analysis versus MODIS coarse DOD at 550 nm for a summer month and a winter month of 2012. As expected, the analysis is closer to the assimilated observations than the first-guess and outperforms a simulation without data assimilation. Timeseries of model coarse DOD at 550 nm for the mean analysis at three AERONET stations are in Fig. 4 together with coarse AOD retrievals at 550 nm (level 2.0, version 2.0) from the spectral de-convolution algorithm (SDA; O'Neill et al., 2003), and assimilated MODIS data, when available. We use the AERONET value closest to the model time step and within a ±30 min interval. These timeseries show that the analysis characterizes well most dust events at sites close and far from emission areas.

Acknowledgements We acknowledge the DustClim project which is part of ERA4CS, an ERA-NET initiated by JPI Climate, and funded by FORMAS (SE), DLR (DE), BMWFW (AT), IFD (DK), MINECO (ES), ANR (FR) with co-funding by the European Union's Horizon 2020 research and innovation programme (Grant n. 690462). BSC co-authors also acknowledge support from the European Research Council under the European Union's Horizon 2020 research and innovation programme (grant n. 773051; FRAGMENT), the AXA Research Fund, the Spanish Ministry of Science, Innovation and Universities (grant n. RYC-2015-18690 and CGL2017-88911-R), the European Union's Horizon 2020 research and innovation programme (grant n. 792103; SOLWARIS). This work has been partially funded by the contribution agreement between AEMET and BSC to carry out development and improvement activities of the products and services supplied by the WMO Sand and Dust Storm Regional Centres. Jerónimo Escribano and Martina Klose have received funding from the European Union's Horizon 2020 research and innovation programme, respectively, under the Marie Skłodowska-Curie grant agreements H2020-MSCA-COFUND-2016-754433 and H2020-MSCA-IF-2017-789630. Martina Klose further acknowledges support through the Helmholtz Association's Initiative and Networking Fund (grant agreement n. VH-NG-1533). We acknowledge PRACE (eDUST, eFRAGMENT1, and eFRAGMENT2) and RES (AECT-2019-3-0001, AECT-2020-1-0007, AECT-2020-3-0013) for awarding access to MareNostrum at the BSC and for providing technical support. The authors thank all the Principal Investigators and their staff for establishing and maintaining the NASA and PHOTONS AERONET sites, and the MODIS mission scientists and associated NASA personnel for the production of the data used in this study.

Questions and Answers

Questioner: Andy Delcloo.

Question: ERA-Interim is still used, can you please elaborate why and when will you switch to ERA-5?

Answer: Indeed we have used in this work the ERA-Interim reanalysis as meteorological initial and boundary conditions of our simulations. We have not used ERA5 (with the exception of ERA5 soil) since, when testing its use, we faced technical issues to simulate some specific dates. Extending the current reanalysis with ERA5 (in particular from the end day of production of ERA-Interim on 31 August 2019) would not be correct for the production of a consistent data set and could introduce spurious trends associated to the change of meteorological initial and boundary conditions. However, we can switch to the use of ERA5 in a new version of our reanalysis. This will allow us to extend our data set to present dates.

Questioner: Andy Delcloo.

Question: Ensemble approach, can you please explain about the introduction of spread in the ensemble, using Kalman filtering.

Answer: We adopted the LETKF scheme for the estimation of our dust reanalysis. Therefore, model uncertainty is estimated from the realizations of the dust fields in MONARCH ensemble simulations. The use of ensemble simulations allows for the estimation of a flow-dependent background uncertainty which would be otherwise difficult to estimate for a dust model due to the highly varying nature of dust concentrations. More specifically, MONARCH ensemble has been generated with the use of different meteorological initial and boundary conditions (MERRA-2 with ERA5 soil and ERA-Interim reanalysis) and different emission schemes, additionally to perturbations of model emission parameters. The latter choice was made on the basis that uncertainties in dust emission is one of the major contributor to model error. Previous studies showed that combined meteorology and aerosol source ensembles are necessary to produce sufficient spread in outflow regions. Similarly, the test experiments presented in this work showed the benefit of using different meteorology and emission schemes for dust data assimilation.

References

Cziczo, D. J., Froyd K. D., Hoose C., Jensen, E. J., Diao, M., Zondlo, M. A., Smith, J. B., Twohy, C. H., & Murphy D. M. (2013). Clarifying the dominant sources and mechanisms of cirrus cloud formation. *Science, 340*(6138). https://doi.org/10.1126/science.1234145

Di Tomaso, E., Escribano, J., Basart, S., Ginoux, P., Macchia, F., Barnaba, F., Benincasa, F., Bretonnière, P.-A., Buñuel, A., Castrillo, M., Cuevas, E., Formenti, P., Gonçalves, M., Jorba, O., Klose, M., Mona, L., Montané, G., Mytilinaios, M., Obiso, V., Olid, M., Schutgens, N., Votsis, A., Werner, E., & Pérez García-Pando, C. (2021). The MONARCH high-resolution reanalysis of desert dust aerosol over Northern Africa, the Middle East and Europe (2007–2016). *Earth System Science Data.* [preprint]. https://doi.org/10.5194/essd-2021-358

Di Tomaso, E., Schutgens, N. A. J., Jorba, O., & Pérez García-Pando, C. (2017). Assimilation of MODIS dark target and deep blue observations in the dust aerosol component of NMMB-MONARCH version 1.0. *Geoscience Model Development, 10,* 1107–1129. https://doi.org/10.5194/gmd-10-1107-2017

Ginoux, P., Prospero, J. M., Gill, T. E., Hsu, N. C., & Zhao, M. (2012). Global-scale attribution of anthropogenic and natural dust sources and their emission rates based on Modis Deep Blue aerosol products. *Reviews of Geophysics, 50.* https://doi.org/10.1029/2012rg000388

Hunt, B. R., Kostelich, E. J., & Szunyogh, I. (2007). Efficient data assimilation for spatiotemporal chaos: A local ensemble transform Kalman filter. *Physica D, 230.*

Jickels, T. D., An, Z. S., Andersen, K. K., Baker, A. R., Bergametti, G., Brooks, N., Cao, J. J., Boyd, P. W., Duce, R. A., Hunter, K. A., Kawahata, H., Kubilay, N., LaRoche, J., Liss, P. S., Mahowald, N., Prospero, J. M., Ridgwell, A. J., Tegen, I., & Torres, R. (2005). Global iron connections between desert dust. *Ocean Biogeochemistry, and Climate, Science, 308,* 67–71. https://doi.org/10.1126/science.1105959

Klose, M., Jorba O., Gonçalves Ageitos, M., Escribano, J., Dawson, M. L., Obiso, V., Di Tomaso, E., Basart, S., Montané Pinto, G., Macchia, F., Ginoux, P., Guerschman, J., Prigent, C., Huang, Y., Kok, J., Miller, R. L., & Pérez García-Pando, C. (2021). Mineral dust cycle in the Multiscale Online NonhydrostaticAtmospheRe CHemistry model (MONARCH) version 2.0. Geoscientific Model Development [preprint].

Miller, R. L., Knippertz, P., García-Pando, C. P., Perlwitz, J. P., & Tegen, I. (2014). Impact of dust radiative forcing upon climate. In *Mineral dust* (pp. 327–357). Springer.

Mytilinaios, M., Ciamprone, S., Trippetta, S., Basart, S., Di Tomaso, E., Jorba, O., Pérez García-Pando, C., Formenti, P., Cuesta, J., Gkikas, A., Kahn, R., & Mona, L. (2022). *Evaluation of the MONARCH reanalysis dust and coarse dust optical depth over Northern Africa, Middle East and Europe, using MODIS.* MISR: IASI and AERONET observations, in preparation.

O'Neill, N. T., Eck, T. F., Smirnov, A., Holben, B. N., & Thulasiraman, S. (2003). Spectral discrimination of coarse and fine model optical depth. *Journal of Geophysical Research: Atmospheres, 108.* https://doi.org/10.1029/2002JD002975

Pérez, C., Haustein, K., Janjic, Z., Jorba, O., Huneeus, N., Baldasano, J. M.,T. Black, T., Basart, S., Nickovic, S., Miller, R. L., Perlwitz, J. P., Schulz, M., & Thomsonet, M. (2011). Atmospheric dust modeling from meso to global scales with the online NMMB/BSC-Dust model - Part 1. *Atmospheric Chemistry and Physics, 11.*

Pu, B., & Ginoux, P. (2016). The impact of the Pacific Decadal Oscillation on springtime dust activity in Syria. *Atmospheric Chemistry and Physics, 16*(21), 13431–13448.

Schutgens N. A. J., Miyoshi, T., Takemura, T., & Nakajima T. (2010) Applying an ensemble Kalman filter to the assimilation of AERONET observations in a global aerosol transport model. *Atmospheric Chemistry and Physics, 10.*

Yu, H., Chin, M., Yuan, T., Bian, H., Remer, L. A., Prospero, J. M., Omar, A., Winker, D., Yang, Y., Zhang, Y., Zhang, Z., & Zhao, C. (2015). The fertilizing role of African dust in the Amazon rainforest: A first multiyear assessment based on data from Cloud-Aerosol Lidar and Infrared Path finder Satellite Observations. *Geophysical Research Letters, 42.*

How Does the Use of Different Soil Mineralogical Atlases Impact Soluble Iron Deposition Estimates?

Elisa Bergas-Massó, María Gonçalves-Ageitos, Stelios Myriokefalitakis, Ron L. Miller, and Carlos Pérez García-Pando

Abstract Atmospheric deposition is the primary input of soluble iron (Fe) in remote regions of the open ocean. Fe constitutes a fundamental micro-nutrient for marine biota, affecting biogeochemical cycles and ultimately carbon uptake. Fe is emitted to the atmosphere primarily from dust sources. Fe solubility at emission is origin-dependent, being mostly insoluble when associated with dust minerals. Earth System Models (ESMs) commonly assume that dust aerosols have a globally uniform composition, neglecting known regional variations in soil mineralogy. This work assesses the implications of soil composition uncertainties on the bio-available Fe delivery to the ocean by using a state-of-the-art ESM with a detailed atmospheric Fe cycle (EC-Earth3-Iron). We run two 1-year simulations considering two different Soil Mineralogy Atlases (SMAs). Our results reveal a non-negligible impact of soil mineralogy uncertainties on soluble Fe estimates. By only changing the source of information for the soil composition and updating the Fe content in minerals, we find differences of 31.5% in mineral Fe emissions and 26.7% in soluble Fe deposition budgets.

Keywords Soluble iron · Soil mineralogy · Iron cycle

E. Bergas-Massó (✉) · M. Gonçalves-Ageitos · C. P. García-Pando
Barcelona Supercomputing Center (BSC), Barcelona, Spain
e-mail: elisa.bergas@bsc.es

E. Bergas-Massó · M. Gonçalves-Ageitos
Universitat Politècnica de Catalunya (UPC), Barcelona, Spain

S. Myriokefalitakis
Institute for Environmental Research and Sustainable Development (IERSD), National Observatory of Athens, Penteli, Greece

R. L. Miller
NASA Goddard Institute for Space Studies, New York, NY, USA

C. P. García-Pando
Catalan Institution for Research and Advanced Studies (ICREA), Barcelona, Spain

C. Mensink and O. Jorba (eds.), *Air Pollution Modeling and its Application XXVIII*, Springer Proceedings in Complexity, https://doi.org/10.1007/978-3-031-12786-1_34

1 Introduction

Soluble Fe acts as an essential micronutrient for marine biota, limiting phytoplankton growth in some regions of the open ocean (i.e., High Nutrient Low-Chlorophyll regions). Ocean productivity and hence CO_2 uptake by the ocean relies on the availability of limiting nutrients. Therefore, quantifying soluble Fe deposition over the ocean is key to estimate ocean primary productivity and ultimately carbon uptake.

Atmospheric mineral dust is the dominant source of Fe deposited over the open ocean. The interaction of airborne mineral dust with the Earth system depends on the mineralogical dust composition itself, particle size distribution (PSD), or its shape, among other characteristics. Hence, the Fe in dust and its ability to dissolve is influenced by dust mineralogy. However, Earth System Models (ESMs) usually consider dust as globally uniform, disregarding the known diversity in the sources' mineralogical composition. In particular, Fe abundance in soils is generally set to a constant ~3.5% (Duce & Tindale, 1991), and its solubility is considered to be around 0.1% (Fung et al., 2000).

Although Fe solubility at emission is low, observations suggest that solubility is boosted downwind of the sources (Zhuang et al., 1992) as a result of different atmospheric chemistry processes. A primary Fe solubilization mechanism is the acidic (proton-promoted) dissolution, as at low pH values, the Fe–O bonds of iron oxides weaken, favoring its dissolution (Johnson & Meskhidze, 2013). Other pathways to soluble Fe production are organic ligand (e.g., oxalate) processing (Myriokefalitakis et al., 2015) and photoreductive processes.

Constraining the dust mineralogical composition in models presents challenges. Soil mineralogy maps are generated by extrapolating a limited number of mineralogical analyses of soil samples, which are particularly rare in arid and semi-arid regions. Furthermore, those mineralogical analyses use the wet sieving technique, which tampers the soil size distribution by breaking coarse particles, making it hard to estimate mineral-dependent PSD at emission. Moreover, the Fe content in some clay minerals can be highly variable depending on their origin: primary mineral characteristics and chemical weathering processes (Shi et al., 2012).

This work assesses the impacts of soil mineralogy uncertainties on soluble Fe deposition over the open ocean using a state-of-the-art ESM, EC-Earth3-Iron (Pehkonen et al., 1993). This model includes a detailed atmospheric Fe cycle and the ability to use two different soil mineralogy data sets (Claquin et al., 1999; Journet et al., 2014).

2 Methodology

EC-Earth3-Iron is used to model soluble Fe deposition. The ESM is composed of several modules representing different components of the Earth system (Noije et al., 2014). The configuration used in this work includes the Integrated Forecast System

(IFS) model, which represents atmospheric dynamics, coupled with the Tracer Model 5 (TM5), which allows for interactive simulation of atmospheric chemistry and transport of aerosols and reactive gas species (Noije et al., 2014). EC-Earth3-Iron further considers:

- The primary emissions of both insoluble and soluble Fe forms, associated with mineral dust (Myriokefalitakis et al., 2015) and combustion aerosols (Ito et al., 2018) which are distributed in two different sizes: accumulation mode (<= 1 μm) and coarse mode (1–10 μm) at emission.
- The atmospheric processing mechanism of Fe accounting for: proton-promoted dissolution, oxalate-promoted Fe dissolution (with oxalate calculated on-line) and photo-reductive dissolution. The emitted Fe is allocated in three different pools that account for different dissolution rates, namely fast, intermediate and slow Fe pools (Ito & Shi, 2016)
- The implementation of two different soil mineralogy datasets (Claquin et al., 1999 and Journet et al., 2014).
- The use of Brittle fragmentation theory (Kok, 2011) to define the PSD of minerals at emission.

The two mineralogical datasets used, differ in various aspects (e.g., number of minerals considered in the different size ranges, spatial coverage, and observations used) The Claquin dataset (Claquin et al., 1999), with the updates of Nickovic et al. (2012), provides mineralogical information for arid dust-source regions based on 239 descriptions of soils, for eight different minerals: illite, kaolinite, smectite, feldspars, hematite, gypsum, quartz, and calcite. The Journet dataset (Journet et al., 2014) is based on more than 700 soil descriptions. It contains information on the relative abundance of 12 minerals: quartz, feldspars, illite, smectite, kaolinite, chlorite, vermiculite, mica, calcite, gypsum, hematite and goethite. In contrast with Claquin, this data set has global coverage. Journet (Journet et al., 2014) also provides updated values for the Fe mineral speciation, which are implemented in the EC-Earth3-Iron model along with the soil map. The Claquin dataset makes use of previously reported Fe content in Nickovic et al. (2012) and Journet et al. (2008).

To assess the differences in soluble iron deposition owed to uncertainties in the soil mineralogical information, we run two identical 1-year simulations (with a 1-year long spin-up) with the two different soil mineralogical datasets and Fe mineral speciation. We use the model configuration above-mentioned (IFS + TM5) for both simulations with observed sea surface temperature and sea ice concentration rather than interactive ocean. We nudge the simulation wind fields towards the ERA-Interim reanalysis (Berrisford et al., 2011), allowing short-term simulations to more closely follow the observed climate.

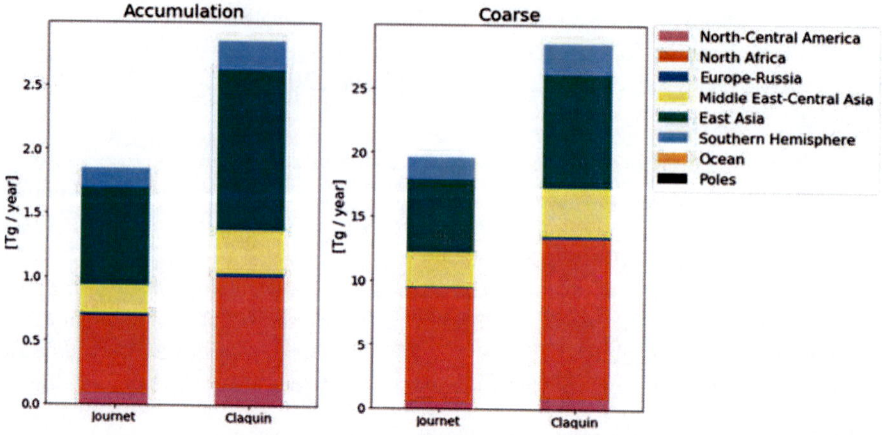

Fig. 1 Annual budget of Fe-dust emission [Tg/yr] for Journet and Claquin simulations with the accumulation mode (left) and coarse mode (right). Colors in the barplots represent the contribution to the total budget of different regions, which are based on HTAP2 regions (Koffi et al., 2016)

3 Results

Fe-dust emissions are higher with Claquin's mineralogy and Fe speciation in all dust source regions. Total budgets of Fe-dust emissions are 31.4 Tg/yr and 21.5 Tg/yr for Claquin and Journet simulations respectively, which is translated to a reduction of 31.5% in Fe-Dust emission with Journet's with respect to Claquin. This relative difference can be considered constant among particular regions and particle size ranges (Fig. 1).

In line with the changes in emission, soluble Fe deposition decreases for the Journet simulation compared with Claquin. This decrease is generalized over the globe but gets accentuated in the Northern Hemisphere (NH), especially in the Atlantic downwind North African dusty regions where the reduction is around 40% (Fig. 2). The sharper decrease in the relative change between Journet and Claquin over dust source regions in the NH could be explained not only by a decrease in Fe emissions, but also by higher concentrations of calcite over those regions with Journet mineralogy (more than double in almost all the NH). Higher calcite concentrations lead to a more basic atmosphere (~5% less acidic), potentially reducing the acidic processing of iron over these areas.

4 Conclusions

In this study, we assess the impact that uncertainties in mineralogical composition of soils have on soluble iron estimates. We do so by using a complex chemistry model

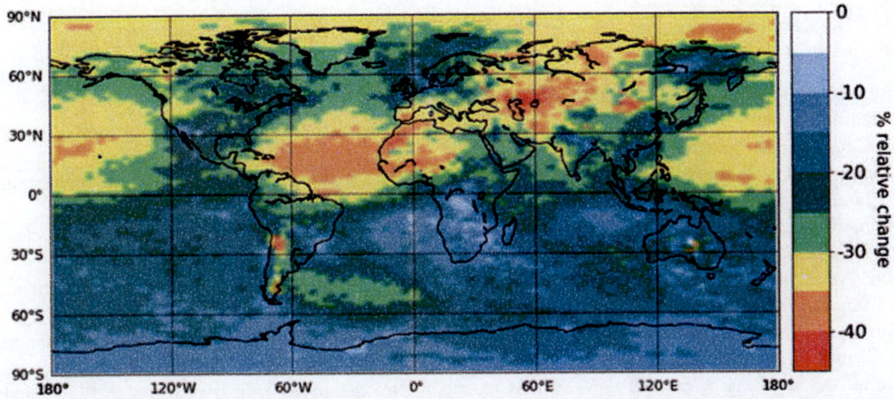

Fig. 2 Relative change [%] in soluble Fe deposition in Journet simulation with respect to Claquin simulation

accounting for a detailed atmospheric processing scheme, which allows us to explore non-linear effects on the resulting soluble Fe deposition budget.

Our results show that current uncertainties regarding soil composition have a significant impact on estimates of Fe-dust emissions and soluble Fe deposition. This relates to the composition of the source regions or parent soils for dust emission, from which little observations have been collected. Besides, dust composition affects aerosol pH, therefore influencing the solubilization of the iron. We detect relative differences of around 30% in Fe emissions from dust sources and about 40% in soluble Fe deposition depending on the assumptions regarding the soil mineralogy used.

Results suggest that more emphasis should be given to constrain soil mineralogy to improve the representation of biogeochemical cycles as the Fe cycle. To keep advancing in our understanding of how soil mineralogy influences the Fe cycle, future works will address the sensitivity of the dissolution processes towards dust mineralogy over longer periods, and with a focus on the regional impacts.

Acknowledgements This work was funded by the Ministerio de Economía y Competitividad (MINECO) as part of the NUTRIENT project (CGL2017-88911-R). The research leading to this work has also received support from the ERC Consolidator Grant "FRAGMENT" (grant agreement No. 773051), the AXA Chair on Sand and Dust Storms and the National Observatory of Athens research grant (n° 5065). Finally, we thankfully acknowledge the RES AECT-2020-3-0020 and PRACE projects to grant this work with computer resources and the technical support provided by the BSC and the Computational Earth Sciences team of the Earth Sciences Department.

References

Berrisford, P., et al. (2011). The ERA-Interim archive. Version 2.0 (1), 23.

Claquin, T., et al. (1999). Modeling the mineralogy of atmospheric dust sources. *Journal of Geophysical Research, 104256*(09), 243–322.

Duce, R. A., & Tindale, N. W. (1991). Atmospheric transport of iron and its deposition in the ocean. *Limnology and Oceanography, 36*(8), 1715–1726.

Fung, I. Y., et al. (2000). Iron supply and demand in the upper ocean. *Global Biogeochemical Cycles, 14*, 281–296.

Ito, A., et al. (2018). Radiative forcing by light-absorbing aerosols of pyrogenetic iron oxides. *Scientific Reports, 8*(1).

Ito, A., & Shi, Z. (2016). Delivery of anthropogenic bioavailable iron from mineral dust and combustion aerosols to the ocean. *Atmospheric Chemistry and Physics, 16*, 85–99.

Johnson, M. S., & Meskhidze, N. (2013). Atmospheric dissolved iron deposition to the global oceans: Effects of oxalate-promoted Fe dissolution, photochemical redox cycling, and dust mineralogy. *Geoscientific Model Development, 6*(4), 1137–1155.

Journet, E., et al. (2014). A new data set of soil mineralogy for dust-cycle modeling. *Atmospheric Chemistry and Physics, 14*(04), 3801–3816.

Koffi, B., et al. (2016). Hemispheric Transport Air Pollution (HTAP): Specification of the HTAP2 experiments.

Kok, J. (2011). A scaling theory for the size distribution of emitted dust aerosols suggests climate models underestimate the size of the global dust cycle. *Proceedings of the National Academy of Sciences of the United States of America, 108*(01), 1016–1021.

Myriokefalitakis, S., et al. (2015). Changes in dissolved iron deposition to the oceans driven by human activity: A 3-D global modelling study. *Biogeosciences, 12*(13), 3973–3992.

Myriokefalitakis, S., et al. (2022). Multiphase processes in the EC-Earth Earth System model and their relevance to the atmospheric oxalate, sulfate, and iron cycles. *Geoscientific Model Development, 15*, 3079–3120.

Nickovic, S., et al. (2012). Technical note: High-resolution mineralogical database of dust-productive soils for atmospheric dust modeling. *Atmospheric Chemistry and Physics, 12*(2), 845–855.

Pehkonen, S. O., et al. (1993). Photoreduction of iron oxyhydroxides in the presence of important atmospheric organic compounds. *Environmental Science & Technology, 27*(10), 2056–2062.

Shi, Z., et al. (2012). Impacts on iron solubility in the mineral dust by processes in the source region and the atmosphere: A review. *Aeolian Research, 5*, 21–42.

van Noije, T. P. C., et al. (2014). Simulation of tropospheric chemistry and aerosols with the climate model EC-Earth. *Geoscientific Model Development, 7*(5), 2435–2475.

Zhuang, G., et al. (1992). Link between iron and sulfur cycles suggested by fe(ii) in remote marine aerosol. *Nature, 355*(02), 537–539.

Modelling the Absorption of Organic Aerosols at Regional and Global Scales

Hector Navarro-Barboza, Vincenzo Obiso, Rubén Sousse, Marco Pandolfi, Carlos Pérez García-Pando, and Oriol Jorba

Abstract Depending on their chemical composition, aerosols cool the atmosphere directly by scattering solar radiation and indirectly through aerosol-cloud interactions or warm the lower atmosphere by absorbing visible solar radiation. The most prevalent types of absorbing aerosols are black carbon (BC) and mineral dust. However, some organic aerosols (OA) with absorbing properties known as brown carbon (BrC) are found to contribute significantly to the total aerosol absorption in the atmosphere. The main sources of primary BrC are biomass burning and residential coal combustion. Still, recent studies have postulated the existence of various secondary sources of BrC resulting from multi-phase reactions of volatile organic compounds exposed to nitrogen oxides and ammonia.

In this study, we evaluate OA's chemistry and optical properties stimulated by the Multiscale Online Nonhydrostatic Atmosphere Chemistry (MONARCH) model. The model is run at the regional (European domain) and global scales perturbing key processes and parameters to better constrain the OA absorbing fraction. The evaluation of the model focuses mostly on the concentrations and optical properties of BC and OA. In-situ surface measurements of PM chemical composition (both off-line and on-line) and optical properties (multi-wavelengths scattering and absorption) provided by IDAEA-CSIC and columnar integrated optical properties (optical depth, single scattering albedo, and asymmetry factor) derived from the Aerosol Robotic Network (AERONET) are used. We discuss the role of emissions, refractive index, and aging processes affecting the BrC burden.

H. Navarro-Barboza (✉) · V. Obiso · R. Sousse · C. P. García-Pando · O. Jorba
Barcelona Supercomputing Center (BSC), Barcelona, Spain
e-mail: hector.navarro@bsc.es

M. Pandolfi
Institute of Environmental Assessment and Water Research, IDAEA-CSIC), Barcelona, Spain

C. P. García-Pando
Catalan Institution for Research and Advanced Studies (ICREA), Barcelona, Spain

C. Mensink and O. Jorba (eds.), *Air Pollution Modeling and its Application XXVIII*, Springer Proceedings in Complexity, https://doi.org/10.1007/978-3-031-12786-1_35

1 Introduction

According to the latest IPCC report, carbonaceous aerosols, in addition to influencing the climate, are also air pollutants and could cause adverse problems for human health and ecosystems (Zhongming et al., 2021). Carbonaceous aerosols include both black carbon (BC) and organic aerosols (OA). On the one hand, BC, produced by incomplete combustion (Wang et al., 2014), is the main absorber of light in the atmosphere. This aerosol has been widely studied for its role in the earth's radiative balance. On the other hand, OA can come directly from natural and anthropogenic sources and form in the atmosphere by multiple heterogeneous and multiphase reactions (secondary organic aerosols, SOA). Generally, these aerosols have been considered as purely scattering. However, recent laboratory analysis and field campaigns have provided strong evidence for the existence of a fraction of OA with absorption properties predominantly at ultraviolet wavelengths, known as "Brown Carbon" (BrC) (Laskin et al., 2015). Although this finding has represented an advance in OA's knowledge, the role of BrC on climate and human health is still uncertain. To better understand the behavior of BrC particles in the atmosphere, it is necessary to use numerical models; this implies an adequate representation of their concentration, chemical composition, and optical properties. However, atmospheric chemistry models have historically shown difficulties in simulating both concentration and chemical composition, particularly OA (Tsigaridis et al., 2014).

This work presents first steps to better constrain the burden and radiative forcing of BrC in the atmosphere. We combine different observational datasets to evaluate OA optical properties simulated by the Multiscale Online Nonhydrostatic AtmospheRe CHemistry (MONARCH) model developed at BSC.

2 Methodology

2.1 MONARCH Model and Simulation Setup

The MONARCH model (Spada, 2015; Badia et al., 2017) couples meteorology and chemistry online and can be run either globally or regionally. The system solves tropospheric chemistry with detailed gas-phase and aerosol mechanisms. The aerosol components considered in the model are: Mineral dust, sea salt, BC, sulfate, nitrate, ammonium, primary organic aerosol (POA), Secondary organic aerosol (SOA), and the unspeciated PM fraction emitted. A simple non-volatile SOA scheme accounts for the contribution of anthropogenic, biomass burning, and biogenic formation. The HERMESv3 emission model provides emissions for regional or global applications (i.e., HTAP, CAMS-REG-anthro). Other natural emissions like desert dust, sea salt and biogenic emissions are computed online by MONARCH.

In this work, a regional experiment is designed with a domain covering Europe at 0.2° horizontal resolution and 24 vertical layers with the top of the atmosphere at 50 hPa. Meteorology and chemistry boundary conditions were obtained from ECMWF IFS and Copernicus CAMS global forecasting systems, respectively. The CAMS-REG-Anthro v3.1 anthropogenic emission inventory was used. An annual simulation of the 2018 year was conducted. Complementary, a global experiment is set up. The domain covers the globe with 0.7° × 0.5° horizontal resolution and 48 vertical layers with the top of the atmosphere at 5 hPa. The study period corresponds to the years from 2013 to 2016. Both runs are used to assess preliminary representations of BrC in the MONARCH model.

2.2 *Observational Data*

We compare MONARCH results with observational data targeting optical properties. Regional results are compared with the data measured by IDAEA-CSIC. This group maintains three strategic sites (urban background Barcelona station (BCN), regional background Montseny station (MSY), and remote background Montsec station (MSA)) with the aim to study the physical, chemical, and optical properties of atmospheric aerosol particles and their climate effects. We compare optical properties (multi-wavelength scattering and absorption). On the other hand, global results are compared with data derived from the Aerosol Robotic Network (AERONET), including columnar integrated optical properties (optical depth, single scattering albedo, and asymmetry factor).

3 Results

We defined six cases in the regional experiment to evaluate the optical properties of OA (Table 1 top), perturbing both the BrC/OA mass fraction and the imaginary refractive index (*ni*) of primary and secondary OA. We consider three fractions assuming that 20/50/100% of the total OA is BrC. We also perturb *ni* by varying from very weakly to strongly OA absorbing properties. In the global experiment, we defined two cases, a base case with no explicit BrC and OA with very weak absorption, and a second case where all primary emissions from biomass burning (POW) were treated as BrC and the rest like the base case (see Table 1).

Figure 1 shows a synthesis of the results for regional (a and b) and global (c). In the regional experiment, periods with a good characterization of BC and OA mass result with an excellent agreement of BrC absorption assuming a large BrC/OA fraction (urban site) and a lower fraction (regional/remote locations) with mixtures of weakly to strongly absorbing BrC. Regarding the global experiment, there is strong sensitivity in areas affected by fires (Fresno site). Clear improvements of single scattering albedo (ω) correlation with a preliminary simple treatment of BrC is observed.

Table 1 Simulation scenarios. Regional cases perturbing BrC/OA fraction and imaginary refractive index [P (primary), S (secondary)]. Global cases tagging biomass burning emissions

REGIONAL	Mass fraction	Imaginary refractive index (ni)		
	BrC/OA	Hydrophobic primary-BrC	Hydrophilic primary-BrC	SOA-BrC anthro./biomass
Case 1—no BrC	0	-	-	-
Case 2—Strongly P and S BrC absorption	100./50./20	1e-1	1e-1	1e-2
Case 3—Moderately P and S BrC absorption	100./50./20	1e-1	1e-2	1e-2
Case 4—Moderately P and weakly S BrC abs	100./50./20	1e-1	1e-2	1e-3
Case 5—Moderately P and very weakly S BrC abs	100./50./20	1e-1	1e-2	1e-4
Case 6—Weakly P and very weakly S BrC abs	100./50./20	1e-2	1e-3	1e-4
Global	Imaginary refractive index (ni)			
	Hydrophobic POA	Hydrophilic POA	Hydrophobic primary-BrC	Hydrophilic primary-BrC
Case 1—no BrC	1e-8	1e-4	-	-
Case 2—Primary biomass burning emissions as BrC	1e-8	1e-4	1e-1	1e-3

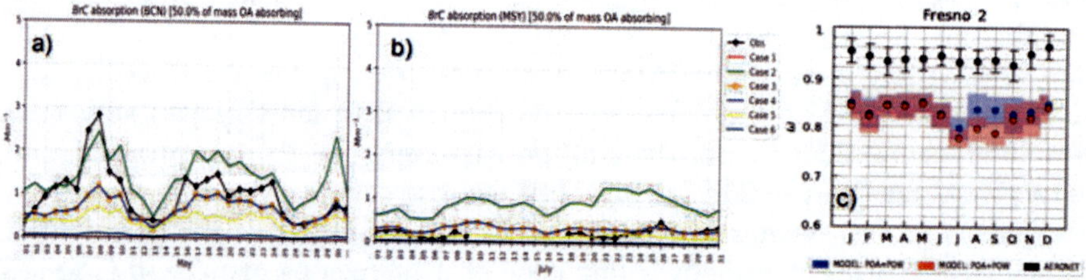

Fig. 1 (a) Regional results of BrC absorption at Barcelona station (BCN) for May 2018 considering 50% of OA as BrC, (b) regional results of BrC absorption at Montseny station (MSY) for July 2018 considering 50% of OA as BrC [black line: observations, color lines: model runs for cases described in Table 1], and (c) single scattering albedo (ω) at Fresno AERONET station from the global run [black dots are observations, blue dots are the model results considering no explicit BrC, and red dots are the model results considering BrC from biomass burning emissions].

4 Conclusions

In this study, we perturbed the mass fraction OA/BrC and their imaginary refractive index. In general, the model reproduces reasonably well the absorption throughout the year. However, in the urban environment, as in BCN, the absorption attributed to OA derived from observations can be explained by the model if most organic is BrC. On the other hand, in regional sites (MSY), the BrC absorption is likely not dominated by a large abundance of BrC, and the signal is more driven by the secondary contribution. For the global run, explicit treatment of strongly absorbing BrC from biomass burning sources has a significant impact in the single scattering albedo compared with classical approaches of an OA with some moderate absorption. The combination of different types of measurements allows identifying which components deserve further refinement in the model.

Acknowledgements Hector Navarro-Barboza holds a PhD fellowship (FPI-SO PRE2018-084988) of the Severo-Ochoa project SEV-2015-0493-18-1 granted to the Barcelona Supercomputing Center. This work was supported by funding from the Ministerio de Ciencia, Innovacíon y Universidades as part of the BROWNING project (RTI2018-099894-BI00) and CAIAC project (PID2019-108990PB-100), and the EU H2020 project FORCES (grant no. 821205). BSC co-authors acknowledge PRACE (eFRAGMENT2) and RES (AECT-2020-1-0007 and AECT-2021-1-0027) for awarding access to MareNostrum at the Barcelona Supercomputing Center and for providing technical support.

References

Badia, A., Jorba, O., Voulgarakis, A., Dabdub, D., Pérez García-Pando, C., Hilboll, A., Gonçalves, M., & Janjic, Z. (2017). Description and evaluation of the multiscale online nonhydrostatic atmospheric chemistry model (nmmb-monarch) version 1.0: gas-phase chemistry at global scale. Geoscientific Model Development, 10(2):609–638.

Laskin, A., Laskin, J., & Nizkorodov, S. A. (2015). Chemistry of atmospheric brown carbon. *Chemical Reviews, 115*(10), 4335–4382.

Spada, M. (2015). Development and evaluation of an atmospheric aerosol module implemented within the nmmb/bsc-ctm. Ph.D. dissertation, Universitat Politècnica de Catalunya. Tsigaridis, K., Daskalakis, N., Kanakidou, M., Adams, P., Artaxo, P., Bahadur, R., Balkanski, Y., Bauer, S., Bellouin, N., Benedetti, A., et al. (2014). The aerocom evaluation and intercomparison of organic aerosol in global models. Atmospheric Chemistry and Physics, *14*(19):10845–10895.

Wang, X., Heald, C., Ridley, D., Schwarz, J., Spackman, J., Perring, A., Coe, H., Liu, D., & Clarke, A. (2014). Exploiting simultaneous observational constraints on mass and absorption to estimate the global direct radiative forcing of black carbon and brown carbon. *Atmospheric Chemistry and Physics, 14*(20), 10989–11010.

Zhongming, Z., Linong, L., Wangqiang, Z., Wei, L., et al. (2021). Ar6 climate change 2021: The physical science basis.

Comparative Study Between the Effects of Autumn and Winter Rainfall on Aerobiological Variables in the NE of the Iberian Peninsula

Ricard Kirchner, M. Carmen Casas-Castillo, Raül Rodríguez-Solà, Marta Alarcón, Cristina Periago, Concepción De Linares, and Jordina Belmonte

Abstract The correlation analysis between precipitation indices and aerobiological variables could lead to different results if seasonal rainfall is considered. The number of significant correlations, as well as the ratio of positive/negative correlations, is higher when previous autumn rainfall is taken into account than for annual rainfall. However, the percentage of significant correlations is only 5.1% of the total possible correlations. The autumn indices showing more correlations are related to rainfall duration and persistence, while those for winter were related to rainfall amount and intensity. Since the influence of moderately extreme rainfall seems to be low, the reported probable increase, due to climate change, of the occurrence of intense and short rainy events in the Mediterranean area in autumn does not seem that will have an appreciable effect by itself on the subsequent pollen production.

Keywords Airborne pollen · Extreme rainfall · Aerobiology · Autumn · Mediterranean area

R. Kirchner · M. C. Casas-Castillo (✉)
Department of Physics, ESEIAAT, Universitat Politècnica de Catalunya·BarcelonaTech, Colom 1, 08222 Terrassa, Spain
e-mail: m.carmen.casas@upc.edu

R. Rodríguez-Solà
Department of Physics, ETSEIB, Universitat Politècnica de Catalunya·BarcelonaTech, Diagonal 647, 08028 Barcelona, Spain

M. Alarcón · C. Periago
Department of Physics, EEBE, Universitat Politècnica de Catalunya·BarcelonaTech, Eduard Maristany 16, 08019 Barcelona, Spain

C. De Linares · J. Belmonte
Department of Animal Biology, Plant Biology and Ecology, Faculty of Bioscience, Universitat Autònoma de Bellaterra, 08193 Bellaterra, Spain

C. De Linares
Department of Botany, Faculty of Sciences, Universidad de Granada, 18071 Bellaterra, Spain

J. Belmonte
Institute of Environmental Sciences and Technology (ICTA-UAB), Universitat Autònoma de Bellaterra, 08193 Bellaterra, Spain

C. Mensink and O. Jorba (eds.), *Air Pollution Modeling and its Application XXVIII*, Springer Proceedings in Complexity, https://doi.org/10.1007/978-3-031-12786-1_36

1 Introduction

Precipitation has a twofold impact on some aerobiological variables: Previous to pollination it has a positive influence in the subsequent pollen production, while during pollination it could cause the airborne pollen removal by a wash-out effect (Majeed et al., 2018). The influence of climate change in the rainfall pattern of the Mediterranean area, probably causing the increase of the frequency and the intensity of the extreme rainfall episodes (Rodríguez et al., 2014), especially in autumn, might affect the following pollen production. To characterize weather and climate extremes, the Expert Team on Climate Change Detection and Indices (ETCCDI) of the World Meteorological Organization proposed a set of descriptive indicators (Klein-Tank et al., 2009). Rodríguez-Solà et al., (2020, 2022) found that the numbers of significant correlations between aerobiological variables and some of the rainfall ETCCDI were higher for winter than for annual rainfall, being the ratio between the positive and negative correlations more than twice for winter. However, the percentage of significant winter correlations of the possible total was still low (5.6%), which seems to indicate a low influence of previous moderately extreme rainfall by itself on the subsequent pollen production, especially compared to that of temperature. The aim of the present study is to investigate how these results would change if previous autumn rainfall is considered.

2 Data and Methods

To explore the impact of seasonal extreme rainfall on airborne pollen/spores in the NE Iberian Peninsula, Spearman correlations between some of the precipitation ETCCDI and the aerobiological variables corresponding to 25 pollen and two fungal spore taxa measured between 1994 and 2018 in six aerobiological stations have been calculated. The precipitation ETCCDI, which characterize moderate to extreme rainfall, have been calculated seasonally from the daily rainfall dataset of the Servei Meteorològic de Catalunya (Llabrés-Brustenga et al., 2019), and are listed in Table 1.

Airborne pollen and spore data were extracted from the Catalan Aerobiological Network database. The studied locations were Barcelona (BCN), Girona (GIC), and Tarragona (TAU), all three urban environments; Lleida (LLE), closer to rural than to urban conditions; and Bellaterra (BTU) and Manresa (MAN), both urban/rural locations. Pollen and spore data (obtained from Hirst traps samples) were analysed according to the standard methodology of the Spanish Aerobiology Network (Galán et al., 2007) and correspond to the fungi, threes, bushes and herbs listed in Table 2. Table 3 shows the description of the pollen/spore variables.

The Spearman correlation has been calculated for every location and throughout the considered temporal periods (25 years from 1994 to 2018 for BCN and BTU, and 23 years from 1996 to 2018 for GIC, LLE, MAN, and TAU) between every pollen/spore taxa, every aerobiological variable, and every rainfall ETCCDI. The

Table 1 Description of the precipitation ETCCDI considered in the present study

Precipitation ETCCDI	Description
RX1day	Maximum daily rainfall amount (mm)
RX5day	Maximum rainfall amount (mm) over 5 consecutive days
SDII	Ratio of total rainfall to the number of wet days
R10mm	Number of days with rainfall greater than 10 mm
R20mm	Number of days with rainfall greater than 20 mm
CWD	Maximum number of consecutive wet days (≥ 1 mm)
R95p	Total rainfall (mm) exceeding the 95th percentile
PRCPTOT	Total wet-day precipitation (mm)

Table 2 Fungal spore and pollen taxa considered in the present study

	Total Pollen (PTOT)	
Fungi	Alternaria (ALTE)	Cladosporium (CLAD)
Trees	Alnus (ALNU)	Betula (BETU)
	Castanea (CAST)	Cupressaceae (CUPR)
	Fagus (FAGU)	Fraxinus (FRAX)
	Olea (OLEA)	Pinus (PINU)
	Platanus (PLAT)	Populus (POPU)
	Quercus deciduous type (QUCA)	Quercus evergreen type (QUPE)
	Quercus Total (QTOT)	Ulmus (ULMU)
Bushes	Corylus (CORY)	Ericaceae (ERIT)
	Pistacia (PIST)	
Herbs	Amaranthaceae (AMAR)	Artemisia (ARTE)
	Mercurialis (MERC)	Plantago (PLAN)
	Poaceae or Gramineae (POAC)	Polygonaceae (POLY)
	Urticaceae (URTI)	

total number of calculated correlations has been 10,368 for autumn rainfall, the same number than for winter in Rodríguez-Solà et al., (2020, 2022). A statistical significance level of 95% (p-value < 0.05) has been considered.

Table 3 Description of the pollen/spore variables

Pollen/spore variables	Description
APIn/ASIn	Annual pollen/spore integral (Pollen * day/m^3, Spore * day/m^3)
MPS/MSS	Main pollen/spore season, between the dates for which 2.5% and 97.5% of the APIn/ASIn is reached
SPIn/SSIn	Seasonal pollen/spore integral (Pollen * day/m^3, Spore * day/m^3) in the MPS/MSS
Days P/S	Number of days with pollen/spores
Max	Maximum daily concentration (P/m^3, S/m^3)
Peak	Date of the maximum daily concentration
Start	Start date of the MPS/MSS
End	End date of the MPS/MSS
Length	Number of days of the MPS/MSS

3 Results and Discussion

Figure 1 shows the total numbers of statistically significant Spearman correlations for the autumn rainfall ETCCDI, compared to the winter and annual numbers found by Rodríguez-Solà et al., (2020, 2022). As it happened for winter, the number of significant correlations, as well as the ratio of positive/negative significant correlations, resulted higher for autumn than for annual rainfall. However, the percentage of significant correlations found is only 5.1% of the possible total.

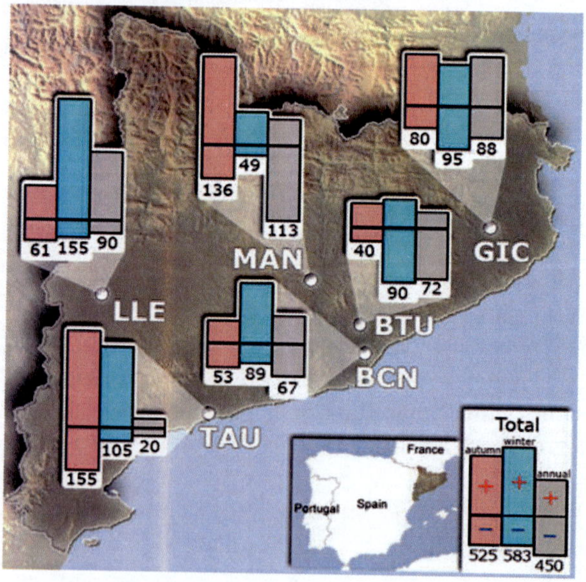

Fig. 1 Total numbers of statistically significant Spearman correlations for the autumn, winter and annual rainfall ETCCDI at every studied location

Regarding to the specific rainfall ETCCDI, while the winter indices with more correlations were those more related in some way to rainfall amount and intensity, for autumn the indices with more correlations have been those more related to rainfall persistence (Fig. 2). For indices RX5day and CWD, both related to rainfall episodes duration, a much higher number of correlations have been found for autumn (90–100) than for winter (60–70). However, for indices related to rainfall amount and intensity, as R10mm and R20mm, the number of autumn correlations decreases by more than a third compared to winter. Therefore, the temporal extension of the autumn rainfall episodes seems to be more important with regard to the subsequent pollen production than rainfall intensity.

As for the pollen/spore parameters, while winter rainfall had more influence on temporal aerobiological variables such as the Start, End and Length of the Main Pollen Season, and the number of Days with pollen/spores in the same yearly period, autumn rainfall affects more on the variables related to pollen amount such as APIn/ASIn, SPIn/SSIn and maximum daily concentration of the next yearly period (Fig. 3). The number of correlations found for these last three variables increase from approximately 60 for winter to almost 90 for autumn rainfall, while the temporal variables decrease by a half.

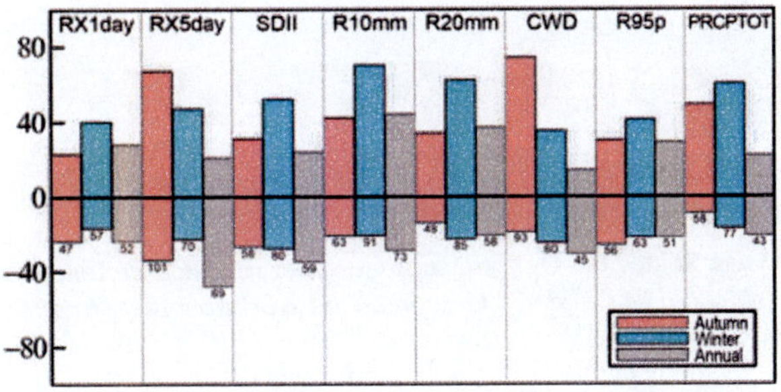

Fig. 2 Numbers of statistically significant Spearman correlations for every autumn, winter and annual ETCCDI

Fig. 3 Numbers of statistically significant Spearman correlations for every pollen/spore parameters for autumn, winter and annual rainfall

4 Conclusions

The correlation analysis between aerobiological variables and some seasonal precipitation ETCCDI leads to different results if autumn or winter rainfall previous to pollination is considered. An important number of negative correlations due to the wash-out effect are obtained when annual rainfall is used in the correlation study. As found for winter rainfall, the number of significant correlations and the ratio of positive/negative correlations are higher if previous autumn rainfall is considered instead of annual rainfall. However, the percentage of significant correlations is only 5.1% of the possible total. The study has also shown differences between the autumn and winter rainfall correlations: The more correlated autumn indices are those related to rainfall persistence and the duration of the rainy episodes, while those for winter are related to rainfall amount and intensity. Within the low influence that previous extreme rainfall seems to have on the subsequent pollen production, especially compared to that of temperature, the probable increase of the occurrence of intense rainfall events in autumn due to the influence of climate change in the Mediterranean area might slightly affect the subsequent pollen production. However, the kind of autumn rainy episodes for which occurrence is expected to increase are intense but short, so no increase is expected for the most correlated rainfall characteristics, that is, duration and persistence.

References

Galán Soldevilla, C., Cariñanos González, P., Alcázar Teno, P., & Domínguez Vilches, E. (2007). *Manual de Calidad y Gestión de la Red Española de Aerobiología*. Servicio de Publicaciones.

Klein-Tank, A. M. G., Zwiers F. W., Zhang X. (2009) Guidelines on Analysis of extremes in a changing climate in support of informed decisions for adaptation. Rep WCDMP-No. 72, WMO-TD no. 1500, World Meteorological Organization, Geneva, 55 pp https://www.ecad.eu/documents/WCDMP_72_TD_1500_en_1.pdf

Llabrés-Brustenga, A., Rius, A., Rodríguez-Solà, R., Casas-Castillo, M. C., & Redaño, À. (2019). Quality control process of the daily rainfall series available in Catalonia from 1855 to the present. *Theoretical and Applied Climatology, 137*(3–4), 2715–2729. https://doi.org/10.1007/s00704-019-02772-5

Majeed, H. T., Periago, C., Alarcón, M., & Belmonte, J. (2018). Airborne pollen pa-rameters and their relationship with meteorological variables in NE Iberian Penin-sula. *Aerobiologia, 34*, 375–388. https://doi.org/10.1007/s10453-018-9520-z

Rodríguez, R., Navarro, X., Casas, M. C., Ribalaygua, J., Russo, B., Pouget, L., & Redaño, À. (2014). Influence of climate change on IDF curves for the metropolitan area of Barcelona (Spain). *International Journal of Climatology, 34*, 643–654. https://doi.org/10.1002/joc.3712

Rodríguez-Solà, R., Casas-Castillo, M. C., Ho Zhang, J. J., Kirchner, R., Alarcón, M., Periago, C., De Linares, C., Belmonte, J. (2020) Study of the correlation between precipitation indices and airborne pollen in the NE of the Iberian Peninsula [Poster session]. Seventieth European Symposium on Aerobiology, November 16 20, 2020, virtual edition. https://doi.org/10.13140/RG.2.2.36428.74886

Rodríguez-Solà, R., Casas-Castillo, M. C., Ho Zhang, J. J., Kirchner, R., Alarcón, M., Periago, C., De Linares, C., Belmonte, J. (2022) A study on correlations between precipitation ETCCDI and airborne pollen/spore parameters in the NE Iberian Peninsula. *International Journal of Biometeorology, 66,* 1173–1187. https://doi.org/10.1007/s00484-022-02267-5

Estimating Particulate Matter Using Remote Sensing Data and Meteorological Variables Over Ahvaz, Iran

Maryam Gharibzadeh and Abbas Ranjbar Saadatabadi

Abstract Increased air pollution in the past two decades in the Middle East has caused several problems. The concentrations of surface-level particulate matter have always been considered as important indicators in evaluation of air pollution. In order to find out the PM_{10} and $PM_{2.5}$ concentration, multiple linear and nonlinear regression models were performed in this study. The relationship between concentrations of PM_{10}, $PM_{2.5}$ and Moderate Resolution Imaging Spectroradiometer (MODIS) aerosol optical depth (AOD) was calculated using several meteorological parameters such as temperature, relative humidity, wind speed, wind direction, K-index and visibility. AOD data and meteorological parameters were used to produce four statistical models including linear and nonlinear multi-regression models for Ahvaz. The higher correlation between measured and estimated PM_{10} and $PM_{2.5}$ concentrations was observed in the calculated nonlinear equations.

Keywords PM_{10} · $PM_{2.5}$ · Aerosol optical depth · Multiple nonlinear regressions

1 Introduction

The increase of aerosol and dust cause many problems such as air pollution, serious health risks and significant environmental effects (Sportisse, 2007). One of the key components in air pollution is PM_{10}, which has many effects, especially on respiratory problems (Zaman et al., 2017). $PM_{2.5}$ can also cause cardiovascular problems (Pope et al., 2002; Seo et al., 2015). Other effects of these particles include breathing premature death and birth defects (Ballester et al., 2010). Particulate matter can also influence the climate and the radiation budget through direct, semi-direct and indirect effects (Jacob, 1999).

M. Gharibzadeh (✉)
Institute of Geophysics, University of Tehran, Tehran, Iran
e-mail: m.gharibzad@ut.ac.ir

A. Ranjbar Saadatabadi
Atmospheric Science and Meteorological Research Center (ASMERC), Tehran, Iran

C. Mensink and O. Jorba (eds.), *Air Pollution Modeling and its Application XXVIII*, Springer Proceedings in Complexity, https://doi.org/10.1007/978-3-031-12786-1_37

To find out the relationship between satellite data, especially AOD, and the concentration of PM_{10} and $PM_{2.5}$ many studies have been conducted. In many studies an experimental linear model using only one parameter (AOD) has been applied to estimate PM_{10} and $PM_{2.5}$ (Gupta & Christopher, 2008; Kanniah et al., 2016). The AOD parameter alone is not sufficient to predict PM, since the AOD corresponds to the aerosols throughout the atmosphere column, while PM refers to the particle concentration at the surface. So PM depends on several meteorological factors in addition to AOD (Chitranshi et al., 2014; Seo et al., 2015; Zaman et al., 2017). More parameters such as temperature, wind speed and relative humidity were added in other works to improve the estimation results (Wang et al., 2010; Xiao et al., 2011; Benas et al., 2013).

In this work to investigate the relationship between PM_{10}, $PM_{2.5}$ and AOD in Ahvaz (one of the most polluted cities in the Middle East), AOD datasets from satellite and various ground-based meteorological parameters were used. First satellite remote sensing data and meteorological in situ measurements and also K-index were obtained in the period of 2009–2019. Then multivariable linear and nonlinear regression models were used to estimate the concentration of PM_{10} and $PM_{2.5}$ over Ahvaz. Finally the models were validated and discussed.

2 Materials/Methodss

2.1 *Site Description*

Iran is situated in the Middle East in the middle of the dust belt (Gharibzadeh et al., 2019). Ahvaz is a city in the southwest of Iran (N 31.33, E 48.66, and 12 m above sea surface). Ahvaz is often affected by dust storms from Iraq, Kuwait and Saudi Arabia dust sources, especially in the spring and summer. Severe dust storms that occur especially in summer led to thousands of hospital admissions for cardiovascular and respiratory diseases (Shahsavani et al., 2012). The study area of Iran, Ahvaz is shown in Fig. 1.

2.2 *Aerosol Optical Depth*

Aerosol can cause extinction of sunlight passing through the atmosphere by absorption or scattering. The Aerosol Optical Depth (AOD) is a measure of the amount of aerosols in the vertical column of the atmosphere from the top of the atmosphere to the ground level. The AOD was obtained from MODIS Daily Level 2 Aerosol Product at spatial resolutions of 10 km × 10 km. The seasonal average of AOD is shown in the Table 1.

Fig. 1 The study area of Iran, Ahvaz

Table 1 Seasonal Average of AOD obtained from MODIS over Ahvaz during 2009–2019

	Spring	Summer	Fall	Winter
AOD	0.49 ± 0.02	0.47 ± 0.02	0.28 ± 0.01	0.27 ± 0.01

2.3 *Meteorological Parameters*

A number of effective meteorological parameters on PM concentration have been used to investigate the relation between PM_{10}, $PM_{2.5}$, and AOD more precisely. In addition to AOD and PM, other meteorological data such as temperature (T), relative humidity (RH), wind speed (WS), wind direction (WD), and horizontal visibility (V) have been obtained from Iran Meteorological Organization. The wind direction is entered into the model using the WD parameter and is obtained from this equation (Sotoudeheian et al., 2014):

$$WD = \cos(2\pi\theta/360) \quad (1)$$

Table 2 shows the seasonal average of meteorological parameters of Ahvaz:

K-index is the factor for measuring the potential for thunderstorms in meteorology and it is calculated by George formula (1960):

Table 2 Seasonal average of meteorological parameters of Ahvaz during 2009–2019

	Spring	Summer	Fall	Winter
T (°C)	30.2	38	22.6	15.6
RH %	30.2	25.7	49.8	57.8
WD	−0.28	−0.36	−0.24	−0.17
WS (m/s)	2.62	2.65	1.68	2.02
V (Km)	7.92	8.28	8.21	7.57

Table 3 Seasonal averages of K-index in Ahvaz during 2019–2009

	Spring	Summer	Fall	Winter
K-index	10.90	9.28	3.74	−1.73

$$Kindex = (T_{850} - T_{500}) + TD_{850} - (T_{700} - TD_{700}) \tag{2}$$

where T_{850}, T_{500} and T_{700} are temperature in Celsius at 850 mb, 500 mb and 700mb, respectively. TD_{850} and TD_{700} is Dew point in Celsius at 850 mb and 700 mb. The first term represents the lapse rate. The moisture between 850 and 700 mb are shown by the second and third terms. This parameter is taken from Wyoming Weather Web, upper air data. The seasonal average of the K-index is given in Table 3.

3 Results

3.1 *Multivariable Linear Regression Model for PM_{10} and $PM_{2.5}$*

Multivariable linear regression model focused on determining relation between PM and AOD as a dependent and an independent variable. To obtain more accurate equation for PM_{10} concentration, AOD and meteorological parameters such as T, RH, WS, WD, V and also K-index used as variables. All statistical analyzes were performed using Statistical Analysis System (SAS) software. This linear equation for PM_{10} can be seen in Eq. 1 (E1)

$$PM_{10} = \alpha_1 + \alpha_{1AOD}AOD + \alpha_{1T}T + \alpha_{1RH}RH + \alpha_{1Kindex}Kindex + \alpha_{1WS}WS + \alpha_{1WD}WD + \alpha_{1V}V \tag{3}$$

Using multivariable linear regression, the following equation was obtained for estimation of $PM_{2.5}$ (E2). Table 4 shows the values of coefficients for different equations. αi is the intercept and αij are the regression coefficients of the dependent variables (j) on equation i.

$$PM_{2.5} = \alpha_2 + \alpha_{2AOD}AOD + \alpha_{2T}T + \alpha_{2RH}RH + \alpha_{2Kindex}Kindex$$

Table 4 Estimated regression coefficients for the linear PM_{10} and $PM_{2.5}$ equations

	α	α_{AOD}	α_T	α_{RH}	α_{Kindex}	α_{WS}	α_{WD}	α_V
E1	850	3.87	−1.06	−2.8	−0.28	18.65	−59	−0.07
E2	81.9	37.59	1.17	0.19	−0.05	−5.3	5.6	$-8*10^{-3}$

$$+\alpha_{2WS}WS+\alpha_{2WD}WD+\alpha_{2V}V \tag{4}$$

3.2 *Multivariable Nonlinear Regression Model for PM₁₀ and PM₂.₅*

Multivariable linear regression models were developed to estimate PM_{10} and $PM_{2.5}$ concentrations using multivariable nonlinear regression. This model was improved by combining five meteorological parameters as well as AOD and K-index. The developed multivariable nonlinear regression model for the PM_{10} estimations over Ahvaz is:

$$PM_{10}=e^{(\alpha_3+\alpha_{3T}T+\alpha_{3RH}RH+\alpha_{3Kindex}Kindex+\alpha_{3WD}WD+\alpha_{3V}V)} \times AOD^{\alpha_{3AOD}} \times WS^{\alpha_{3WS}} \tag{5}$$

The log function is used to obtain the linear form (E3) of the equation on both sides of the above equation.

$$LNPM_{10}=\alpha_3+\alpha_{3AOD}LNAOD+\alpha_{3T}T+\alpha_{3RH}RH+\alpha_{3Kindex}Kindex+\alpha_{3WS}LNWS+\alpha_{3WD}WD+\alpha_{3V}V \tag{6}$$

Multivariable nonlinear regression model was used to find out $PM_{2.5}$ concentrations as a function of seven different parameters (E4).

$$LNPM_{2.5}=\alpha_4+\alpha_{4AOD}LNAOD+\alpha_{4T}T+\alpha_{4RH}RH+\alpha_{4Kindex}Kindex+\alpha_{4WS}LNWS+\alpha_{4WD}WD+\alpha_{4V}V \tag{7}$$

The calculated values for the coefficients (E3 and E4) are shown in Table 5.

Table 5 Estimated regression coefficients for the nonlinear PM_{10} and $PM_{2.5}$ equations

	α	α_{AOD}	α_T	α_{RH}	α_{Kindex}	α_{WS}	α_{WD}	α_V
E3	7.7	0.03	−0.004	−0.015	$-9*10^{-4}$	0.01	−0.05	$-2*10^{-4}$
E4	4.75	0.07	0.01	−0.006	$-2*10^{-4}$	−0.15	−0.01	$-1*10^{-4}$

4 Discussion

The monthly mean values of PM_{10} estimated by multivariable linear regression model (E1) and multivariable nonlinear regression model (E3) compared with the monthly mean values of observed PM_{10} over Ahvaz are shown in Fig. 2.

The monthly mean values of $PM_{2.5}$ estimated by multivariable linear regression model (E2) and multivariable nonlinear regression model (E4) compared with the monthly mean observed $PM_{2.5}$ over Ahvaz are shown in Fig. 3.

The positive sign for all AOD coefficients was observed as expected. These positive values indicate the positive relationship between PM_{10} concentration and AOD. Temperature has different effects in different seasons, therefore both negative and positive signs are observed. For example, higher PM levels can occur during the colder seasons of the year due to the inversion phenomenon (Sotoudeheian & Arhami, 2014). Mostly a negative sign was obtained for relative humidity, indicating an inverse effect of RH on PM. In conditions of high relative humidity, the hygroscopic aerosols can grow several times their normal size and increase extinction efficiency of sun lights due to particle. Small RH values are expected to increase PM_{10}

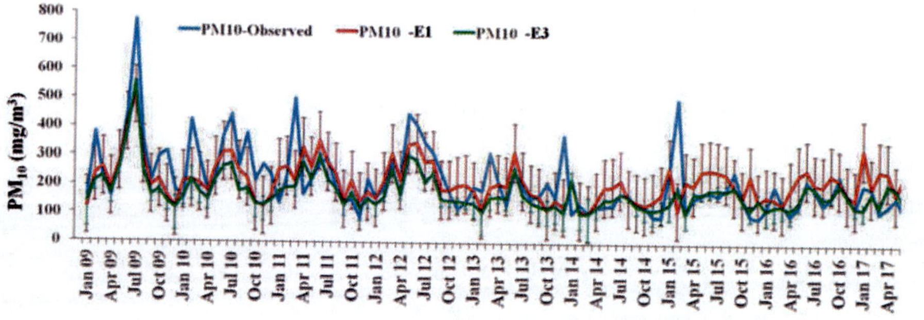

Fig. 2 The monthly mean PM_{10} estimated by the multivariable linear and nonlinear regression model (E1 and E3) compared with the observed PM_{10} over Ahvaz during 2008–2017

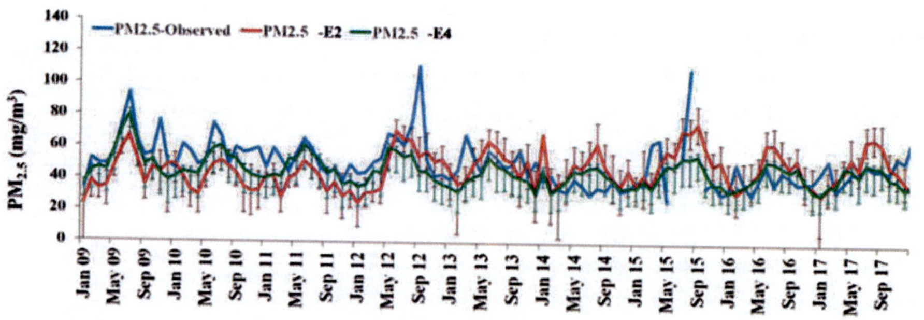

Fig. 3 The monthly mean $PM_{2.5}$ estimated by the multivariable linear and nonlinear regression model (E2 and E4) compared with observed $PM_{2.5}$ over Ahvaz during 2009–2017

Table 6 The prediction accuracy of PM_{10} and $PM_{2.5}$ concentration with linear and nonlinear regression

	E1	E2	E3	E4
Prediction accuracy (%)	62	60	64	70

concentration (Benas et al., 2013). A negative correlation was found between PM_{10} and K-index. K-index is associated to atmospheric stability and consequently to the vertical mixing. Increasing in the boundary layer height relate to increased aerosol vertical mixing. K-Index values increased because of the turbulent movements and atmospheric instability (Benas et al., 2013; Gupta & Christopher, 2009).

Measured PM concentrations were used to validate estimated values of PM_{10} and $PM_{2.5}$ concentrations obtained from multivariable linear and nonlinear regression models. The percentage of correct predictions of PM concentration changes for all linear and nonlinear regression models are shown in Table 6.

5 Conclusions

The MODIS 10 km AOD was used to estimate the concentrations of $PM_{2.5}$ and PM_{10} over Ahvaz, Iran during 10 years. In order to improve this estimation, various meteorological parameters were used. Then, several equations for estimating $PM_{2.5}$ and PM_{10} using Statistical Analysis System (SAS) were calculated. The results of multivariable linear regression model show that this model has good ability to predict 62% of PM_{10} and 60% of $PM_{2.5}$ changes in Ahvaz. The prediction accuracy of PM_{10} and $PM_{2.5}$ concentration with nonlinear equations was 64% and 70%, respectively.

Acknowledgements The authors would like to express appreciation to Wyoming Weather Web for providing the data. Iran Meteorological Organization is acknowledged for providing the data at Ahvaz site. The team of MODIS is also acknowledged for providing available data products.

References

Ballester, F., Estarlich, M., Iniguez, C., Llop, S., Ramon, R., Esplugues, A., Lacasana, M., & Rebagliato, M. (2010). Air pollution exposure during pregnancy and reduced birth size: a prospective birth cohort study in valencia. *Spain. Environmental Health, 9*, 6.

Benas, N., Beloconi, A., & Chrysoulakis, N. (2013). Estimation of urban PM10 concentration, based on MODIS and MERIS/AATSR synergistic observations. *Atmospheric Environment, 79*, 448.

Chitranshi, S., Sharma, S. P., & Dey, S. (2014). Satellite-based estimates of outdoor particulate pollution (PM10) for Agra City in northern India. *Air Quality, Atmosphere and Health, 8*, 55–56.

George, J. J. (1960). *Weather forecasting for aeronautics*. Academic Press.

Gharibzadeh, M., & Alam kh. Abedini Y. Bidokhti A.A Masoumi A. Bibi H. and Zeb B. (2019). Climatological analysis of the optical properties of aerosols and their direct radiative forcing in the Middle East. *Journal of Atmospheric and Solar-Terrestrial Physics, 183*, 86–98.

Gupta, P., & Christopher, S. A. (2008). Seven-year particulate matter air quality assessment from surface and satellite measurements. *Atmospheric Chemistry and Physics, 8*, 3311–3324.

Gupta, P., & Christopher, S. A. (2009). Particulate matter air quality assessment using integrated surface, satellite, and meteorological products: Multiple regression approach. *Journal of Geophysical Research-Atmospheres, 114*, 1–13.

Jacob, D. J. (1999). Introduction to atmospheric chemistry. Princeton University Press, Chapter 8.

Kanniah, K. D., Kaskaoutis, D. G., San, L. H., Latif, M. T., Zaman, N. A. F. K., & Liew, J. (2016). Overview of Atmospheric Aerosol Studies in Malaysia: Known and unknown. *Atmospheric Research, 182*, 302–318.

Pope, C. A., Burnett, R. T., Thun, M.J., Calle, E. E., Krewski, D., Ito, K., & Thurston, G. D. (2002). Lung cancer, cardiopulmonary mortality, and long-term exposure to fine particulate air pollution, *Journal of the American Medical Association, 287*, 1132–1141.

Seo, S., Kim, J., Lee, H., Jeong, U., Kim, W., Holben, B. N., Kim, S. W., Song, C. H., & Lim, J. H. (2015). Estimation of PM10 concentrations over Seoul using multiple empirical models with AERONET and MODIS data collected during the DRAGON-Asia campaign. *Atmospheric Chemistry and Physics, 15*, 319–334.

Shahsavani, A., Naddafi, K., Haghighifard, N. J., Mesdaghinia, A., Yunesian, M., Nabizadeh, R., Arahami, M., Sowlat, M. H., Yarahmadi, M., Saki, H., Alimohamadi, M., Nazmara, S., Motevalian, S. A., & Goudarzi, G. (2012). The evaluation of PM10, PM2.5, and PM1 concentrations during the Middle Eastern Dust (MED) events in Ahvaz, Iran, from april through september 2010. *Journal of Arid Environments, 77*, 72–83.

Sotoudeheian, S., & Arhami, M. (2014). Estimating ground-level PM10 using satellite remote sensing and ground-based meteorological measurements over Tehran. *Journal of Environmental Health Science & Engineering, 12*, 122.

Sportisse, B. (2007). *Fundamentals in Air Pollution, from Processes to Modelling*. Springer.

Wang, Z., Chen, L., Tao, J., Zhang, Y., & Su, L. (2010). Satellite-based estimation of regional particulate matter (PM) in Beijing using Vertical-and-RH correcting method. *Remote Sensing of Environment, 114*, 50–63.

Xiao, Z. (2011). Estimation of the main factors influencing haze, based on a long-term monitoring campaign in Hangzhou, China. *Aerosol and Air Quality Research*, 873–882.

Zaman, N. A. F. K., Kanniah, K. D., & Kaskaoutis, D. G. (2017). Estimating Particulate Matter Using Satellite Based Aerosol Optical Depth and Meteorological Variables in Malaysia. *Atmospheric Research, 193*, 142–162.

Emission Modelling and Processing

Modelling of Biogenic Volatile Organic Compounds Emissions Using a Detailed Vegetation Inventory Over a Southern Italy Region

C. Silibello, S. Finardi, N. Pepe, R. Baraldi, P. Ciccioli, M. Mircea, and P. Ciccioli

Abstract Biogenic Volatile Organic Compounds (BVOC) play a key role in ozone-forming reactions and are precursors of secondary organic aerosols that absorb and scatter the atmospheric radiation, affect precipitations by acting as cloud condensation nuclei and increase PM concentrations affecting thus the human health. Therefore, it is important to develop models that can accurately estimate such emissions. BVOC emissions show wide variability depending on the distributions of plant species, their emission potentials and seasonal behavior. This work presents further developments of Plant-Specific Emission Model (PSEM) and its application over Campania region (Southern Italy) considering a detailed recognition of tree species, natural vegetation, and agricultural land performed at a very high spatial resolution (200 m). Its BVOC emissions were compared with those provided by Model of Emissions of Gases and Aerosol from Nature (MEGAN, v2.04) evidencing significant differences, particularly for isoprene. Such emissions were further used to feed a Chemical Transport Model (CTM) applied over the Vesuvius area, where vertical profiles of BVOC concentrations were measured by tethered balloon soundings. The comparison between the observed and predicted BVOC concentrations evidenced the capability of the developed model to estimate the biogenic emissions and highlighted the relevance of an accurate inventory of the vegetation.

C. Silibello (✉) · S. Finardi · N. Pepe
ARIANET, Palermo, Italy
e-mail: c.silibello@aria-net.it

R. Baraldi
Istituto Per La Bioeconomia (CNR), Bologna, Italy

P. Ciccioli
Istituto Per I Sistemi Agricoli E Forestali del Mediterraneo (CNR), Perugia, Italy

M. Mircea
Laboratory of Atmospheric Pollution, Italian National Agency for New Technologies, Energy and Sustainable Economic Development—ENEA, 40129 Bologna, Italy

P. Ciccioli
Istituto Per I Sistemi Biologici (CNR), ADR CNR, 1, Montelibretti, Monterorotondo Scalo, Italy

C. Mensink and O. Jorba (eds.), *Air Pollution Modeling and its Application XXVIII*, Springer Proceedings in Complexity, https://doi.org/10.1007/978-3-031-12786-1_38

1 Introduction

BVOC are organic trace gases mainly composed by isoprenoids, but also alcohols, carbonyls, esters and acids, that are released in the atmosphere by above- and below-ground plant organs. Plants synthesize different blends of BVOC and emit them with different mechanisms. The short-term environmental factors triggering the production the BVOC in plants are the photosynthetically active photon flux density (PPFD) reaching the leaf surface, the leaf temperature and the soil moisture. In some plants, isoprenoids are stored in specialized organs, such as resin ducts or glands, from which they slowly reach the atmosphere by diffusion. This type of emission, which is typical of mono- and sesquiterpenes produced by conifers and some scented plants, is simply indicated as the "T" mechanism because the release into the air of these BVOC depends mostly on their partial pressure, which is known to exponentially increase with temperature. Plants in which the BVOC emission is concurrent with their photosynthetic production follow instead the so called "L + T" mechanism, because the emission is triggered by both light and temperature. Frequent is also the case of plants that do not produce and emit isoprenoids (e.g. *Quercus cerris*, *Ostrya caronifolia* and *Fraxinus ornus*). Such complexity in the BVOC emission process supports the use of species-specific emission models as indicated by Jiang et al. (2019) and Silibello et al. (2017) who concluded that the MEGAN (v.2) emission model could largely overestimate the isoprene emissions in Southern Europe. In this work we present a new version of the Plant-Specific Emission Model (PSEM, Pacheco et al., 2014; Silibello et al. 2017) that implements updates to better simulate the vegetation emissions seasonality and canopy vertical distribution characteristics. In the following sections we briefly describe the improvements and show the results of preliminary results from simulations over Campania region.

2 Materials and Methods

We refer to Silibello et al. (2017) for a description of PSEM, that requires species-specific data for BVOC emission rates, areal coverage, and leaf mass area. The emission rate E^a [μg h^{-1}] of the BVOC a from a grid cell is estimated as:

$$E^a = \sum_{k=0}^{N} E_k^a \cdot A_c \cdot LMA_k \cdot LAI_k \cdot \gamma_{s,k} \cdot \gamma_{c,k} \cdot (1 + \gamma_{L,k})$$

where A_c is the surface area of the grid cell in m^2 and, for each vegetation species k, LMA_k [g_{dw} m_{leaf}^{-2}] is the leaf mass area, LAI_k [m_{leaf}^2 m_{ground}^2] its contribution to the overall LAI, $E_k{}^a$ the BVOC a emission factor [μg g_{dw}^{-1} h^{-1}], $\gamma_{S,k}$, $\gamma_{c,k}$ and $\gamma_{L,k}$ are adimensional correction factors accounting respectively for the seasonal variation of the emissions, the different light and temperature levels experienced by leaves or needles emitting inside a canopy and the contribution from dead biomass accumulated on the soil (litter). The approach proposed by Karl et al. (2009) has been used to estimate LAI_k: $LAI_k = LAI_{MODIS}\frac{f_k \cdot S_k}{\sum_{i=1}^{n} f_i \cdot S_i}$ where S_k is the seasonality factor of the vegetation species k (vegetation growing/senescense). S_k and $\gamma_{S,k}$, can be described by the product of two logistic functions:

$$S_k or \gamma_{s,k}(M) = \left[\beta \cdot \left(\frac{1}{1+e^{-K_g(M-M_g)}}\right) \cdot \left(1 - \frac{1}{1+e^{-K_d(M-M_d)}}\right)\right] + \alpha$$

where K_g and K_d are the growth and decay rates of S_k and $\gamma_{S,k}$ through the year, M, is the time in months of the calendar year (i.e. $1 \div 12$), M_g and M_d are the inflection points of the growth and decay curves, still expressed in months and referred to a central month of the year, usually July; α and β are adjusting terms to fit the curves to a maximum value of 1 and to a minimum value that can be different from zero. S_k and $\gamma_{S,k}$ are generally shifted in time and described by different M_g and M_d. Figure 1 shows an example of the application of such approach to reproduce *Castanea Sativa* LAI seasonality. A set of 21 logistic functions has been de-rived to describe the seasonality terms S_k and $\gamma_{S,k}$ for the vegetation species on the basis of a detailed recognition of natural and agricultural vegetation species (at 200 m resolution) over the Campania region, and MODIS LAI data. The MEGAN3 (https://bai.ess.uci.edu/megan) canopy environment model has been used to compute the γ_C term. This model considers six different canopy types: Needleleaf trees, tropical forest trees, temperate broadleaf trees, shrubs, herbaceous and crops and splits out the vegetation canopy into 5-layer.

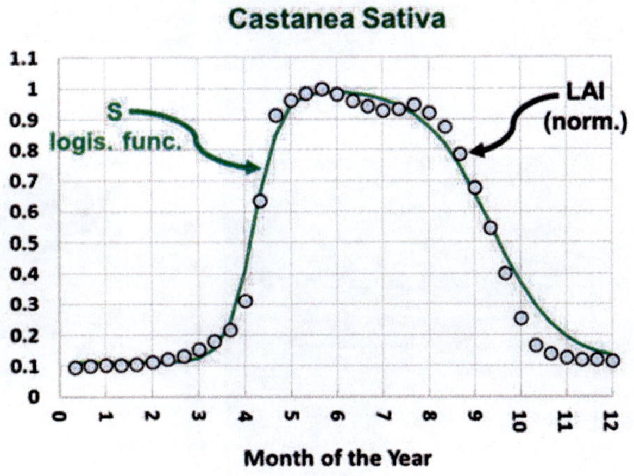

Fig. 1 Use of logistic functions to reproduce Castanea Sativa LAI seasonality

3 Results

Figure 2 shows the yearly isoprene and monoterpenes emissions over Campania region for the year 2018 estimated by PSEM and MEGAN (v 2.04) models. It can be observed that MEGAN application could largely overestimate the isoprene emissions and underestimate monoterpenes ones, with respect to PSEM prediction. The spatial distribution of such emissions significantly changes according to the different BVOC emission rates and spatial distribution of the vegetation species. The availability of BVOCs, ozone and other VOCs experimental data, collected during a field campaign performed in July 2015 (14–16) over a forest site located on the Mount Vesuvius (Naples, Italy), allowed a further evaluation of PSEM and MEGAN emissions estimate by means of CTM simulations. We refer to Finardi et al. (2018) for a description of field measurements and CTM set-up. Figure 3 illustrates the comparison between observed and predicted BVOC vertical profiles, during the second day of the experimental campaign at 13:00 UTC.

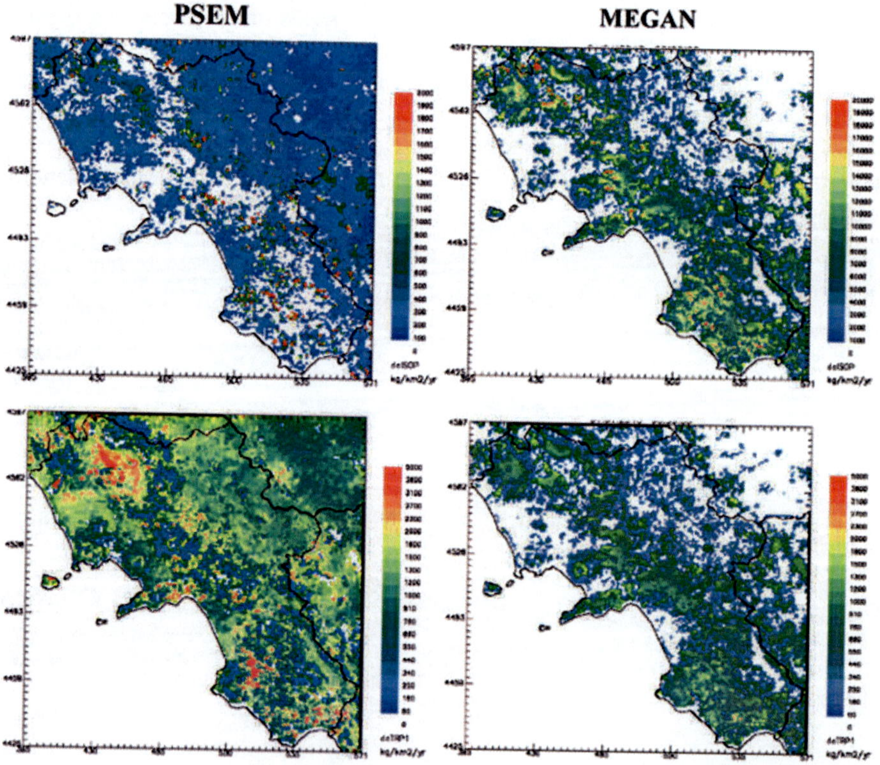

Fig. 2 Isoprene (up) and Monoterpenes (down) yearly emissions [kg km^{-2} yr^{-1}] estimated by PSEM and MEGAN emission models

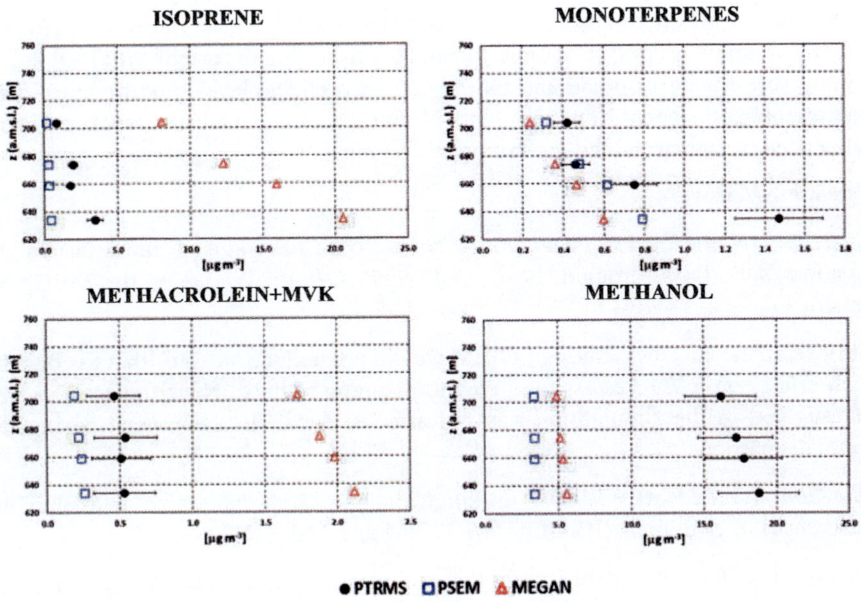

Fig. 3 Observed and predicted BVOCs vertical profiles using PSEM and MEGAN models during July 15 at 13:00 UTC

The comparison of predicted and observed profiles evidences a good agreement using PSEM emissions and confirms the significant overestimation of isoprene concentrations, and consequently of its ozonolysis products (methacrolein and methyl-vynil-ketones -MVK-), using MEGAN emissions (similar results have been obtained considering the other available vertical profiles). As for monoterpenes, a better agreement is obtained using PSEM emissions. Both models underestimate methanol levels, probably due to anthropogenic sources not fully estimated.

4 Conclusions

In this study, the updated version of PSEM and the largely used BVOC emission model MEGAN have been used to estimate yearly emissions over Campania region. MEGAN model provided significant larger isoprene emissions and lower monoterpenes ones than PSEM. Moreover, the spatial distribution of the emissions was different according to the different BVOC emission rates and vegetation species mapping considered by the two models. The availability of measured BVOC vertical profiles, collected during an experimental campaign performed over Mount Vesuvius, allowed the application of a CTM to further evaluate the two emission models. The experimental campaign was performed during the summer, when the maximum emission is reached because LAI, PPFD and temperature rise to their yearly maxima.

The comparison between observed and predicted BVOCs concentrations evidences a better agreement using PSEM emission and confirms that the use of MEGAN model could overestimate isoprene and its ozonolysis products levels and underestimate monoterpenes concentrations. Both models underestimate methanol levels probably influenced by undetermined anthropogenic sources.

Questions/Answers

Question: (Name unclear) Did you account for stress-induced emissions in the creation of BVOCs (drought, heat, cold, wind, O_3 stress) Like in the MEGAN3 version?

Answer: In the current version of PSEM we did not include an algorithm to consider such effects on BVOC emissions. We are planning to do it. Nevertheless, the area investigated in the simulations was not affected by such phenomena during the simulated period.

Question: (Rafael Borge) What is the single one most important factor to explain the differences of emissions (BVOC) from PSEM and MEGAN?

Answer: The most important factor is the different way in which the basal emission factors are considered in the two models. As an example, in PSEM the contribution of different types of oaks on isoprene emissions is explicitly considered according to their emission factors: In areas where non-isoprene emitting oaks, such as *Quercus cerris*, are present no isoprene emissions are estimated while in MEGAN, since all oaks belong to the same Plant Functional Type (PFT), in these areas the emissions are estimated according to the emission factor associated to the related PFT.

Acknowledgements The development and verification of PSEM model is supported by Life VEG-GAP project (https://www.lifeveggap.eu/) which was c-funded by European Union Life Program in 2018, Grant Number LIFE18 PRE IT 003.

References

Jiang, J., Aksoyoglu, S., Ciarelli, G., Oikonomakis, E., et al. (2019). Effects of two different biogenic emission models on modelled ozone and aerosol concentrations in Europe. *Atmospheric Chemistry and Physics, 19*, 3747–3768.

Karl, M., Guenther, A., Köoble, R., Leip, A., & Seufert, G. (2009). A new European plant specific emission inventory of biogenic volatile organic compounds for use in atmospheric transport models. *Biogeosciences, 6*, 1059–1087.

Finardi, S., Agrillo, G., Baraldi, R., Calori, G., Carlucci, P., Ciccioli, P., D'Allura, A., Gasbarra, D., Gioli, B., Magliulo, V., Radice, P., Toscano, P., & Zaldei, A. (2018). Atmospheric dynamics and ozone cycle during sea breeze in a Mediterranean complex urbanized coastal site. *Journal of Applied Meteorology and Climatology, 57*(5), 1083–1099.

Pacheco, C. K., Fares, S., & Ciccioli, P. (2014). A highly spatially resolved GIS-based model to assess the isoprenoid emissions from key Italian ecosystems. *Atmospheric Environment, 96*, 50–60.

Silibello, C., Baraldi, R., Rapparini, F., Facini, O., et al., (2017) Modelling of biogenic volatile organic compounds emissions over Italy. HARMO 2017 Proceedings, 2017-October, 14–18.

Modelling Pollutant Concentrations in Streets: A Sensitivity Analysis to Asphalt and Traffic Related Emissions

T. Sarica, K. Sartelet, Y. Roustan, Y. Kim, L. Lugon, M. André, B. Marques, B. D'Anna, C. Chaillou, and C. Larrieu

Abstract Higher concentrations of nitrogen dioxide and particles (organics and black carbon) are often observed in streets compared to urban background. Road traffic is generally seen as one of the dominant sources of emissions in streets. The main goal of this work is to evaluate the impact of uncertainties of traffic emissions (speciation of VOC at exhaust and non-exhaust emissions) on pollutant concentrations in streets, and to evaluate the potential impact of asphalt-related emissions, which are usually ignored.

Keywords Sensitivity · Emissions · Traffic · Pollutant concentrations · Street network

T. Sarica (✉) · K. Sartelet · Y. Roustan · Y. Kim · L. Lugon
CEREA, École des Ponts, EDF R&D, Île-de-France, France
e-mail: thibaud.sarica@enpc.fr

K. Sartelet
e-mail: karine.sartelet@enpc.fr

M. André
COSYS Univ Gustave Eiffel, IFSTTAR, Lyon, France

B. Marques · B. D'Anna
Aix-Marseille Univ, CNRS, LCE, UMR 7376, Marseille 13331, France

B. Marques
French Agency for Ecological Transition, ADEME, Angers 49000, France

C. Chaillou · C. Larrieu
Aramco Fuel Research Center, Aramco Overseas Company, Rueil-Malmaison 92500, France

C. Mensink and O. Jorba (eds.), *Air Pollution Modeling and its Application XXVIII*, Springer Proceedings in Complexity, https://doi.org/10.1007/978-3-031-12786-1_39

1 Introduction

Despite increasingly stringent regulations, road transport remains a sources of air pollution (nitrogen oxides and particles to a lesser extend) in urban areas. The European emission standards (Euro norms) drive in Europe the reduction of automobile-related emissions through several regulations, from Euro 1 in 1992 to the Euro 7 regulation expected for 2025. However, only emissions at exhaust are considered by these regulations and non-exhaust emissions are omitted. The COPERT methodology, described in the EMEP/EEA air pollution emission inventory guidebook (EMEP/EEA, 2019), lists traffic-related pollutant emission factors for the different Euro norms and vehicle technologies. It is often used in Europe to build emission inventories. However, uncertainties remain high for some emission factors, such as those linked to non-exhaust emissions, and the methodology lacks a detailed speciation of some emitted pollutants such as volatile organic compounds (VOC). Non-exhaust emissions and VOC speciation may influence particle concentrations (black carbon BC and organic respectively) (Lugon et al., 2020b; Sartelet et al., 2018). Depending on the speciation, VOC may contain precursors of organic aerosols (intermediate and semi volatile organic compounds IVOC and SVOC, toluene, xylenes etc.).

Numerical modelling is an effective tool to evaluate the impacts of emission changes but also the importance of urban topology on pollutant concentrations at street scale. In this work, the Model of Urban Network of Intersection Canyons and Highways (MUNICH) is used to compute concentrations of pollutants at street scale (Kim et al., 2018; Lugon et al., 2020a). MUNICH is coupled with the modular box model SSH-aerosol (Sartelet et al., 2020) to represent formation and ageing of primary and secondary gases and particles in streets. Four sensitivity scenarios are designed to study uncertainties in the COPERT methodology linked to VOC speciation at the exhaust, non-exhaust emissions and asphalt emissions.

2 Simulated Scenarios

Simulations are performed over the month of April 2014 using the street network presented in Kim et al. (2018). This street network is composed of 577 street segments representing a district of Le Perreux-sur-Marne, a suburb 13km east of Paris. Meteorological parameters and background concentrations originate from Lugon et al. (2020b). Deposition and scavenging of both gas and particle phases are taken into account. The size distribution of particulate species are represented in the model by 6 size bins of diameters ranging between 0.01 and 10 μm (Lugon et al., 2020b).

Fleet composition and traffic flows and speeds have been determined for a typical weekday and a typical weekend day from simulations using the traffic simulator SymuVia André et al. (2020). The fleet is composed on average by 77% of passenger cars (PC) and 14% of light commercial vehicles (LCV). Diesel is the most widespread fuel in the fleet representing respectively 65.2% and 97.5% of PC and LCV. Traffic flows of PC are higher during the weekday than weekend day with peaks at up to 1 400 vehicles per hour at morning and afternoon rush hours over the main street section of interest for this study. During the weekend day, the maximal flow is about 200 vehicles per hour. For both days the PC speed profiles are similar, and vary between 24 and 32 km h^{-1}.

Table 1 presents the reference simulation S0 and the four sensitivity scenarios, S1 to S4, performed in this study. The reference simulation S0 is based on the COPERT methodology described in EMEP/EEA (2019) for exhaust emissions of CO, NO_x ($NO+NO_2$), VOC, CH_4 and $PM_{2.5}$, and non-exhaust emissions of PM_{10}. NO_2 fractions, VOC speciations and particulate matter (PM) decomposition into BC, organic matter and dust are also provided by the COPERT methodology. At exhaust, the "cold-start" emissions when the operational temperature of the internal combustion engine and after-treatment devices have not yet been reached are taken into account. In all scenarios, only the street emissions for the local-scale simulation with MUNICH are modified. The impact of the scenarios on the background concentrations is not taken into account.

Scenario S1 evaluates the impacts of using finer and more recent Non-Methane VOC (NMVOC) speciations for petrol PC and LCV. These new speciations originate from laboratory measurements performed during the MAESTRO project.[1] With these new speciations, the fractions of IVOC and SVOC are reduced from 6.0% to less than 2.0% and from 4.9% to less than 0.2% compared to COPERT respectively. Scenario S2 investigates the implications of estimating IVOC, SVOC and low volatility volatile organic compounds (LVOC) emissions from the NMVOC emissions as presented in Sartelet et al. (2018), with formulae based on chamber measurements performed in the United-States for petrol and diesel vehicles. The primary organic aerosols emitted in S0 are not considered in this scenario, as they are assumed to be LVOC. Using this method, the averaged total emission of I/S/LVOC is reduced by 71% compared to the reference S0, 78 $\mu g\ s^{-1}$ against 272 $\mu g\ s^{-1}$. The volatility distribution is also modified with a larger fraction of IVOC and a lower fraction of LVOC. Non-exhaust emissions originating from tyre, brake and road wears are bound to high uncertainties partly due to the difficulties to differentiate the origin of the measured materials. Following Lugon et al. (2020b), the tyre emission factors are multiplied by 9 in scenario S3 as the simulation S0 underestimates the concentrations of BC compared to observations. Furthermore, the chemical speciation of tyre-wear emissions assumes 25% of BC (compared to 15.3% in simulation S0). Finally, scenario S4 considers emissions of Secondary Organic Aerosols (SOA) precursors in the gas phase from road asphalt due to pavement heating and exposure to sunlight, as detailed in Khare et al. (2020). As a first assessment, these emissions are assumed to occur during daytime at a constant rate of 0.025 mg/min/kg asphalt. This rate is derived assuming a temperature of the road asphalt of 20 °C during the simulation period.

[1] Modélisation/caractérisation des précurseurs gazeux et des AErosols Secondaires du Transport Routier—joint project between LCE-AMU, IRCELYON, UGE-EASE and CEREA.

Table 1 List of the simulations performed

Scenario	Description
S0	Reference simulation—COPERT methodology only
S1	More recent NMVOC speciations for petrol PC and LCV
S2	IVOC, SVOC and LVOC are estimated from Sartelet et al. (2018)
S3	Higher tyre-wear emissions following Lugon et al. (2020b)
S4	Emissions from road asphalt derived from Khare et al. (2020)

3 Concentration Levels

The reference S0 is compared to observations, performed on one segment of the street network, using the statistical criteria presented in Hanna and Chang (2012). Although NO_x and NO concentrations are underestimated, NO_2 concentrations are well modelled and satisfy the strictest criteria. Particles are also well modelled: PM_{10} and $PM_{2.5}$ both comply with the strictest criteria, although PM_{10} are slightly underestimated and $PM_{2.5}$ are slightly overestimated. However, BC concentrations are significantly underestimated and none of the criteria are satisfied. These results and comparisons to observations are in accordance with previous works (Kim et al., 2018; Lugon et al., 2020b).

Scenario S1 with the use of more recent speciations for NMVOC emissions of petrol PC and LCV has low impacts on pollutant concentrations. PM_{10} and $PM_{2.5}$ average concentrations are decreased by about 0.12% compared to the reference S0. Organics in the particle phase are impacted with a reduction of the average concentration of 0.6%. The gas-phase concentrations of IVOC and SVOC are also reduced by 12.6% and 5.2% respectively. This low sensitivity was expected as petrol is not the preponderant fuel in the fleet. 30% and 2% of the PC and LCV respectively are running on petrol. Scenario S2 PM_{10} and $PM_{2.5}$ average concentrations are 1.6% and 1.8% lower compared to the reference S0 respectively. Organics are the most impacted compounds of PM with a reduction of 8.7%, showing that the IVOC and SVOC speciation of VOC may have a large influence on organic concentrations in the particles.

There is a high increase in the PM concentrations in scenario S3, due to higher emission factors and modified chemical speciation of tyre wear. PM_{10} and $PM_{2.5}$ average concentrations are 10.9% and 1.1% higher compared to the reference S0 respectively. BC and organics are the particle compounds that increase the most by 41.9% and 23.8% respectively. In scenario 4, taking into account road asphalt emissions results in a large increase of the concentrations of gas-phase IVOC and SVOC by as much as 492.0% and 321.5% respectively. The impacts on particle-

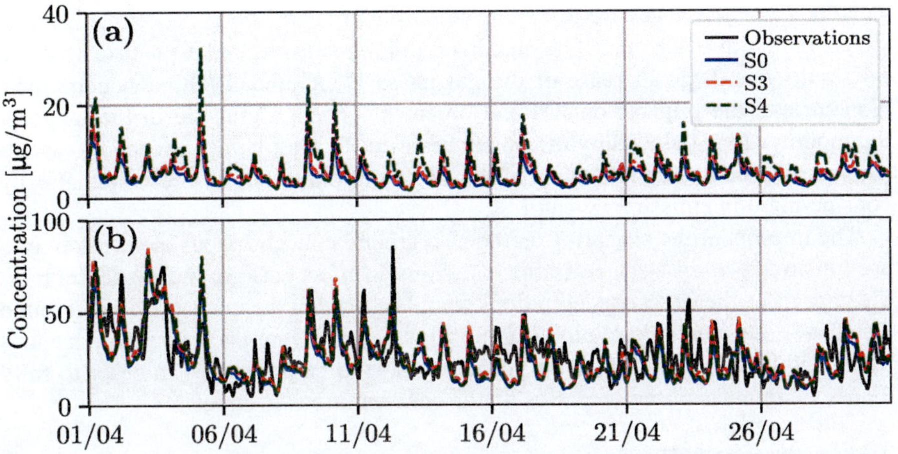

Fig. 1 Hourly simulated concentrations of organics (**a**) and PM_{10} (**b**) at a segment of the street network for the reference simulation S0 and scenarios S3 and S4. PM_{10} concentrations are compared to observations [μgm^3]

phase organics are not as high, 36.0% for the organics. The PM_{10} and $PM_{2.5}$ average concentrations increase by 6.6% and 7.4% respectively, compared to the reference S0. These results are plausible and confirm the conclusion of Khare et al. (2020) that road asphalt can be a major missing source of SOA precursor emissions (Fig. 1).

4 Conclusion

The impacts of different sources of uncertainties linked to traffic emissions on pollutant concentrations in streets have been evaluated using the street network model MUNICH, coupled with the complex chemistry module SSH-aerosol. The reference simulation S0 presents concentrations that compare well to the observed concentrations and previous works.

The impacts of considering more recent NMVOC speciations for petrol PC and LCV in scenario S1 are limited as petrol is not the preponderant fuel in the fleet. Nevertheless, slight reductions of PM and organics are observed, as the fraction of IVOC and SVOC are lower in the new speciations than in COPERT. Computing IVOC and SVOC emissions using the method presented in Sartelet et al. (2018), which is not specific to European cars, in scenario S2 results in larger modifications of organic concentrations in streets compared to the reference S0. Having access to more recent NMVOC speciations for diesel European vehicles is decisive to fully assess the impacts of NMVOC speciation on pollutant concentrations.

The simulated BC concentrations are closer to observation in the scenario S3, which has higher emission factors and modified chemical speciation for tyre wear.

This indicates the necessity to better measure and characterize non-exhaust emissions. Taking into account SOA precursor emissions from road asphalt in scenario S4 results in a huge increase of the gas-phase IVOC and SVOC concentrations. Nonetheless, the impacts on PM and organics concentrations are in the range of the modifications induced by the other scenarios. Further work is needed to assess the significance and uncertainties of this source of emissions before systematically considering it in emission inventories.

The impacts of the scenarios on the background concentrations is not taken into account in this study. Because traffic emissions do affect background concentrations, the impacts on the street concentrations would be larger if the impacts on background concentrations were taken into account. Simulations using background concentrations modified accordingly to the scenarios have to be performed in order to fully assess the impacts of the scenarios on the concentrations in streets.

Acknowledgement This work was partially funded by the MAESTRO (Modelling/characterisation of gaseous precursors and secondary aerosols in road transport) project of the ADEME program CORTEA.

References

André, M., Sartelet, K., Moukhtar, S., André, J. M., & Redaelli, M. (2020). Diesel, petrol or electric vehicles: What choices to improve urban air quality in the Ile-de-France region? A simulation platform and case study. *The Journal of Atmospheric Environment*. https://doi.org/10.1016/j.atmosenv.2020.117752

EMEP/EEA. (2019). *EMEP/EEA air pollutant emission inventory guidebook 2019*. EEA report no. 13/2019.

Hanna, S., & Chang, J. (2012). Acceptance criteria for urban dispersion model evaluation. *The Journal of Meteorology and Atmospheric Physics*. https://doi.org/10.1007/s00703-011-0177-1.

Khare, P., Machesky, J., Soto, R., He, M., Presto, A. A., & Gentner, D. R. (2020). Asphalt-related emissions are a major missing nontraditional source of secondary organic aerosol precursors. *The Journal of Science Advances*. https://doi.org/10.1126/sciadv.abb9785.

Kim, Y., Wu, Y., Seigneur, C., & Roustan, Y. (2018). Multi-scale modeling of urban air pollution: Development and application of a street-in-grid model (v1.0) by coupling MUNICH (v1.0) and Polair3D (v1.8.1). *The Journal of Geoscientific Model Development*. https://doi.org/10.5194/gmd-11-611-2018

Lugon, L., Sartelet, K., Kim, Y., Vigneron, J., & Chrétien, O. (2020a). Nonstationary modeling of NO2, NO and NOx in Paris using the Street-in-Grid model: Coupling local and regional scales with a two-way dynamic approach. *The Journal of Atmospheric Chemistry*. https://doi.org/10.5194/acp-20-7717-2020

Lugon, L., Vigneron, J., Debert, C., Chrétien, O., & Sartelet, K. (2020b). Black carbon modelling in urban areas: Investigating the influence of resuspension and non-exhaust emissions in streets using the Street-in-Grid (SinG) model. *The Journal of Geoscientific Model Development*. https://doi.org/10.5194/gmd-2020-386.

Sartelet, K., Couvidat, F., Wang, Z., Flageul, C., & Kim, Y. (2020). SSH-Aerosol v1.1: A modular box model to simulate the evolution of primary and secondary aerosols. *The Journal of Atmosphere-Basel*. https://doi.org/10.3390/atmos11050525

Sartelet, K., Zhu, S., Moukhtar, S., André, M., André, J. M., Gros, V., Favez, O., & BrasseurA and Redaelli M. (2018). Emission of intermediate, semi and low volatile organic compounds from traffic and their impact on secondary organic aerosol concentrations over Greater Paris. *The Journal of Atmospheric Environment*. https://doi.org/10.1016/j.atmosenv.2018.02.031

Questions and Answers

Questioner: Janot Tokaya.

Question: Are the non-vehicle/traffic road emissions, which you showed can be quite high, dependent on road/asphalt type and age of road?

Answer: Emissions of particles from road abrasion due to the vehicles motions are dependent on asphalt type and age of the road. However, methodologies to compute emission inventories, like the COPERT methodology, generally do not represent that level of detail. The quantification of road-abrasion emissions is difficult as road asphalt presents a complex chemical composition and as emissions are mixed with road dust and re-suspended materials. Furthermore, from an operational point of view, it is complicated, and in some cases impossible, to have access to the type of asphalt used and the age of the road.

An Updated Agriculture Emissions Inventory and Contribution Estimation Using WRF-CMAQ Model for Turkey

Ezgi Akyuz, Fulya Cingiroglu, Alper Unal, and Burcak Kaynak

Abstract In this study, NH_3 and NMVOC emissions from agricultural activities were updated and compared with EMEP inventory for a region known for agricultural activities in Turkey. An advanced air quality model, CMAQ, was used to determine the spatio-temporal distribution of NH_3 and NMVOC, which have seasonal patterns. Model was performed for base (EMEP 2018 emissions) and scenario (updated 2019 emissions) cases and 2018 summer months with high-resolution grid size. Moreover, secondary particle formation over the domain was identified. This study is significant due to the agriculture emissions to be updated and re-gridded using high-resolution CORINE land use data and examined of pollutants which are climate dependent. EMEP agriculture emissions were 955.92 kt/annual and 486.60 kt/annual (Turkey totals), and 284.35 kt/annual and 227.54 kt/annual (Turkey emissions in the domain) for NH_3 and NMVOCs, respectively; updated agriculture emissions were found as 579.58 kt/annual and 289.97 kt/annual (Turkey totals), and 195.01 kt/annual and 98.66 kt/annual (Turkey emissions in the domain) for NH_3 and NMVOC. Not only the emissions were different in overall quantities, but also there were spatial differences in concentrations between cases. For NH_3, the largest source of which is agriculture, monthly averaged concentration differences were found up to 60% in some areas of the domain due to the distribution of total emissions over agricultural areas unlike the EMEP spatial distribution methodology. VOC emissions were overestimated in EMEP inventory; thus, a difference had also become in concentrations between two

E. Akyuz (✉) · F. Cingiroglu · A. Unal
Eurasia Institute of Earth Sciences, Istanbul Technical University, Istanbul, Turkey
e-mail: akyuzezg@itu.edu.tr

F. Cingiroglu
e-mail: cingiroglu@itu.edu.tr

A. Unal
e-mail: aunal@itu.edu.tr

B. Kaynak
Department of Environmental Engineering, School of Civil Engineering, Istanbul Technical University, Istanbul, Turkey
e-mail: kaynakbu@itu.edu.tr

C. Mensink and O. Jorba (eds.), *Air Pollution Modeling and its Application XXVIII*, Springer Proceedings in Complexity, https://doi.org/10.1007/978-3-031-12786-1_40

cases. Highest concentrations were found in July for $PM_{2.5}$, $PM_{2.5}(OC)$, $PM_{2.5}(SO_4)$, and $PM_{2.5}(NH_4)$.

1 Introduction

Agricultural activities mainly provide livestock and crops for human population along with the industrial use. In recent centuries, especially in the last century, extended agricultural lands are needed to provide more products to sustain the needs of the growing world population. Agriculture sector, which provides the basic needs of people, is one of the most important anthropogenic sources of air pollution. It is essential to evaluate agricultural emissions, which increase daily due to the expanding population, together with high spatio-temporal resolution in order to determine how much agriculture contributes to air pollution levels.

Ammonia (NH_3) emissions, which the agriculture sector is the most contributor, reached three times of its value in the last 50 years (Hoesly et al., 2018). Non-methane volatile organic compounds (NMVOCs) emissions, which were around 5 Tg in the 1925s, have now exceeded 150 Tg/yr worldwide. The NMVOC contribution of global agriculture emissions is around 5%. Almost 90% of NH_3 emissions are emitted from the agriculture sector in Turkey (EMEP, 2020). Moreover, this sector, which comprises manure management, crop production, and field burning activities, contributes to particulate matter (PM), sulfur dioxide (SO_2), nitrogen oxides (NO_x), and carbon monoxide (CO) emissions (EMEP/EEA, 2019). NH_3 leads to formation of secondary inorganic particles reacting with sulfuric acid (H_2SO_4), nitric acid (HNO_3), hydrochloric acid (HCl) and some other gases (Chen et al., 2014). Along with their primary impact on air pollution, NH_3 and NMVOCs from agriculture lead to the production of secondary particles in the atmosphere, providing positive or negative feedback to climate change as well. Known as an agricultural country, Turkey is among the countries that will be significantly affected by the climate change (Türkeş, 2019). One of the sectors that will be most affected by the climate change is the agriculture sector (Burney & Ramanathan, 2014). Therefore, this study focuses on agricultural emissions, and their contribution to secondary particles. Along with the studies based on emission calculations, improvements and reduction of uncertainties (Chang et al., 2019; Harizanova-Bartos & Stoyanova, 2018; Szuba-Barańska & Mrówczyńska-Kamińska, 2017; Skjøth et al., 2004), modelling studies are also implemented for agricultural activities (Wu et al., 2008, Battye et al., 2019; Li et al., 2020). Furthermore, air pollution and health related studies were also implemented in the US, the number of deaths from agriculture caused NH_3 emissions are 69% of total agriculture pollution (mainly from livestock waste and fertilizer application) and 27% of the deaths are related to primary $PM_{2.5}$ (dust from tillage, livestock dust, field burning, and fuel combustion in agricultural equipment use) (Domingo et al., 2021).

The main objective of this study is the comparison of EMEP and an updated national emissions inventory with high spatial resolution for agriculture sector. In this scope, NH_3, NMVOCs and other agriculture related emissions were calculated using EMEP methodology. Then, the WRF-CMAQ model was used to simulate two cases with EMEP and updated emission inventory. Secondary pollutant formation caused from these agricultural emissions was also compared for these cases.

2 Methodology

EMEP 2018 emissions were used for all sectors in base case (EMEP, 2020). Agricultural emissions were distributed to model grids using area weighted allocation. Thus, agricultural areas were not taken into account. In scenario case, emissions from manure management, crop production, and field burning were recalculated using one of the EMEP methodologies; Tier 1 or Tier 2 (EMEP/EEA, 2019). For scenario case emission calculations, livestock originated NH_3 and NO_x emissions were calculated using Tier 2 methodology (N-flow tool) while NMVOC and PM emissions were calculated using Tier 1 methodology due to lack of detailed activity data. Activity data (number of livestock) were obtained from Turkish Statistical Institute (TUIK, 2019). Emissions from crop production were calculated using Tier 1 methodology. Fertilizer data were obtained from Environmental Province Status Report of each province (MoEU, 2020). Unlike EMEP emission inventory, in this study field burning emissions were added to the updated emissions, and Tier 2 methodology was used for calculation. All emission factors were obtained from EMEP 2019 Guidebook (EMEP/EEA, 2019). All agriculture related emissions estimated for provinces were spatially distributed to model grids by intersecting agricultural areas of CORINE land use data (CORINE, 2018). A speciation profile for VOC and PM speciation was also implemented (Huang et al., 2017). Finally, spatio-temporal distribution of all emissions for all sectors was prepared using Turkey-specific python code as an input to the CMAQ model.

The WRF-CMAQ (v5.3.2 with CB6-AERO7) platform was used to simulate the concentration levels for two emission cases (EPA, 2020). Outer domains' coverages were $36 \times 36\,km^2$ and $12 \times 12\,km^2$. Main study domain covers mainly Mediterranean part of Turkey and the surrounding area with the $4 \times 4\,km^2$ grid size for focusing on agriculturally intense regions. Model simulations were performed for summer months (June, July, and August) of 2018 for two different cases: base case with EMEP emissions and scenario case with updated agricultural emissions.

3 Results

For the main $4 \times 4\,km^2$ gridded domain, EMEP and updated emission inventory were used in two different cases. Manure management (MM) corresponds to K_Livestock sector; the sum of crop production (CP) and field burning (FB) corresponds to

L_AgriOther sector in EMEP. In EMEP emission inventory, CO, SO_x, and NO_x emissions did not exist for agriculture sector. For updated emission inventory, calculated CO and NO_x emissions were in small quantities; however, NO_x emissions were found 89.30 and 31.97 kton/annual for total Turkey and Turkey emissions in the domain, respectively (Table 1). PM_{10} emissions in EMEP and updated inventories were very similar, while $PM_{2.5}$ emissions were calculated lower than EMEP inventory. EMEP NMVOC emissions were more than two times compared to the updated NMVOC emissions for Turkey in the domain. NH_3 emissions were calculated as 955.92 and 579.58 kton/annual for EMEP and updated inventories, respectively (Table 1).

While EMEP emissions are spread over wider areas, updated emissions are distributed over whole agricultural areas using CORINE high resolution land use data, resulting in regional emission differences and will subsequently lead to differences in the concentration distribution (Fig. 1). EMEP emissions were explicitly higher in the southeastern part of the domain (Syria) and it is pointing out a probable issue in emission estimation for the region.

Table 1 EMEP and updated emissions for Turkey (kt/year), first part of the table indicates the total emissions for Turkey while the second part shows the total emissions for the domain

	CO	NO_X	NH_3	$PM_{2.5}$	PM_{10}	NMVOC	SO_X
K_Livestock	0.00	0.00	778.03	6.13	27.57	236.65	0.00
L_AgriOther*	0.00	0.00	177.89	32.02	37.36	249.95	0.00
EMEP TR (K + L)	***0.00***	***0.00***	***955.92***	***38.15***	***64.94***	***486.60***	***0.00***
EMEP TR National Totals	***1619.71***	***923.60***	***996.82***	***384.03***	***553.34***	***1078.14***	***2527.70***
MM	0.00	5.41	261.44	7.14	21.07	269.95	0.00
CP + FB	2.67	83.89	318.14	1.53	36.29	20.02	0.01
Updated TR (MM + CP + FB)	***2.67***	***89.30***	***579.58***	***8.67***	***57.36***	***289.97***	***0.01***
Updated TR National Totals	***1622.38***	***1012.90***	***620.47***	***354.55***	***545.76***	***881.52***	***2527.72***
K_Livestock	0.00	0.00	202.28	1.40	6.29	112.23	0.00
L_AgriOther*	0.00	0.00	82.07	14.77	17.24	115.31	0.00
EMEP TR (K + L)	***0.00***	***0.00***	***284.35***	***16.17***	***23.53***	***227.54***	***0.00***
EMEP TR National Totals	***574.26***	***355.65***	***295.78***	***152.15***	***250.40***	***413.64***	***1249.10***
MM	0.00	1.81	84.97	2.38	7.03	88.92	0.00
CP + FB	1.22	30.16	110.05	0.74	17.64	9.74	0.01
Updated TR (MM + CP + FB)	***1.22***	***31.97***	***195.01***	***3.12***	***24.67***	***98.66***	***0.01***
Updated TR National Totals	***578.27***	***387.92***	***208.18***	***139.57***	***252.06***	***285.20***	***1249.14***

* Field burning calculations were not performed because it is illegal to burn agricultural waste

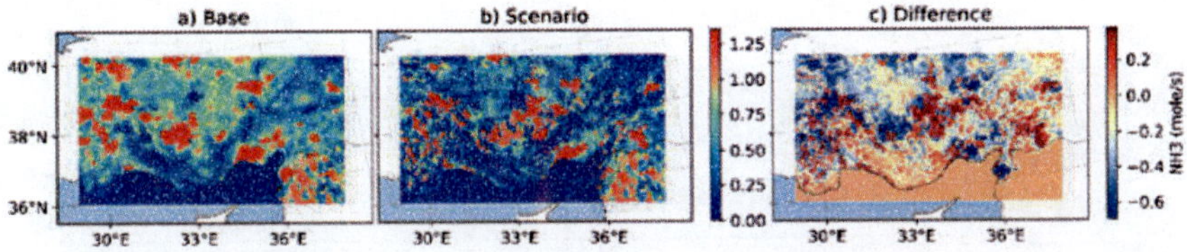

Fig. 1 Daily emission difference of NH_3 for base and scenario cases

Because the fate and transport of air pollutants are significantly affected by meteorological parameters, such as solar radiation (SOL_RAD), relative humidity (RH), and precipitation (P), monthly averages of these parameters were investigated for selected summer months. Solar radiation is prominently higher in July compared to the other two months. Average relative humidity is the highest in August; however, July and August are significantly higher than June. The average rainfall is zero indicating no precipitation in July and August. Simulation results for July were selected for the detailed investigation of the selected pollutants due to highest solar radiation, relatively high relative humidity and lower precipitation.

Spatial distributions of some agricultural originated and other formed pollutants point out differences between the base and scenario cases (Fig. 2). In general, similar concentrations were estimated for July and August while June concentrations were estimated lower for both cases except NH_3. Higher RH and lowest P may cause the lower concentrations in these two months. In June, the month with the lowest humidity and the highest precipitation, NH_3 has the highest concentrations among the summer months. The lowest concentrations were found in August, the month with the highest humidity and the lowest precipitation. Although the NH_3 concentration distributions were similar regions in both cases, the pollution spreads to much farther and wider areas from the agricultural areas due to the different distribution of emissions in the base case.

Concentration differences were found up to 60% between the cases. High concentrations were estimated in the base case in the provinces of the southeastern region (Adana) of the domain. The maximum difference for NMVOCs was 20%, and a higher and spatially extended concentration estimate was made in the base case. Total $PM_{2.5}$ concentrations were the highest in July and the lowest in June and base case was indicating similar concentrations along with a slight difference in the southern region (Adana) of the domain.

$PM_{2.5}(NH_4)$ was found lowest in June and highest in July for two cases. $PM_{2.5}(NH_4)$ concentrations were generally higher across the domain in the base case, and the difference was estimated to be a maximum of 10%. There were domain-wide differences for $PM_{2.5}(OC)$. Significantly, the base case was higher than the scenario case across the domain. $PM_{2.5}(OC)$ and $PM_{2.5}(SO_4)$ concentrations were estimated as the highest in July and the lowest in June for two cases. Base case had higher concentrations than scenario case for both pollutants.

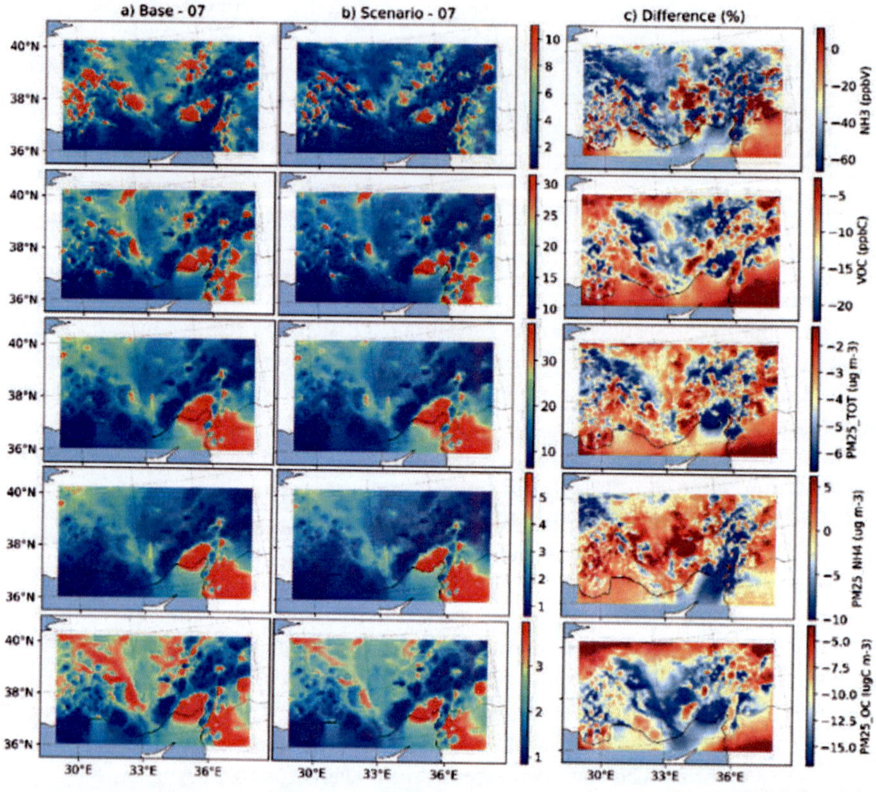

Fig. 2 Monthly average NH_3, VOC, Total $PM_{2.5}$, $PM_{2.5}$ (NH_4), and $PM_{2.5}$ (OC) concentrations for base, scenario and difference (scenario-base) for July 2018

4 Conclusions

In addition to the possible uncertainties of the model simulations, the uncertainties in the emission inventories are one of the most important sources that cause differences in model results. In this study, a comprehensive emission inventory was prepared for Turkey, and simulations using these updated emissions were performed for the Mediterranean Region and its surrounding area, a region where agricultural practices are intensified. Some of the emissions were directly estimated for county level and also high-resolution land use data was used to spatially distribute these emissions to the model grids. Spatial and quantitative differences were observed between the updated inventory and EMEP emissions inventory.

Another purpose of the study is to investigate the spatio-temporal distribution of NH_3 and NMVOC concentrations and contribution of NH_3 and NMVOCs to secondary inorganic and organic aerosols. EMEP inventory overestimates NMVOC emissions. There are no ground observations for NH_3, but NH_3 monitoring with satellites can be used for this region, and model results can be compared with these observations and correlations can be estimated for agricultural areas for these two cases. As NH_3 and NMVOC are pollutants dependent on meteorological parameters such as temperature and precipitation, they are also among the pollutants that will be most affected by climate change. For this reason, the accurate estimation of these emissions and their spatial distribution to the relevant regions are very important for improving model performance.

References

Battye, W. H., Bray, C. D., Aneja, V. P., Tong, D., Lee, P., & Tang, Y. (2019). Evaluating ammonia (NH_3) predictions in the NOAA NAQFC for Eastern North Carolina using ground level and satellite measurements. *Journal of Geophysical Research: Atmospheres, 124*(14), 8242–8259.

Burney, J., & Ramanathan, V. (2014). Recent climate and air pollution impacts on Indian agriculture. *Proceedings of the National Academy of Sciences of the United States of America, 111*(46), 16319–16324.

Chang, Y., Zou, Z., Zhang, Y., Deng, C., Hu, J., Shi, Z., Dore, A. J., & Collett, J. L. (2019). Assessing contributions of agricultural and nonagricultural emissions to atmospheric ammonia in a Chinese megacity. *Environmental Science and Technology.*

Chen, X., Day, D., Schichtel, B., Malm, W., Matzoll, A. K., Mojica, J., McDade, C. E., Hardison, E. D., Hardison, D. L., Walters, S., Van De Water, M., & Collett, J. L. (2014). Seasonal ambient ammonia and ammonium concentrations in a pilot IMPROVE NHx monitoring network in the western United States. *Atmospheric Environment, 91*, 118–126.

CORINE. (2018). *CLC CORINE land use 2018 data.* https://land.copernicus.eu/pan-european/corine-land-cover/clc2018

Domingo, N. G. G., Balasubramanian, S., Thakrar, S. K., Clark, M. A., Adams, P. J., Marshall, J. D., Muller, N. Z., Pandis, S. N., Polasky, S., Robinson, A. L., Tessum, C. W., Tilman, D., Tschofen, P., & Hill, J. D. (2021). Air quality-related health damages of food. *Proceedings of the National Academy of Sciences of the United States of America, 118*(20).

EMEP/EEA. (2019). *Air pollutant emission inventory guidebook 2019. Technical guidance to prepare national emission inventories.* https://www.eea.europa.eu/publications/emep-eea-guidebook-2019/

EMEP. (2020). *Officially reported emission data.* https://www.ceip.at/data-viewer

EPA. (2020). *CMAQ (Version 5.3.2) [Software].*

Harizanova-Bartos, H., & Stoyanova, Z. (2018). Impact of agriculture on air pollution. *CBU International Conference Proceedings, 6*, 1071–1076.

Hoesly, R. M., Smith, S. J., Feng, L., Klimont, Z., Janssens-Maenhout, G., Pitkanen, T., Seibert, J. J., Vu, L., Andres, R. J., Bolt, R. M., Bond, T. C., Dawidowski, L., Kholod, N., Kurokawa, J. I., Li, M., Liu, L., Lu, Z., Moura, M. C. P., O'Rourke, P. R., & Zhang, Q. (2018). Historical (1750–2014) anthropogenic emissions of reactive gases and aerosols from the Community Emissions Data System (CEDS). *Geoscientific Model Development, 11*(1), 369–408.

Huang, G., Brook, R., Crippa, M., Janssens-Maenhout, G., Schieberle, C., Dore, C., Guizzardi, D., Muntean, M., Schaaf, E., & Friedrich, R. (2017). Speciation of anthropogenic emissions of non-methane volatile organic compounds: A global gridded data set for 1970–2012. *Atmospheric Chemistry and Physics, 17*(12), 7683–7701.

Li, L., Wang, K., Chen, W., Zhao, Q., Liu, L., Liu, W., Liu, Y., Jiang, J., Liu, J., & Zhang, M. (2020). Atmospheric pollution of agriculture-oriented cities in Northeast China: A case in Suihua. *Journal of Environmental Sciences (china), 97*, 85–95.

MoEU. (2020). *Environmental province status reports*. https://ced.csb.gov.tr/il-cevre-durum-raporl ari-i-82671
Skjøth, C. A., Hertel, O., Gyldenkærne, S., & Ellermann, T. (2004). Implementing a dynamical ammonia emission parameterization in the large-scale air pollution model ACDEP *109*, 1–13.
Szuba-Barańska, E., & Mrówczyńska-Kamińska, A. (2017). An attempt to assess the impact of agriculture on the environment in the Countries of Central-Eastern Europe. *Journal of Agribusiness and Rural Development, 10*(3).
TUIK. (2019). *Livestock statistics Turkish Statistical Institute, Ankara*. http://www.turkstat.gov.tr/
Türkeş, M. (2019). *Climate change training module series 1: Scientific basis of climate change and impacts on Turkey*.
Wu, S., Krishnan, S., Zhang, Y., & Aneja, V. (2008). *Modeling atmospheric transport and fate of ammonia in North Carolina—Part I : Evaluation of meteorological and chemical predictions, 42*, 3419–3436.

Evaluation of Satellite Vegetation Indices for BVOCs Emission Modelling. Case Study: Basque Country

Iñaki Zuazo, Eduardo Torre-Pascual, and Jose Antonio García

Abstract Vegetation Indices generated from satellite observations are an alternative for assessing vegetation condition, which is necessary for modelling BVOCs emissions. In this work, we compared following vegetation indices: LAI, FPAR, NDVI and EVI obtained from the MODIS sensor of the US Geological Survey Database. Twelve locations with a single species and 100% density were selected for the most typical tree species of the Basque Country: Fagus sylvatica and Quercus faginea as deciduous trees, and Pinus radiata and Quercus ilex as evergreen trees. The results indicate that the stability and data quality of the LAI V6 product reprocessed by Sun Yat-sun University make it a valid alternative for BVOCs emission modelling.

Keywords Vegetation indices · Remote sensing · BVOCs modelling · NDVI · LAI · FPAR · EVI

1 Introduction

Biogenic Volatile Organic Compounds (BVOCs) are well-known precursors of ozone and secondary organic aerosols which affect the gas phase and heterogeneous chemistry of the troposphere, being isoprene and monoterpenes the most abundant species (Atkinson & Arey, 2003). The estimation of their emissions is crucial for air quality modelling, and specifically, for chemical transport models (Curci et al., 2009). Biogenic emissions are often estimated based on the (Guenther, 1997) approach, which considers the type of vegetation, the leaf density, a short-term activity factor

I. Zuazo (✉) · E. Torre-Pascual · J. A. García
Faculty of Engineering Bilbao, University of the Basque Country (UPV/EHU), Plaza Ingeniero Torres Quevedo 1, 48013 Bilbao, Spain
e-mail: joseignacio.zuazo@ehu.eus

E. Torre-Pascual
e-mail: eduardo.delatorre@ehu.eus

J. A. García
e-mail: joseantonio.garciaf@ehu.eus

C. Mensink and O. Jorba (eds.), *Air Pollution Modeling and its Application XXVIII*, Springer Proceedings in Complexity, https://doi.org/10.1007/978-3-031-12786-1_41

that depends on temperature and radiation, and a long-term activity factor that depends on the seasonality and the greenness of the vegetation.

An alternative for the determination of the long-term activity factor are satellite observations. The use of vegetation indices, generated with satellite sensors, provide a set of biophysical variables, describing the state, the dynamism and the disturbances of the terrestrial vegetation (Huete et al., 2011). In this study, we have analysed the indices that provide a vegetation greenness optical measure, a property that describes the leaf chlorophyll content, the leaf area, the canopy cover, and the canopy structure. Most of BVOCs emission models implement the Leaf Area Index (LAI) as an input for the calculation of the long-term activity factor, that is the case for the Model of Emissions of Gases and Aerosols from Nature (MEGAN) (Guenther et al., 2012), while other authors use other indices, such as (Oderbolz et al., 2013) which implements the Normalised Difference Vegetation Index (NDVI).

In this work, we compared the most common vegetation indices used in the literature: LAI, Fraction of Photosynthetically Active Radiation (FPAR), NDVI and Enhanced Vegetation Index (EVI) in the Basque Country, a highly forest vegetated mid-latitude region in the north of Spain, and we analysed their seasonal behaviour.

2 Methodology

The study was carried out in the Basque Country, a Spanish Autonomous Community in the north of the Iberian Peninsula, located between the Cantabrian Range and the Pyrenees. The Forest Map of the Basque Country of 2010 (https://opendata.euskadi.eus/catalogo/-/mapa-forestal-del-pais-vasco/) together with orthophotos of 2009 (https://www.geo.euskadi.eus/ortofotografia-de-25cm-de-la-comunidad-autonoma-del-pais-vasco-ano-2009/s69-geodir/es/) were used for this analysis. We selected a total of 12 forest locations based on the fact that only one species is present in each of them, with a 100% tree density, and that the stand condition should be pole stage or timber stage. These conditions minimize possible misinterpretations due to a mixture of species and low tree density areas. The selected areas contain four of the main tree species of the region: Fagus sylvatica (beech) and Quercus faginea (oak), which are deciduous trees; and Pinus radiata (pine) and Quercus ilex (holm oak), which are evergreen trees (Fig. 1).

We downloaded the LAI, FPAR, NDVI and EVI indices from the U.S. Geological Survey website (https://earthexplorer.usgs.gov/) for each location shown in Fig. 1 for 2009. The LAI and FPAR indices were obtained from the MCD15A2 product of the MODIS platform, a composite image for every 8 days, with a horizontal resolution of 1 km, obtained by combining images from the Terra and Aqua satellites. In addition to MODIS LAI, we also downloaded the MODIS LAI Version 6 product MCD15A2H (LAI V6), developed by the "Land–Atmosphere Interaction Research Group at Sun Yat-sen University", which increases data quality and results with images every 8 days with a horizontal resolution of 500 m (Yuan et al., 2011, 2020). The NDVI and EVI images were obtained from the MYD13Q1 product of the MODIS-Aqua

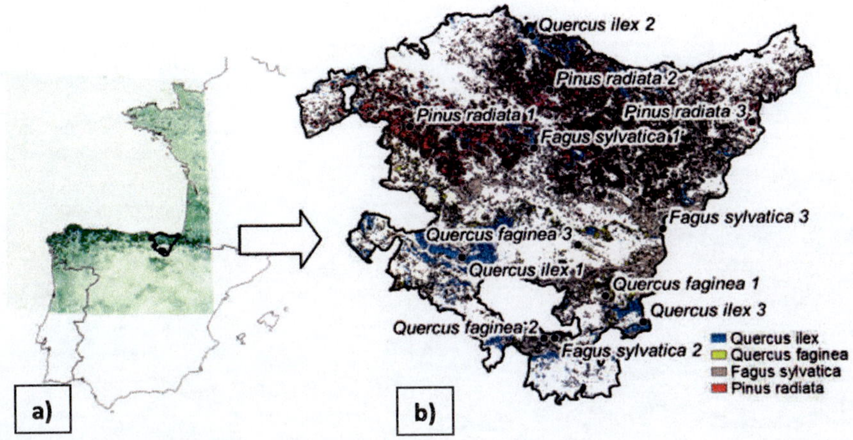

Fig. 1 **a** Location of the Basque Country in the north of Spain with example image of Vegetation Index in green. **b** Location of the 12 study areas, three for each tree species

platform, which corresponds to a composite image every 16 days, with a horizontal resolution of 250 m. All the imagery was processed with the QGIS software.

3 Results

The LAI and FPAR values obtained from the MODIS product MCD15A2 show high variability and do not represent adequately the annual cycle of the selected species. Values of these indices for Fagus Sylvatica and Pinus Radiata are shown in Fig. 2 as an example. The values plotted do not distinguish the different phases of the year for a deciduous species (leafy and leafless). Furthermore, the variability of the images for consecutive dates is very high when they should be similar. For these reasons, their use for the calculation of biogenic emissions is discouraged and they have been discarded in this study.

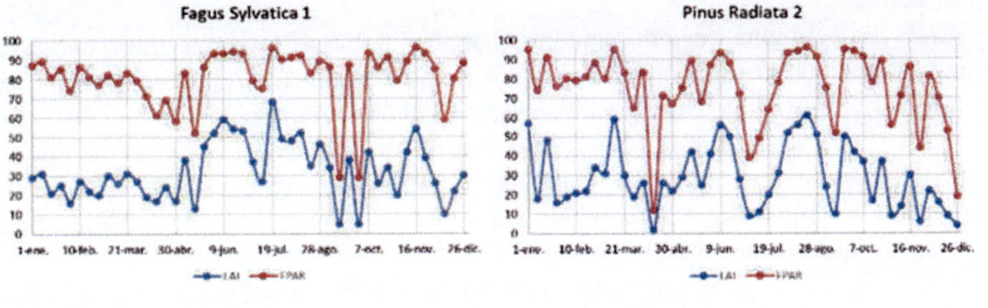

Fig. 2 FPAR and LAI indices in Fagus Sylvatica 1 and Pinus Radiata 2

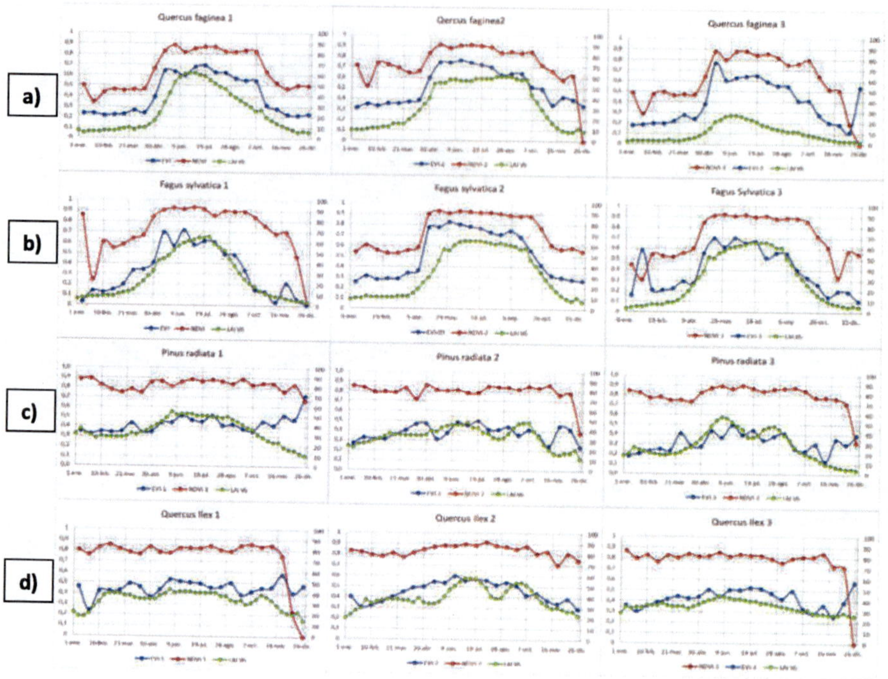

Fig. 3 Results of vegetation indices EVI, NDVI, LAI V6 reprocessed for **a** Quercus faginea, **b** Fagus sylvatica, **c** Pinus radiata and **d** Quercus ilex

In Fig. 3, the evolution of the reprocessed NDVI, EVI and LAI V6 indices values for the 12 selected locations is shown. The NDVI and EVI values are represented on a scale of 0–1, and the LAI V6 values on a scale of 0–100. From the results obtained, the following discussion can be made for each of the indices:

NDVI. In deciduous trees this index is between 0.5–0.6 in winter and 0.8–0.9 in summer. The presence and absence of leaves due to the seasonality of this type of trees is clearly visible. In evergreen trees, the index is between 0.8–0.9 throughout the year showing no seasonality. Anomalous values, mainly in January and December, are due to a lack of quality in the images, which can infer higher cloudiness during these dates.

EVI. In deciduous trees this index is between 0.2–0.4 in winter and 0.6–0.8 in summer showing seasonal differences as well. In evergreen trees, the index shows an annual variability, with ranging values from 0.2–0.3 in winter to a maximum of 0.5–0.6 in summer. This variability indicates that this index detects seasonal variations also in evergreen trees. As in the NDVI index, outliers are detected in the EVI index, especially in January and December. This is to be expected, as both indices are derived from the same satellite product.

LAI V6. The quality and stability of this index for the selected locations are remarkable, no outliers are detected, and the curves obtained have hardly any noise.

In the case of Quercus Faginea 3, index values are not coherent with respect to the values obtained in EVI and NDVI. The vegetation of the site could have not been correctly detected. In deciduous trees, values of 0–10 in winter and 60–70 in summer are shown, showing once again the seasonality of this type of vegetation. Evergreen trees show values between 10–30 in winter and 40–60 in summer, indicating that the reprocessed LAI is able to distinguish seasonal variations in evergreen trees.

In summary, the NDVI index detects adequately the presence of tree leaves. The EVI and reprocessed LAI V6 indices, in addition to detecting the presence of leaves, are able to detect the variation of enzymatic activity that is responsible for BVOCs emissions. This behaviour is in line with the research of (Zhang et al., 2005) concluding that NDVI better represents the total canopy, while EVI is more associated with the presence of the chlorophyll used in photosynthesis. The LAI V6 product presents stable results and can detect both canopy variations characteristic of deciduous trees and variations at the enzymatic level of evergreen trees, so it is presented as an alternative to be considered when used as a long-term activity factor in BVOCs emission models.

4 Conclusions

Satellite observation can be used to assess the long-term activity factor of vegetation in BVOCs emission modelling. The selection of a suitable index that adequately measures the phenology and seasonality of vegetation was found to be relevant. MODIS NDVI index adequately characterises the total canopy, while the MODIS EVI and LAI V6 indices are also able to detect the presence of the active chlorophyll responsible for photosynthesis. The stability and data quality of the LAI V6 product make it a valid alternative for BVOCs emission modelling.

References

Atkinson, R., & Arey, J. (2003). Gas-phase tropospheric chemistry of biogenic volatile organic compounds: A review. *Atmospheric Environment*, *37*(Suppl. 2). https://doi.org/10.1016/S1352-2310(03)00391-1

Curci, G., Beekmann, M., Vautard, R., Smiatek, G., Steinbrecher, R., Theloke, J., & Friedrich, R. (2009). Modelling study of the impact of isoprene and terpene biogenic emissions on European ozone levels. *Atmospheric Environment*, *43*(7). https://doi.org/10.1016/j.atmosenv.2008.02.070

Guenther, A. (1997). Seasonal and spatial variations in natural volatile organic compound emissions. *Ecological Applications*, *7*(1). https://doi.org/10.2307/2269405

Guenther, A. B., Jiang, X., Heald, C. L., Sakulyanontvittaya, T., Duhl, T., Emmons, L. K., & Wang, X. (2012). The model of emissions of gases and aerosols from nature version 2.1 (MEGAN2.1): An extended and updated framework for modeling biogenic emissions. *Geoscientific Model Development*, *5*(6). https://doi.org/10.5194/gmd-5-1471-2012

Huete, A. R., Solano-Barajas, R., Glenn, E. P., & Restrepo-Coupe, N. (2011). Monitoreo de propiedades y procesos ecosistémicos con índices de vegetación MODIS. In *Aplicaciones del sensor MODIS para el monitoreo del territorio* (pp. 195–230). Secretaría del Medio Ambiente y Recursos Naturales. https://www.ciga.unam.mx/publicaciones/images/abook_file/aplicaciones MODIS.pdf

Oderbolz, D. C., Aksoyoglu, S., Keller, J., Barmpadimos, I., Steinbrecher, R., Skjøth, C. A., Plaß-Dülmer, C., & Prévôt, A. S. H. (2013). A comprehensive emission inventory of biogenic volatile organic compounds in Europe: Improved seasonality and land-cover. *Atmospheric Chemistry and Physics, 13*(4). https://doi.org/10.5194/acp-13-1689-2013

Yuan, H., Dai, Y., Xiao, Z., Ji, D., & Shangguan, W. (2011). Reprocessing the MODIS Leaf Area Index products for land surface and climate modelling. *Remote Sensing of Environment, 115*(5), 1171–1187. https://doi.org/10.1016/j.rse.2011.01.001

Yuan, H., Dai, Y., & Li, S. (2020). *Reprocessed MODIS Version 6 Leaf Area Index data sets for land surface and climate modelling.* Sun Yat-sun University. http://globalchange.bnu.edu.cn/res earch/laiv6

Zhang, Q., Xiao, X., Braswell, B., Linder, E., Baret, F., & Moore, B. (2005). Estimating light absorption by chlorophyll, leaf and canopy in a deciduous broadleaf forest using MODIS data and a radiative transfer model. *Remote Sensing of Environment, 99*(3), 357–371. https://doi.org/10.1016/J.RSE.2005.09.009

Impact of Facility-Specific Temporal Profiles for Public Power Sector on WRF-CMAQ Simulations

Fulya Cingiroglu, Ezgi Akyuz, Alper Unal, and Burcak Kaynak

Abstract The energy demand is increasing day by day, although the installation and operation of coal-fired power plants slowly fade out throughout the world, it is still increasing in Turkey. Air pollutant emissions from public power sector, mainly composed of power plants contribute significantly to Turkey national totals (70.4% SO_2, 38.9% NO_x, 9.8% $PM_{2.5}$). Along with the emission estimation, the temporal and spatial distribution of these emissions are also crucial for accurate simulations with low uncertainty in air quality modeling. Sector-specific profiles are widely used as temporal profiles to input hourly emissions for modeling. In this study, a new temporal profile was established using hourly electricity generation of a large-capacity coal-fired power plant: Afsin Elbistan Power Plant (AEPP). The WRF/CMAQ model was run for two cases; one with sector-specific temporal profile (base case) and another with facility-specific temporal profile (scenario case). EMEP 2018 emissions were used and model was run for 2018 summer months (JJA) with high-resolution 4 $\times$ 4 km^2 grid size. Temporal profile comparisons for the summer period indicated maximum differences between 00.00 and 07.00 h with higher ratios for scenario case. Day of the week comparison showed consistently high differences on the weekends, especially on Sundays (average of +40.4% for JJA) with base case always allocating lower emissions on Saturdays and Sundays than scenario case. The largest daily differences were observed between 14 and 22 July where the highest was on 22 July (+91%) which were again weekend days. Hourly concentrations over AEPP showed the highest differences in July with scenario case being usually higher. The

F. Cingiroglu (✉) · E. Akyuz · A. Unal
Eurasia Institute of Earth Sciences, Istanbul Technical University, Istanbul, Turkey
e-mail: cingiroglu@itu.edu.tr

E. Akyuz
e-mail: akyuzezg@itu.edu.tr

A. Unal
e-mail: aunal@itu.edu.tr

B. Kaynak
Department of Environmental Engineering, School of Civil Engineering, Istanbul Technical University, Istanbul, Turkey
e-mail: kaynakbu@itu.edu.tr

C. Mensink and O. Jorba (eds.), *Air Pollution Modeling and its Application XXVIII*, Springer Proceedings in Complexity, https://doi.org/10.1007/978-3-031-12786-1_42

peak concentrations and concentration differences usually occur around midnight (00.00). The days with the highest hourly differences were on 18 July (1614 ppbV for SO_2, 26 ppbV for NO_2, and 9 ppbV for VOC) for Layer 7, which is around midnight.

Keywords Temporal profile · Power plant · WRF-CMAQ

1 Introduction

Releasing of air pollutants into the atmosphere is detrimental to human health and also harms the environment. Fossil fuel power plants, especially coal-fired power plants (CPPs) whose environmental pollution burden is high, are one of the largest sources of air pollutant and greenhouse gas emissions. Thus, continuous emission monitoring has been established for some power plants worldwide such as EPA Continuous Emission Monitoring Systems (CEMS) applied in the US for accurate emission estimation.

Emissions should be estimated as accurately as possible in order to identify the pollution caused by CPPs with the most accurate spatio-temporal distribution for modeling studies. Temporal profiles such as EDGAR/TNO, CAMS-TEMPO, EMEP are readily available for some countries (Guevara et al., 2021). Situations such as temporary shutdowns at CPPs due to technical or legislative issues have not been considered in these sector-specific profiles resulting in non-representative emissions distributions. Considering the large amount of emissions of CPPs, this issue may have significant uncertainties in model simulations. In addition to these sector profiles, newly developed temporal profile studies have been conducted for CPPs using region-specific real-time data. CO_2 and SO_2 emission profiles were created using monthly electricity generation profiles for PPs in China for 1990–2010 (Liu et al., 2015). Pham et al. (2008) created monthly, daily, and hourly temporal profiles and updated their emission allocation using real production data.

Air pollutant emissions from public power sector, mainly composed of power plants contribute significantly more to Turkey national totals (70.4% SO_2, 38.9% NO_x, 9.8% $PM_{2.5}$) compared to total EMEP region (51.0% SO_2, 20.0% NO_x, 6.0% $PM_{2.5}$) (EMEP, 2020). PP originated air pollution is still a problem for Turkey and an important issue that causes environmental and health concerns. The electricity productions of the facilities have diurnal, seasonal, and annual variations, and thus changing annual productions. Changes in electricity demand during the day can cause peaks in energy demand and pollution levels. In addition, some facilities in Turkey were closed in some periods due to technical deficiencies. For these reasons, a representative temporal emission distribution is critical. High-resolution temporal profiles are required in estimation of the emissions and pollution impacts of such large point sources.

This study aims to investigate the impacts of the different temporal profiles on the pollutant concentrations along with meteorological parameters. Real time electricity generation data was used to establish a facility-specific temporal profile for a large capacity coal-fired power plant in Turkey: Afsin Elbistan Power Plant (AEPP). A recent study showed very high SO_2 and NO_2 signals over AEPP with TROPOMI satellite retrievals (Cingiroglu et al., 2021). Thus, this study focuses on the impact of the differences in the temporal profiles on air pollutant concentrations over AEPP for two cases for the summer of 2018.

2 Methodology

Annual emissions from EMEP 2018 inventory were used (EMEP, 2020). Hourly fractions used in facility-specific temporal profile were calculated using hourly electricity productions divided by annual electricity production. Real-time hourly electricity productions of AEPP in Turkey were used for 2018. AEPP which is the highest capacity lignite-fired power plant with 1440 MW in Turkey, produced a total of 5680 GWh in 2018. With its current installed capacity, AEPP has 2% of Turkey's electricity production.

To investigate temporal variability impact of emissions for public power sector, temporal profiles were compared. First one is sector-specific temporal profile that was prepared using EDGAR temporal profile (Crippa et al., 2020). Second one is facility-specific temporal profile that was calculated according to hourly electricity production of AEPP (Fig. 1).

WRF (v4.1.1)-CMAQ (v5.3.2 with CB6-AERO7 mechanism) model simulations were run with grid size 36 × 36 km^2 for Europe (D01), 12 × 12 km^2 for Turkey (D02), and lastly 4 × 4 km^2 for central Anatolia (D03). Simulations over D01 and D02 were performed for boundary conditions for D03 to account for the pollutant

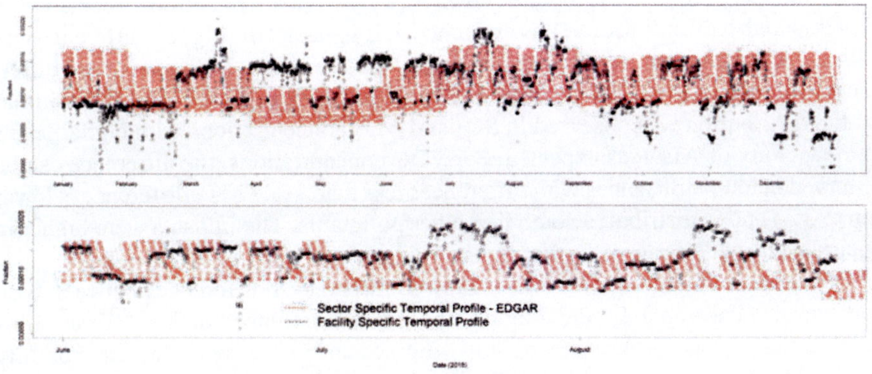

Fig. 1 Comparison of sector-specific and facility-specific temporal profiles for AEPP for the year 2018 (top) and simulation period (June–August 2018) (bottom)

transport from outside the study region. Two simulations were performed for D03: base case and scenario case with changed temporal profile for AEPP for June–August 2018.

3 Results and Discussion

WRF-CMAQ simulations were performed for two cases: base case with standard temporal profile (EDGAR profile) which does not vary with respect to facility, and scenario case with facility-specific temporal profile derived from hourly electricity production. The episode was selected as the summer of 2018 (JJA) in order to eliminate the impact of residential heating.

Day-to-day electricity production variability may not be represented properly in EDGAR temporal profile for older power plants in Turkey because of shutdowns and/or repairments of the facilities. A good example for that is AEPP, which is one of the largest facilities in Turkey, where the electricity production significantly differs from the standard temporal profile (Fig. 1). When the time series and differences for the 3-month summer period were examined, the maximum differences were between 00.00 and 07.00 h (with an hourly maximum difference of 64% in July). The scenario case has higher hourly fractions than the base case. Between 09.00 and 13.00 h, the differences were minimum. Day of the week comparison showed the highest differences on weekends, especially on Sundays (average of +40.4% for JJA and 54% for July). The base case with standard profile always has less emissions on Saturdays and Sundays than scenario case. The largest daily differences were observed between 14 and 22 July such as 14–15 July (71–87%) and 21–22 July (82–91%) which are again on weekends.

The base, scenario and differences for monthly average ground-level SO_2, NO_2, VOC, and $PM_{2.5}$ concentrations are shown for July from top to bottom order (Fig. 2). When spatial distribution of SO_2 is examined, it is seen that the highest difference was with a maximum over 10% mainly transported to the south. Maximum NO_2 differences where the emissions are dominantly released were higher than 5%. Similar spatial distribution was observed in SO_2 and NO_2 concentrations with higher levels in the vicinity of AEPP as expected. For VOC concentrations, the difference can be observed in longer distances with differences less than 1%. $PM_{2.5}$ difference is lower (up to 3%) and is distributed south like other pollutants. The impacts were higher in upper levels, which can be further transported.

Hourly concentrations over AEPP showed the highest differences between base and scenario case on July and August, and lower differences in June. Time series showed the highest differences in July with scenario case being higher. For July 17–23, the concentrations at Layer 7 where emissions were dominantly introduced increased with peaks in both cases and the differences were maximum on July 23 (Fig. 3). The peaks and concentration differences usually occur around midnight (00.00). The upper-level concentrations are usually higher in the scenario with large differences, however opposite is observed on incidences such as on July 25. The

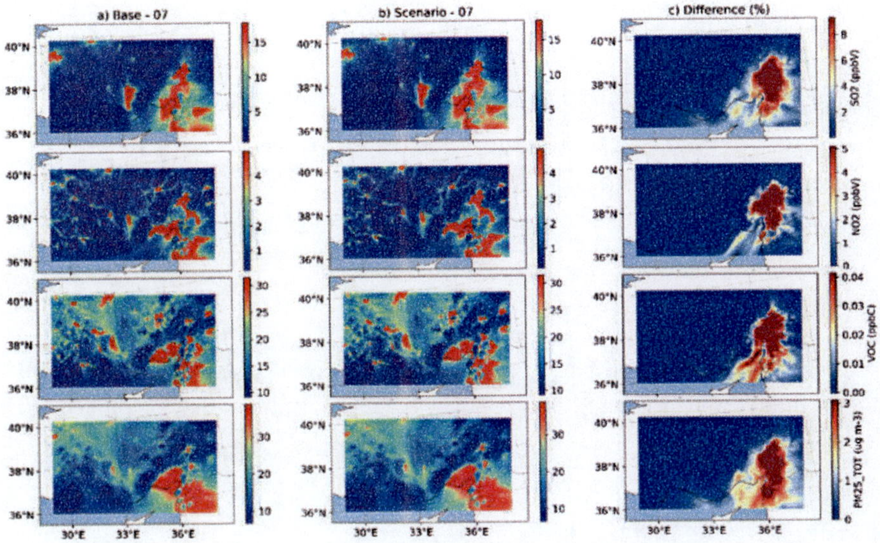

Fig. 2 Monthly averaged SO_2, NO_2, VOC, and $PM_{2.5}$ concentrations for July 2018 (from top to bottom) for base (left), scenario (middle) case, and differences (scenario—base, %)

differences were highest in SO_2 with similar profile in VOC. The temporal profile of NO_2 was showing lower peaks and extreme differences. The maximum hourly differences are 1614 ppbV (18 July) for SO_2, 26 ppbV (18 July) for NO_2 and 9 ppbV (18 July) for VOC, all of which are around midnight with scenario being higher.

4 Conclusions

The impact of temporal profile change for AEPP on model simulations was investigated. Since this study focused on only one PP, the impact on monthly average concentrations can be seen small; however, hourly and daily estimations showed significant differences in the vicinity of AEPP. Updating all PPs with this updated facility-specific temporal profile would have significant impact considering the differences between sector-specific profile summarized in this study.

The evaluation of these improvements in temporal profiles cannot be fully assessed by ground measurements considering the elevated nature of these emissions and meteorological effects on these plumes. However, incorporating satellite retrievals in air quality modeling practices will help in simulation performances. In addition to temporal profiling, an average vertical distribution profile was used for emission allocation (Bieser et al., 2011). The vertical allocation of these emissions would use real stack height information which is facility-specific. This change would result in

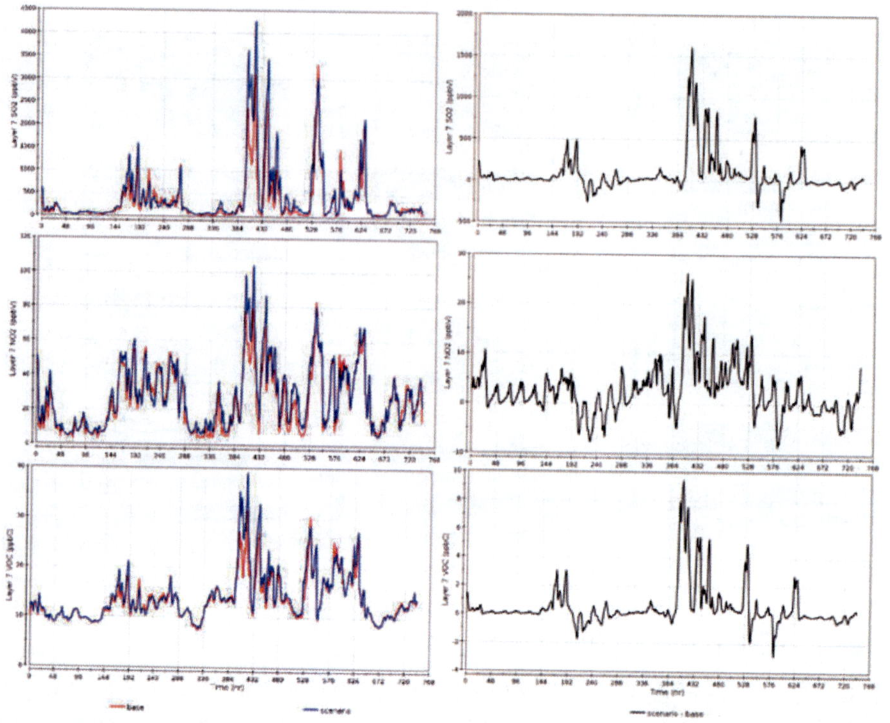

Fig. 3 Timeseries of hourly concentrations over AEPP for base and scenario case (left) and the differences (right): layer 7 SO_2 (top), NO_2 (middle) and VOC (bottom) concentrations for July

allocation of these emissions into one layer, which amplifies the errors in temporal profiling.

When compared with EMEP emissions, the location of AEPP is found slightly off the real location. This problem can be solved by introducing large point sources separately according to exact locations than other emissions. Spatial allocation of these large point sources in grids overmixes these emissions. A better representation can be achieved by introducing these emissions to CMAQ with plume-in-grid treatment.

References

Bieser, J., Aulinger, A., Matthias, V., Quante, M., & Denier Van Der Gon, H. A. C. (2011). Vertical emission profiles for Europe based on plume rise calculations. *Environmental Pollution, 159*(10), 2935–2946.

Cingiroglu, F., Deger, S. S., Ceker, A. O., Akyuz, E., & Kaynak, B. (2021). TROPOMI NO_2 and SO_2 Temporal changes over large point sources and their relationship with electricity production. In

17th International Conference on Environmental Science and Technology (CEST2021), Athens-GREECE, 1–4 September 2021.

Crippa, M., Solazzo, E., Huang, G., Guizzardi, D., Koffi, E., Muntean, M., Schieberle, C., Friedrich, R., & Janssens-Maenhout, G. (2020). High resolution temporal profiles in the emissions database for global atmospheric research. *Scientific Data, 7*(1), 1–17.

EMEP. (2020). *Gridded EMEP 2018 emissions.* https://www.ceip.at/the-emep-grid/gridded-emissions

Guevara, M., Jorba, O., Tena, C., Denier Van Der Gon, H., Kuenen, J., Elguindi, N., Darras, S., Granier, C., & Pérez García-Pando, C. (2021). Copernicus Atmosphere Monitoring Service TEMPOral profiles (CAMS-TEMPO): Global and European emission temporal profile maps for atmospheric chemistry modelling. *Earth System Science Data, 13*(2), 367–404.

Liu, F., Zhang, Q., Tong, D., Zheng, B., Li, M., Huo, H., & He, K. B. (2015). High-resolution inventory of technologies, activities, and emissions of coal-fired power plants in China from 1990 to 2010. *Atmospheric Chemistry and Physics, 15*(23), 13299–13317.

Pham, T. B. T., Manomaiphiboon, K., & Vongmahadlek, C. (2008). Development of an inventory and temporal allocation profiles of emissions from power plants and industrial facilities in Thailand. *Science of the Total Environment, 397*(1–3), 103–118.

Air Quality Effects on Human Health and Ecology

Urban Population Exposure to Air Pollution Under COVID-19 Lockdown Conditions—Combined Effects of Emissions and Population Activity

Martin Otto Paul Ramacher, Volker Matthias, Ronny Badeke, Ronny Petrik, Markus Quante, Jan Arndt, Lea Fink, Josefine Feldner, Daniel Schwarzkopf, Eliza-Maria Link, and Ralf Wedemann

Abstract The aim of this study is to quantify the BIAS in air pollution ($PM_{2.5}$, NO_2) exposure estimates that arise from neglecting population activity under COVID-19 lockdown conditions. We applied mobility data as derived from different sources (Google, Eurostat, Automatic Identification System, etc.) to model the impact of (1) changing emissions and (2) the change in population activity patterns in a European multi-city (Hamburg, Liège, Marseille) exposure study. Our results show significant underestimations of exposure estimates when activity profiles are either neglected or not adjusted for lockdown conditions.

M. O. P. Ramacher (✉) · V. Matthias · R. Badeke · R. Petrik · M. Quante · J. Arndt · L. Fink · J. Feldner · D. Schwarzkopf · E.-M. Link · R. Wedemann
Department of Chemical Transport Modeling, Helmholtz-Zentrum hereon GmbH, Institute of Coastal Research, Max-Planck-Strasse 1, 21502 Geesthacht, Germany
e-mail: martin.ramacher@hereon.de

V. Matthias
e-mail: volker.matthias@hereon.de

R. Badeke
e-mail: ronny.badeke@hereon.de

R. Petrik
e-mail: ronny.petrik@hereon.de

M. Quante
e-mail: markus.quante@hereon.de

J. Arndt
e-mail: jan.arndt@hereon.de

L. Fink
e-mail: lea.fink@hereon.de

J. Feldner
e-mail: josefine.feldner@hereon.de

D. Schwarzkopf
e-mail: daniel.schwarzkopf@hereon.de

R. Wedemann
e-mail: ralf.wedemann@hereon.de

C. Mensink and O. Jorba (eds.), *Air Pollution Modeling and its Application XXVIII*,
Springer Proceedings in Complexity, https://doi.org/10.1007/978-3-031-12786-1_43

Keywords Urban air quality modeling · Urban population exposure modeling · COVID-19 lockdown · Population activity · Mobility reports

1 Introduction

The lockdown response to the coronavirus disease 2019 (COVID-19) has caused an exceptional reduction in global economic and transport activity. Many recent measurement and modelling studies tested the hypothesis that this has reduced ground-level air pollution concentrations (Gkatzelis et al., 2021) as well as the associated population exposure and health effects on different scales (EEA, 2021). Although Google and Apple mobility data is utilized in air quality modelling studies to derive changes in emissions during lockdown periods, the mobility data is not used to reflect changes in population activity patterns; except for very few studies (Kazakos et al., 2021; Shen et al., 2021). But, neglecting the mobility of populations in exposure estimates is known to introduce substantial BIAS, especially on urban-scales (Kukkonen et al., 2016; Ramacher et al., 2019; Soares et al., 2014).

Taking into account the exceptional changes in population activity during lockdown periods due to home office regulations or curfews, the hypothesis of this study is as follows: Taking into account changes in population activity during lockdown will increase the impact of lockdown measures on urban population exposure to air pollutants (NO_2 and $PM_{2.5}$) beyond changes in concentration only.

To test this hypothesis, we applied different emission and population activity scenarios in an air quality modeling system (AQMS) in combination with a dynamic exposure model to three different cities in Central Europe during the first lockdown in 2020. The aim is to quantify the BIAS in air pollution ($PM_{2.5}$, NO_2) exposure estimates that arise from neglecting population activity under COVID-19 lockdown conditions.

2 Urban-Scale AQMS and Exposure Model

We applied the urban-scale chemistry transport model EPISODE-CityChem (Karl et al., 2019) and the urban dynamic exposure model UNDYNE (Ramacher et al., 2019; Ramacher and Karl, 2020) in the European cities Marseille (FR), Liège (BE) and Hamburg (DE) in January-June 2020 (Fig. 1).

The EPISODE-CityChem model is driven by 1 km^2 meteorological fields from model simulations for 2020 with TAPM (Hurley et al., 2005) driven by synoptic ERA5 reanalysis. The regional boundary conditions for 2020 are processed from regional CMAQ simulations by Matthias et al. (2021), which also took into account two different emission scenarios: with and without lockdown conditions. The regional concentrations by Matthias et al. (2021) were corrected for identified monthly BIAS when compared against measurements, before processing to serve

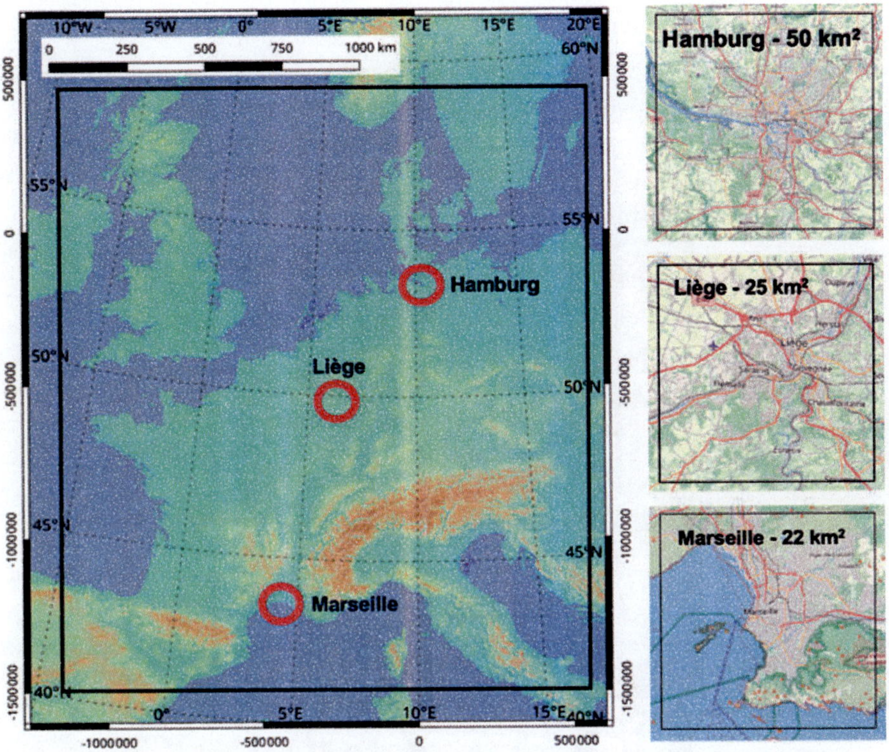

Fig. 1 The urban research domains Hamburg, Liège and Marseille with 50 km^2, 25 km^2 and 22 km^2 extent respectively, represent a North–South gradient in central Europe. Maps are created with © ArcGIS Pro and are based on Open Street Map

as input for boundary conditions in EPISODE-CityChem simulations. The emissions applied for each urban domain in EPISODE-CityChem simulations are based on the CAMS-REG-APv3.1 emission inventory, also following the regional study by Matthias et al. (2021). The regional emissions have been scaled from 2016 to 2020, distributed to point and line sources as well as downscaled to 1 km^2 area sources for each urban domain (Fig. 1). This was done for each sector individually under application of the UrbEm framework (Ramacher et al., 2021). Thus, the applied boundary conditions stem from the emission inventories on the regional and urban-scale, leading to a consistent AQMS setup. The emissions on the local and regional scale were adjusted with proxies from Google, Apple, AIS and Eurostat to derive lockdown conditions in 2020. For a detailed description we refer to Matthias et al. (2021). With this AQMS we simulated two scenarios (Fig. 2, Table 1): (1) a business-as-usual scenario (bau2020) neglecting changes during lockdown and (2) a scenario that takes into account changes in emissions derived with different proxies to account for lockdown regulations (pandemic).

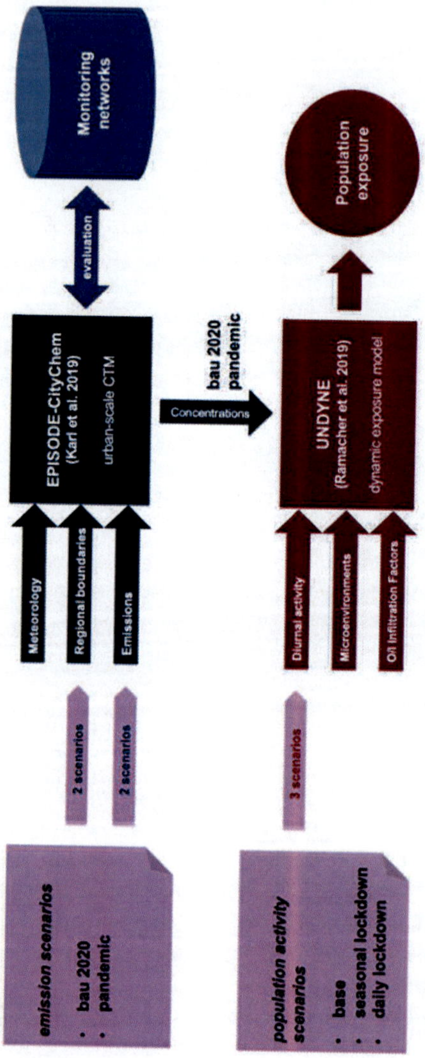

Fig. 2 Study design to estimate population exposure with different population activity and emission scenarios in the urban domains Hamburg, Liège and Marseille

Table 1 Concentration and activity profile scenarios for exposure estimates during the heavy lockdown period 16.–31. March 2020

2020 scenarios	Concentrations	Activity profiles
base_bau2020	Bau (no lockdown adj.)	Base (standard profile, no lockdown adj.)
base_pandemic	Pandemic (lockdown adj.)	Base (standard profile, no lockdown adj.)
lockdown_season_pandemic	Pandemic (lockdown adj.)	Seasonal lockdown (seasonal profile)
lockdown_daily_pandemic	Pandemic (lockdown adj.)	Daily lockdown (daily avg. profiles)

To identify the population exposure in Hamburg, Liège and Marseille, we applied the urban dynamic exposure model UNDYNE (Ramacher & Karl, 2020; Ramacher et al., 2019). In dynamic approaches to exposure estimations, the population activity data is used to account for the diurnal variation in population numbers in different locations of the urban area. This is realized with a time–microenvironment–activity approach in UNDYNE to assess the time-variant exposure of the population in a microenvironment (ME). A microenvironment is defined as the location or area with a relatively uniform pollutant concentration, such as the home or the workplace. Thus, based on microenvironment definitions, UNDYNE allows for a flexible application of population activity in European urban areas. This feature was used to apply different activity profile scenarios (Table 1, Fig. 2) to identify the impact of changes in population activity during lockdown conditions. Therefore, the basic profile for population activity in European cities (Ramacher et al., 2019) was adjusted with city-specific Google and Apple mobility data for two different scenarios that reflect lockdown conditions:

(1) A profile that reflects seasonal averaged change in population activity January–June 2020.
(2) Daily profiles that reflect daily changes in population activity for each day January–June 2020.

With the combination of the AQMS and the urban dynamic exposure model, it is possible to test a series of scenarios (Table 1) with the aim to quantify the BIAS in air pollution ($PM_{2.5}$, NO_2) exposure estimates that arise from neglecting population activity under lockdown conditions.

3 Lockdown Impacts on Urban NO_2 and $PM_{2.5}$ Concentrations

In a first step, we performed the evaluation of modelled results against measurements for $PM_{2.5}$ and NO_2 in all urban domains under investigation. Modelled NO_2

performance slightly improved taking into account the lockdown emissions and lockdown boundary concentrations (incl. offset correction) in the pandemic scenario for concentrations, with very low averaged BIAS of up to 3 $\mu g/m^3$. Modelled $PM_{2.5}$ resulted in indifferent results between the bau2020 and the pandemic scenario, with a less significant signal for better agreement of modeled and measured values in the pandemic scenario. In general, the simulated concentrations for $PM_{2.5}$ overestimate measured values by 3–5 $\mu g/m^3$. The results of simulated $PM_{2.5}$ in both scenarios are closer.

The difference between bau2020 and the pandemic scenario shows an average (January to June 2020) lockdown signal at urban sites of -4 $\mu g/m^3$ in Hamburg, over -8 $\mu g/m^3$ in Marseille to -9 $\mu g/m^3$ in Liège for NO_2, and -1 $\mu g/m^3$ in Hamburg and Marseille to -2 $\mu g/m^3$ in Liège for $PM_{2.5}$ respectively. At road traffic sites the lockdown signal (reduction) is about twice as high. Compared to reductions in concentrations due to lockdown conditions in regional studies, the reduction signal estimated with an urban AQMS leads to higher concentration reductions.

4 Urban Exposure During Lockdown

The simulated concentrations from both scenarios have been applied in combination with different activity profile scenarios (Table 1), for the period of heavy lockdown 16th till 31st March 2020. In Table 2 the results for relative changes in estimated exposure are displayed. Taking into account the changes in concentrations during lockdown but with standard activity profiles shows -30 to -60% reduced NO_2 exposure, and -5 to -15% reduced $PM_{2.5}$ exposure. Taking into account seasonally averaged activity profiles adjusted for lockdown activity there are additional 3–4% reductions in exposure to NO_2 and -3%, -24%, -5% for $PM_{2.5}$ in Hamburg, Liège and Marseille respectively. When daily profiles for lockdown activity are considered, there is an additional 3–5% reduction in exposure for NO_2 and $PM_{2.5}$ estimated.

5 Conclusion

The results of the combined AQMS and exposure model scenario simulation confirm the hypothesis of this study: Taking into account changes in population activity during lockdown period increase the impact of lockdown measures on urban population exposure estimates to air pollution beyond changes in concentration only. Moreover, the reduction in concentrations due to lockdown conditions in urban areas is significantly higher than in regional areas, as indicated by regional-scale CTM studies. Thus, we conclude, that future emission, concentration and exposure modelling studies can benefit from high-resolution activity data. Especially exposure modeling studies in urban areas need to take into account the dynamic activity of populations compared to commonly used static approaches based on residential addresses only.

Table 2 Relative changes in population exposure to NO_2 and $PM_{2.5}$ during the period of heavy lockdown (16.–31. March 2020) in Hamburg, Liège and Marseille in different scenarios

Urban domain	base_pandemic compared to base_bau2020 (%)	Additional impact lockdown_season_pandemic compared to base_pandemic (%)	Additional impact lockdown_daily_impact compared to base_pandemic (%)
Hamburg NO_2	−30	−4	−7
Liège NO_2	−46	−3	−8
Marseille NO_2	−60	−4	−7
Hamburg $PM_{2.5}$	−6	−3	−6
Liège $PM_{2.5}$	−10	−24	−29
Marseille $PM_{2.5}$	−15	−5	−8

Acknowledgements We thank Fabian Lenartz and Marie Dury from ISSeP (Liège, BE), AtmoSud (Marseille, FR), and Hamburg Ministry of Environment and Energy (BUE) for providing AQ monitoring data.

References

EEA. (2021). *COVID-19 and Europe's environment: Impacts of a global pandemic*. EEA. Retrieved January 25, 2020, from https://www.eea.europa.eu/publications/covid-19-and-europe-s

Gkatzelis, G. I., Gilman, J. B., Brown, S. S., Eskes, H., Gomes, A. R., Lange, A. C., et al. (2021). The global impacts of COVID-19 lockdowns on urban air pollution. *Elementa: Science of the Anthropocene, 9*(1).

Hurley, P. J., Physick, W. L., & Luhar, A. K. (2005). TAPM: A practical approach to prognostic meteorological and air pollution modelling. *Environmental Modelling & Software, 20*(6), 737–752.

Karl, M., Walker, S.-E., Solberg, S., & Ramacher, M. O. P. (2019). The Eulerian urban dispersion model EPISODE—Part 2: Extensions to the source dispersion and photochemistry for EPISODE–CityChem v1.2 and its application to the city of Hamburg. *Geoscientific Model Development, 12*(8), 3357–3399.

Kazakos, V., Taylor, J., & Luo, Z. (2021). Impact of COVID-19 lockdown on NO_2 and $PM_{2.5}$ exposure inequalities in London, UK. *Environmental Research, 198*, 111236.

Kukkonen, J., Singh, V., Sokhi, R. S., Soares, J., Kousa, A., Matilainen, L., et al. (2016). Assessment of population exposure to particulate matter for London and Helsinki. In D. G. Steyn & N. Chaumerliac (Eds.), *Air pollution modelling and its application* (pp. 99–105). Springer.

Matthias, V., Quante, M., Arndt, J. A., Badeke, R., Fink, L., Petrik, R., et al. (2021). The role of emission reductions and the meteorological situation for air quality improvements during the COVID-19 lockdown period in central Europe. *Atmospheric Chemistry and Physics, 21*(18), 13931–13971.

Ramacher, M. O. P., Kakouri, A., Speyer, O., Feldner, J., Karl, M., Timmermans, R., et al. (2021). The UrbEm hybrid method to derive high-resolution emissions for city-scale air quality modeling. *Atmosphere, 12*(11), 1404.

Ramacher, M. O. P., & Karl, M. (2020). Integrating modes of transport in a dynamic modelling approach to evaluate population exposure to ambient NO_2 and $PM_{2.5}$ pollution in urban areas. *International Journal of Environmental Research and Public Health, 17*(6), 2099.

Ramacher, M. O. P., Karl, M., Bieser, J., Jalkanen, J.-P., & Johansson, L. (2019). Urban population exposure to NOx emissions from local shipping in three Baltic Sea harbour cities—A generic approach. *Atmospheric Chemistry and Physics, 19*(14), 9153–9179.

Shen, H., Shen, G., Chen, Y., Russell, A. G., Hu, Y., Duan, X., et al. (2021). Increased air pollution exposure among the Chinese population during the national quarantine in 2020. *Nature Human Behaviour, 5*(2), 239–246.

Soares, J., Kousa, A., Kukkonen, J., Matilainen, L., Kangas, L., Kauhaniemi, M., et al. (2014). Refinement of a model for evaluating the population exposure in an urban area. *Geoscientific Model Development, 7*(5), 1855–1872.

Questions and Answers

Questioner: Zhaoyue Chen.

Question: Thanks for wonderful presentation. Can you tell more details about how to change daily activity profile?

Answer: The base profiles for population activity under standard conditions (non-lockdown, in detail explained in Ramacher et al., 2019) are adjusted with data from the Google mobility reports (https://www.google.com/covid19/mobility/) for the non-transport environments (home, work, other) and with Apple mobility trends (https://covid19.apple.com/mobility) for transport environments (road traffic, walking, public transport and cycling). The mobility reports and trends show relative changes for regions and cities compared to a fixed start date (before lockdown). These relative changes are used to adjust the standard profiles, leading to either seasonal profiles or daily profiles (due to the daily availability of mobility reports).

Questioner: Gilles Forêt

Question: Is it more important to consider mobility exposure or indoor exposure? How can we tackle the issue of indoor air exposure?

Answer: In terms of outdoor population exposure to air pollutants mobility (road transport, shipping, aviation) can be considered as the most important source. Due to the fact, that most exposure estimates and connected health effect estimates do only account for outdoor air pollution, mobility is the most important sector to be considered. Nevertheless, in terms of total exposure to air pollution, indoor exposure needs to be considered in the future. But, for indoor air pollution measurements, modeling approaches and the general scientific knowledge are rather scarce. Thus, I vote for both, putting an emphasis right now on possibilities to reduce air pollution by mobility and extending the knowledge by more research and measurements on indoor air pollution.

Towards a Comprehensive Evaluation of the Environmental and Health Impacts of Shipping Emissions

J. Kukkonen, E. Fridell, J.-P. Jalkanen, J. Moldanova, L. Ntziachristos, A. Grigoriadis, F. Barmpas, G. Tsegas, A. Maragkidou, Mikhail Sofiev, T. Grönholm, E. Majamäki, J. Borken-Kleefeld, R. S. Sokhi, P. R. Tiwari, U. A. Ozdemir, V. Zervakis, E. Krasakopoulou, I.-M. Hassellöv, E. Ytreberg, I. Williams, M. Hudson, L. Zapata-Restrepo, L. R. Hole, M. Aghito, O. Breivik, M. Petrovic, S. Rodriguez-Mozaz, A. Ktoris, M. Neophytou, A. Monteiro, M. A. Russo, F. Oikonomou, P. Arampatzi, A. Gondikas, A. Marcomini, E. Giubilato, L. Calgaro, J. J. K. Jaakkola, S.-P. Kiihamäki, R. Aittamaa, G. Broström, M. Hassellöv, J. Tamminen, F. Nicolas, J. Kaitaranta, M. Granberg, and K. Magnusson

Abstract We present a new concept for marine research, applied in the EU-funded project EMERGE, "Evaluation, control and Mitigation of the EnviRonmental impacts of shippinG Emissions" (2020–2024; https://emerge-h2020.eu/). For the first time, both the various marine and atmospheric impacts of the shipping sector have been and will be comprehensively analyzed, using a concerted modelling and measurements framework. The experimental part of the project focuses on five

J. Kukkonen (✉) · J.-P. Jalkanen · A. Maragkidou · M. Sofiev · T. Grönholm · E. Majamäki · J. J. K. Jaakkola · J. Tamminen
Finnish Meteorological Institute (FMI), Helsinki, Finland
e-mail: Jaakko.Kukkonen@fmi.fi

E. Fridell · J. Moldanova · M. Granberg · K. Magnusson
Swedish Environmental Research Institute, Stockholm, Sweden

L. Ntziachristos · A. Grigoriadis · F. Barmpas · G. Tsegas
Aristotle University of Thessaloniki, Thessaloniki, Greece

J. Borken-Kleefeld
International Institute of Applied Systems Analysis, Laxenburg, Austria

J. Kukkonen · R. S. Sokhi · P. R. Tiwari · U. A. Ozdemir
University of Hertfordshire, Hatfield, UK

V. Zervakis · E. Krasakopoulou
University of the Aegean, Mytilene, Greece

I.-M. Hassellöv · E. Ytreberg
Chalmers University of Technology, Gothenburg, Sweden

I. Williams · M. Hudson · L. Zapata-Restrepo
University of Southampton, Southampton, UK

L. R. Hole · M. Aghito · O. Breivik
Norwegian Meteorological Institute, Oslo, Norway

C. Mensink and O. Jorba (eds.), *Air Pollution Modeling and its Application XXVIII*, Springer Proceedings in Complexity, https://doi.org/10.1007/978-3-031-12786-1_44

European geographical case studies in different ecologically vulnerable regions, and a mobile onboard case study. The EMERGE consortium has also developed a harmonised and integrated modelling framework to assess the combined impacts of shipping emissions, both (i) on the marine ecosystems and (ii) the atmospheric environment. The first results include substantial refinements of a range of models to be applied, especially those for the STEAM and OpenDrift models. In particular, the STEAM (Ship Traffic Emission Assessment Model) model has been extended to allow for the effects of atmospheric and oceanographic factors on the fuel consumption and emissions of the ships. The OpenDrift model has been improved to take into account the partitioning, degradation, and volatilization of pollutants in water. The predicted emission and discharge values have been used as input for both regional scale atmospheric dispersion models, such as WRF-CMAQ (Weather Research and Forecasting—Community Multiscale Air Quality Model) and SILAM (System for Integrated modeLling of Atmospheric composition), and water quality and circulation models, such as OpenDrift (Open source model for the drifting of substances in the ocean) and Delft3D (oceanographic model). The case study regions are Eastern Mediterranean, Northern Adriatic Sea, the Lagoon of Aveiro, the Solent Strait and the Öresund Strait. We have also conducted a substantial part of the experimental campaigns scheduled in the project. The final assessment will include the benefits and costs of control and mitigation options affecting water quality, air pollution exposure, health impacts, climate forcing, and ecotoxicological effects and bioaccumulation of pollutants in marine biota.

M. Petrovic · S. Rodriguez-Mozaz
Catalan Institute for Water Research, Girona, Spain

A. Ktoris · M. Neophytou
Maritime Institute of Eastern Mediterranean, Larnaca, Cyprus

A. Monteiro · M. A. Russo
University of Aveiro, Aveiro, Portugal

F. Oikonomou · P. Arampatzi
DANAOS, Limassol, Cyprus

A. Gondikas
Creative Nano, Athens, Greece

A. Marcomini · E. Giubilato · L. Calgaro
University of Venice, Venice, Italy

J. J. K. Jaakkola · S.-P. Kiihamäki · R. Aittamaa
University of Oulu, Oulu, Finland

G. Broström · M. Hassellöv
University of Gothenburg, Gothenburg, Sweden

F. Nicolas · J. Kaitaranta
HELCOM, Helsinki, Finland

Keywords Shipping · Impacts · Environmental impacts · Health impacts · EMERGE · STEAM · WRF-CMAQ · SILAM · OpenDrift · Eastern Mediterranean · Adriatic Sea · Aveiro · Solent Strait · Öresund Strait

1 Introduction

New global standards have been enforced in 2020 for shipping emissions, as a consequence of the potentially significant health and environmental effects, especially from SO_x and sulphate aerosol. In so-called Sulphur Emission Control Areas (SECAs), a stricter regulation of 0.1% Fuel Sulphur Content has been in place since 2015. In Europe, these comprise the Baltic and North Seas.

In complying with the limit values within SECAs, ships are currently mandated to use fuel oil with fuel sulphur content within the limits. Alternatively, vessels may be equipped with abatement systems—SOx scrubbers—that decrease SO_2 in the exhaust to within the limits. However, the seawater scrubbing process produces large volumes of acidic and contaminated exhaust scrubber effluent. This could potentially result in a substantial deterioration of marine water quality. However, there is limited data on the environmental toxicity of scrubber washwater.

Combustion in ship engines results in a range of primary and secondary airborne pollutants that have important environmental, health, economic and climatic impacts. Particulate matter (PM), SO_x, NO_x and O_3 exposures from shipping have been associated with an increase in morbidity and premature mortality rates of human populations (e.g., Nunes et al., 2021).

The EMERGE project has up to date focused on establishing the overall framework, through the combination of laboratory and field measurements and modelling activities. The overarching objectives of EMERGE are: (i) to comprehensively quantify and evaluate the effects of potential emission reduction solutions for shipping in Europe for several scenarios, and (ii) to develop effective strategies and measures to reduce the environmental impacts of shipping.

2 Methods

We have updated emission factors for shipping, especially regarding the effects of advanced aftertreatment technologies, based on an extensive review of previous literature. These values were collected in a structured database, which contains emission rate values for the most relevant pollutants (SO_x, NO_x, CO, HC, PM), and for various emission abatement options.

The STEAM model (e.g., Jalkanen et al., 2021; Johansson et al., 2017) has been and will be used to provide the detailed emission values to air and discharge values to water, on a high spatial and temporal resolution. The model utilizes automatic identification system (AIS) data from individual ships and evaluates emissions using ship data, modelling of fuel consumption and emission factors. The emission and discharge results have been and will be used in the modelling of dispersion and transport of pollutants in the atmosphere and the seas, and for the impact assessments.

Several chemical transport models have been refined and applied for the evaluation of air quality on regional and continental scales. The model predictions will be evaluated against the measured concentrations and deposition of pollutants within the European Monitoring and Evaluation Programme (EMEP), and selected satellite observations. The regional scale models will target a spatial resolution of 5 km at European scale and zoom down to 1 km for the case study areas. In addition, computations on an urban scale will also be done in part of the case studies.

The output of the STEAM model consists of the volumetric fluxes of discharges per area. The fluxes of each pollutant to the sea will subsequently be computed by using the fractions of pollutants in the scrubber discharge effluent; the latter have been evaluated based on the existing literature. The pollutant fluxes per area have then been used as input values in the chemical fate and transport models.

In particular, we have used and will use the Lagrangian OpenDrift model and the Eulerian 3D water quality model Delft3D-WAQ. In addition, we will evaluate the fluxes of pollutants to the sea attributed to atmospheric deposition. The partitioning, degradation, and volatilization of chemicals have been included in a new module (ChemicalDrift) of the OpenDrift model (Dagestad et al., 2018), thus facilitating the modelling of the different environmental behaviour of heavy metals and organic pollutants (e.g., polyaromatic hydrocarbons, PAHs).

Several scenarios have been constructed for estimating the impacts of the application of various emission control technologies in the future. The project will focus on the evaluation of the emission changes resulting from (i) the different developments of marine traffic, (ii) the potential uses of emission control technologies, especially scrubbers, and (iii) the use of various alternative fuels.

We aim to eventually provide for harmonized information on the ecotoxicological effects of scrubber washwater in various European sea regions. The toxicity of scrubber effluents has been tested for a wide range of marine species in different life stages, using different health end points. The consortium has developed a common protocol for one species, the blue mussel larvae, to intercompare the ecotoxicological effects of scrubber water in various European sea regions. The chemical analyses of scrubber water contaminants will be used in water modelling, to estimate whether the seawater concentrations will pose a threat to marine ecosystems.

Several full-scale measurement campaigns have already been conducted.

3 Results

We have compiled the data available in the literature on the content of scrubber effluents, inlet water and sludge. A comprehensive database was compiled, which includes (i) the concentrations of contaminants, nutrients and acidifying substances (e.g., Lunde Hermansson et al., 2021), and (ii) the available information regarding the mode of operation of a ship. The emission factors for a range of compounds were subsequently evaluated corresponding to various assumptions on the scrubber effluent discharge rates for open and closed loop scrubbers. The corresponding information was also evaluated for other shipping waste streams (e.g., grey water, bilge water and anti-fouling paint).

Untargeted high resolution mass spectrometry analysis was used for a detailed chemical characterization of scrubber water. We found that the substances of major concern were several PAHs, together with their alkylated derivatives, and metals.

Experimental studies have been conducted using both open loop effluents from an operating ship and samples obtained from an open-loop pilot scrubber system operated at a laboratory in Chalmers University of Technology. Repeated experimental studies have found statistically significant effects on marine zooplankton after exposure to concentrations as low as 0.01% scrubber effluent (Thor et al., 2021). However, the data also indicates that the sensitivity to scrubber effluent varies greatly both between species and between different life stages of species (Ytreberg et al., 2019).

The STEAM emission model has been updated to use improved emission factors. In particular, the model has been refined to include the effects of the resistance of sea ice, wind, waves and sea currents on the passage of the ships and on the emissions. These improvements have used the data provided by Copernicus Marine and the meteorological data of ERA5. These updates will substantially improve the capability to accurately evaluate shipping emissions and discharges; the refined model will be used globally in the evaluation of shipping emissions in the future.

The magnitude of these effects has been illustrated in Fig. 1 in case of the Baltic Sea, during one year. The Baltic Sea is commonly covered with ice from December to May. During these months, sea ice cover significantly increases the fuel consumption. Wind and waves are dominant environmental effects in September and October. The effects of sea currents almost totally cancel out in this sea region. The model subsequently evaluates the impacts of such environmental effects on the emissions of various pollutants.

Grönholm et al. (2021) presented for the first time a quantitative assessment of the climatic impacts of various liquid natural gas (LNG)-fuelled ships in real-life marine conditions. They found out that the unintentional release of methane (called also as the methane slip) was substantial for part of the LNG-fuelled ships. They evaluated that these releases will actually cause larger impacts on the climate, compared with the corresponding releases originating from ships that use conventional marine fuels.

The predicted discharges to the sea were used as input values in the computations on water quality, using the ChemicalDrift module of the OpenDrift model (Dagestad et al., 2018). Preliminary results have been presented in Fig. 2.

Meteorological datasets have been computed using the Weather Research and Forecasting (WRF) model. These datasets have been evaluated against experimental meteorological data on a European scale. The results will be used in the atmospheric dispersion computations by the partners in their regional and partly also in urban scale computations.

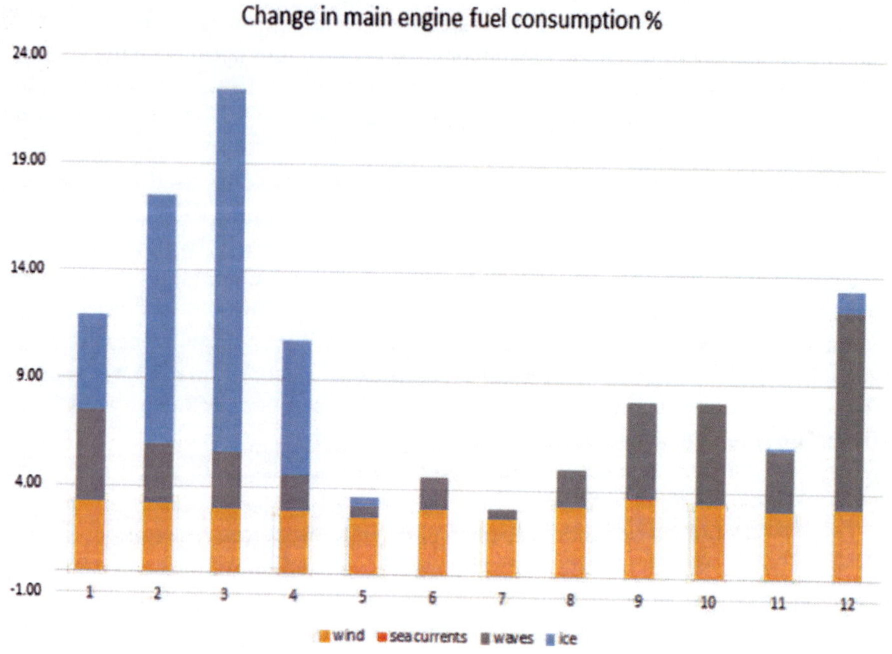

Fig. 1 The impact of ambient conditions, including wind, sea currents, waves and ice cover, to monthly totals of fuel used in the Baltic Sea in 2018. The results were computed using a refined version of the STEAM model. The colours refer to the influence of sea ice (blue), waves (grey), sea currents (red) and wind (yellow)

4 Conclusions

Finally, we will analyze the impacts of ship emissions and effluents on both the atmospheric and marine environments. The cost-efficiency will also be analyzed for various abatement methods, fuels and technological solutions. The project adopts a holistic approach: for the first time, both the various marine and atmospheric impacts of the shipping sector will be comprehensively analyzed, using a concerted modelling and measurements framework.

The project is expected to provide new harmonised knowledge on water and air pollution and their impacts, both on a European scale, and within the vulnerable case study regions. The assessment of scenarios for the future will highlight the optimal procedures towards the mitigation of the environmental and health impacts. The scenarios address various emission abatement technologies, fuel switch and different operation of the ships. The project will provide for recommendations on policy actions for decision-makers and ship owners in the EU and globally.

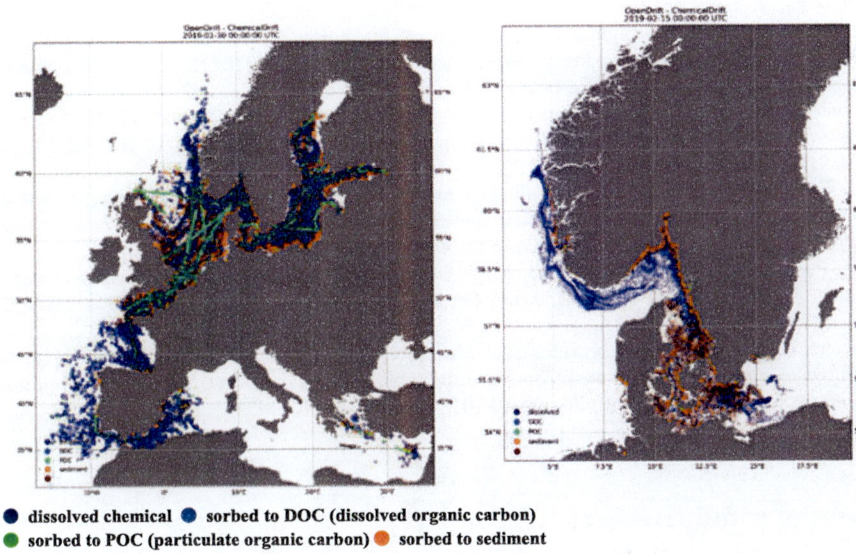

Fig. 2 Preliminary model results on the transport and fate of phenantrene emissions from shipping activity in the European sea regions (panel on the left) and in the Öresund area and its surroundings (panel on the right), computed using the OpenDrift-ChemicalDrift model. The legend (colours) indicates the chemical species. The shipping emission data was provided by the STEAM model (data from January–February 2019). The forcing data was extracted from the global models HYCOM (HYbrid Coordinate Ocean model) and CMEMS (Copernicus Marine Environment Monitoring Service)

Acknowledgements This work has received funding from the European Union's Horizon 2020 research and innovation program under grant agreement No. 874990 (EMERGE project). This work reflects only the authors' view and the Innovation and European Climate, Infrastructure and Environment Executive Agency (CINEA) is not responsible for any use that may be made of the information it contains.

References

Dagestad, K.-F., Röhrs, J., Breivik, Ø., & Aadlandsvik, B. (2018). OpenDrift v1.0: A generic framework for trajectory modelling. *Geoscientific Model Development, 11*(4), 1405–1420. https://doi.org/10.5194/gmd-11-1405-2018

Grönholm, T., Mäkelä, T., Jalkanen, J.-P., Kuula, J., Hatakka, J., Laurila, T., Laakso, L., & Kukkonen, J. (2021). Evaluation of methane emissions originating from LNG ships based on the measurements at a remote marine station. *Environmental Science & Technology*. https://doi.org/10.1021/acs.est.1c03293. Published by American Chemical Society.

Jalkanen et al. (2021). Modeling of discharges from Baltic Sea shipping. *Ocean Science, 17*, 699–728. https://doi.org/10.5194/os-17-699-2021

Johansson, L., Jalkanen, J.-P., & Kukkonen, J. (2017, October). Global assessment of shipping emissions in 2015 on a high spatial and temporal resolution. *Atmospheric Environment, 167*, 403–415. http://www.sciencedirect.com/science/article/pii/S1352231017305563?via%3Dihub

Lunde Hermansson, A., Hassellöv, I.-M., Moldanová, J., & Ytreberg, E. (2021). Comparing emissions of polyaromatic hydrocarbons and metals from marine fuels and scrubbers. *Transportation Research Part D: Transport and Environment, 97*, 102912. https://doi.org/10.1016/j.trd.2021.102912

Nunes, et al. (2021, November). Estimating the health and economic burden of shipping related air pollution in the Iberian Peninsula. *Environment International, 156*, 106763. https://www.sciencedirect.com/science/article/pii/S0160412021003883

Thor, P., Granberg, M., Winnes, H., & Magnusson, K. (2021). Severe toxic effects on pelagic copepods from maritime exhaust gas scrubber effluents. *Environmental Science and Technology, 55*, 5826–5835.

Ytreberg, E., Hassellöv, I.-M., Nylund, A. T., Hedblom, M., Al-Handal, A. Y., & Wulff, A. (2019). Effects of scrubber washwater discharge on microplankton in the Baltic Sea. *Marine Pollution Bulletin, 145*, 316–324. https://doi.org/10.1016/j.marpolbul.2019.05.023

Questions and Answers

Questioner: Lafafi Borge

Question: The increase of shipping emissions due to the explicit consideration of wind and surface ice has an impact on consumption, and therefore CO_2/SO_2 emissions. Is that taken into account to estimate emission of other pollutants, such as NOx?

Answer: Yes. The ambient effects will change the power need and engine load of vessels and therefore these will also affect fuel consumption and the emissions of all atmospheric pollutants. In addition, the ambient effects will influence also the discharges to water (such as, e.g., the scrubber discharges), as these have been modeled as a function of the instantaneous engine power.

Modelling Short-Term Health Effects in Milan Area Due to Lockdown Reduced Emissions: Combined Uncertainty Analysis from Estimated NO_2 Levels and Exposure–Response Functions

Andrea Piccoli, Valentina Agresti, Giovanni Lonati, and Guido Pirovano

Abstract To contain the spread of the COVID-19 pandemic, several governments declared lockdowns. The reduction of human activities linked with mobility restriction caused an unprecedented drop in emissions, especially in the road transport sector. This study aims to evaluate the uncertainty of short-term health effects (i.e. avoided hospital admission (AHA) associated to NO_2 ambient concentrations) derived from the change in air quality (AQ) due to lockdown. The CAMx-WRF modelling suite is applied for a series of nested domains using EMEP and Lombardy region emission inventories. The health impact analysis is focused on a 70×70 km domain centered on Milan metropolitan area with 1-km resolution, from February 24th 2020 to April 30th 2020. Two simulations, Business as usual (BAU) and lockdown scenario (LOCK), are carried out and results are compared with air quality monitoring data to assess the model uncertainty. Health effects for the difference between LOCK and BAU simulation are evaluated for NO_2 following the WHO Health risks of air pollution in Europe (HRAPIE) project recommendation. The combined effect of both modelled concentrations and exposure–response functions (ERF) on the uncertainty of calculated AHA is then evaluated for different air quality station types. We find that the ERF is the major cause for uncertainty for Urban and suburban AQ stations, measured as the size of the 95% confidence interval (CI) of the estimated number of AHA. The AQ uncertainty of the BAU scenario has a lower impact on AHA CI than for the LOCK scenario. When comparing results according to station type, the lowest AHA uncertainty is obtained for urban background stations while the highest for rural background stations.

A. Piccoli (✉) · G. Lonati
Department of Civil and Environmental Engineering, Politecnico di Milano, Milano, Italy
e-mail: andrea.piccoli@polimi.it

G. Lonati
e-mail: giovanni.lonati@polimi.it

A. Piccoli · V. Agresti · G. Pirovano
Sustainable Development and Energy Sources Department, RSE Spa, Milano, Italy
e-mail: valentina.agresti@rse-web.it

G. Pirovano
e-mail: guido.pirovano@rse-web.it

C. Mensink and O. Jorba (eds.), *Air Pollution Modeling and its Application XXVIII*, Springer Proceedings in Complexity, https://doi.org/10.1007/978-3-031-12786-1_45

Keywords Air quality · Health impacts assessment · Air pollution · Uncertainty analysis · Po valley

1 Introduction

To contain the spread of the COVID-19 pandemic, starting from March 2020 several governments declared lockdowns. The reduction of human activities linked with mobility restrictions caused an unprecedented drop in atmospheric emissions, especially in the road transport sector. The Italian region of Lombardy, where Milan and the study area are located, was one of the first European areas affected by the COVID19 pandemic. Restrictions were first implemented on February 24th, with soft measures such as school closure; then, a national lockdown was declared on March 9th. Several studies (Collivignarelli et al., 2020; Menut et al., 2020; Sicard et al., 2020) showed a significant drop in ambient concentrations of NO_2, and to a lesser extent also of particulate matter. This study aims to evaluate the uncertainty of short-term health effects (i.e. avoided hospital admission (AHA) associated to NO_2 ambient concentrations) derived from the change in air quality (AQ) due to lockdown in the city of Milan. For this purpose, we evaluate the AHA by comparing two AQ simulation, a Business As Usual scenario (BAU), where we left the emissions unchanged, and a lockdown scenario (LOCK), in which emissions were reduced to simulate the effects of the real world decrease in human activity. In the Methods section we present the modeling chain, the two scenario set-up and the uncertainty calculation, while results and conclusion are presented in the last paragraph.

2 Methods

2.1 Modeling Set-Up and Data

We used the Comprehensive Air Quality Model with Extension (CAMx) version 7.0 (Ramboll, 2020) to simulate the atmospheric concentration of NO_2. Meteorological input data for CAMx came from the Weather Regional Forecast (WRF) model (Skamarock et al., 2008). CAMx simulations were carried out over a set of two nested domains (Fig. 1). The master domain (ITA) covers the entire Italian peninsula with a resolution of 4 km and a size of 284×362 cells, the nested domain (MIL) is centered on the city of Milan, has a 1 km resolution and covers an area of 70×70 km.

The WRF model was applied over the ITA domain, while the meteorological condition for the nested MIL domain were obtained by downscaling the Italian results using the WRFCAMX processor. The emission for the ITA domain were obtained from the 2017 EMEP inventory (Mareckova et al., 2019), using the High-Elective Resolution Modelling Emission System version 3—Global_Regional (HERMESv3_GR) model for spatialization (Guevara et al., 2019), speciation and

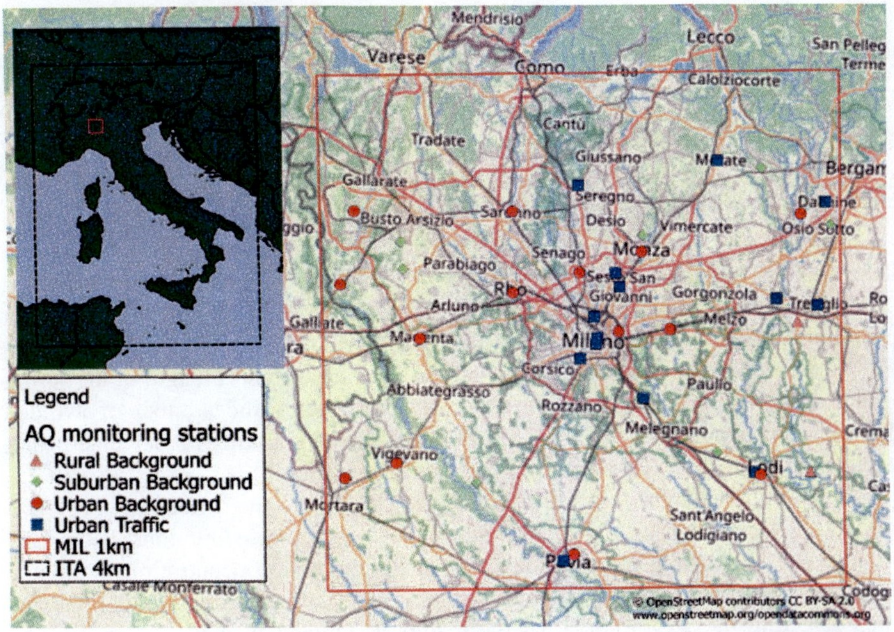

Fig. 1 Small image: Italian (ITA) and Milan (MIL) domain. Main image: MIL domain ad ARPA Lombardia air quality monitoring sites used for uncertainty analysis

temporal disaggregation of the original EMEP inventory into CAMx-ready emission. For the MIL domain we used the Sparse Matrix Operator Kernel Emissions model (SMOKE v3.5) (UNC, 2013) to process the regional INEMAR emission inventory, for year 2014 (Lombardia https://www.inemar.eu/xwiki/bin/view/Inemar/HomeLombardia). We estimated biogenic and sea salt emission for the master domain using the Model of Emissions of Gases and Aerosols from Nature (MEGAN v2.03) (Guenther et al., 2006) and SEASALT model (Environ, 2015), respectively. Biogenic emission for the MIL model domain are a downscaled result of the MEGAN output. For the purpose of this study, we will focus only on the MIL domain.

2.2 *Scenario Set-Up*

Weekly emission reduction coefficients were applied to the BAU emission for both domains. We applied the reduction to the emission only for the road transport sector, as it is the main source of NOx (EEA, 2018) and also because it is the sector that registered the greatest reduction in emission. This approach was adopted in a previous paper and showed negligible differences in NO_2 concentration when considering other sectors (Piccoli et al., 2020). For the MIL domain we used emission reduction coefficients produced by the Environmental Agency of Regione Lombardia (ARPA

Lombardia) (Marongiu et al., 2020), while for the ITA domain the coefficients were based on the "COVID-19 mobility trends" published by Apple (https://covid19.apple.com/mobility).

2.3 *Effects on Human Health and Uncertainty Estimation*

Health effects for the difference between LOCK and BAU simulation are evaluated for NO_2 following the WHO Health risks of air pollution in Europe (HRAPIE) project recommendations (World Health Organization, 2013).

We decided to focus on short terms effects because the short length of the spring lockdown (Mar. 9th to April 30th), namely considering the number of avoided hospital admissions (AHA), based on NO_2 daily mean, among the adverse health effects indicators for NO_2 suggested by HRAPIE project.

There are three sources of uncertainties in computing the differences in health effects between scenarios, one linked to the exposure–response functions (ERF) used in the AHA calculation, and two linked to the NO_2 ambient concentrations estimated for each scenario. The uncertainty is estimated using Monte Carlo simulations and then selecting the 95% interval of the results as upper and lower limits and the median as the point value. Uncertainty in the ERF was evaluated starting from the 95% confidence intervals for the Relative Risk (RR) as reported by the HRAPIE project (RR = 1.0180 [1.0115–1.0245] per 10 $\mu g\ m^{-3}$) and considering a normal distribution. Uncertainty for NO_2 concentrations is assessed at 38 sites within the MIL domain (Table 1, Fig. 1), corresponding to the locations of air quality monitoring stations (AQMS) for NO_2 of the ARPA Lombardia monitoring network (Lombardia https://www.dati.lombardia.it/Ambiente/Dati-sensori-aria/nicp-bhqi).

The uncertainty range is calculated considering the natural logarithm of the modelled NO_2 concentration (MOD) at each AQMS as the sole predictor of the natural logarithm of the observed NO_2 (OBS), in a linear regression with the regression coefficient equal to 1 an intercept equal to 0.

$$ln(OBS) = \alpha + \gamma \cdot ln(MOD) + \varepsilon \; with \; \alpha = 0 \; and \; \gamma = 1$$

Calculations were performed, grouping AQMS sites by station type (Rural Background, Suburban background, Urban background Urban traffic) and relying on data from February 10th to March 8th for the BAU scenario and from March 9th to April 30th for the LOCK scenario.

Table 1 Number of ARPA Lombardia AQMS considered in this study

Type	Rural background (RB)	Sub-URBAN background (SB)	Urban background (UB)	Urban traffic (UT)
N° of AQMS	2	8	14	14

Using the natural logarithm is necessary to comply with the assumption of Gaussianity and homoscedasticity of the standardized residuals. The residuals $\varepsilon_{i,t}$ (*i* indicates the AQMS and *t* the day) for NO_2 daily mean concentrations at the sites of the monitoring stations are given by:

$$\varepsilon_{i,t} = Y_{i,t} - X_{i,t} = ln(OBS_{i,t}) - ln(MOD_{i,t})$$

Standard deviation for point predictions can be therefore estimated as:

$$\sigma_{i,t} = S_e \cdot \sqrt{1 + \frac{1}{n} + \frac{(X_{i,t} - \overline{X})}{\sum (X_{i,t} - \overline{X})^2}}$$

where

$$S_e^2 = \frac{\sum (Y_{i,t} - X_{i,t})^2}{n - 2}$$

is the unbiased estimator of the variance, n is the number of observation for each station type group (number of AQ station in the group multiplied by the number of days with valid AQ data). During the Monte Carlo simulation a value for the residuals $res_{i,t}$ are obtained considering ε as a normally distributed with mean 0 and standard deviation equal to $\sigma_{i,t}$. The model concentration is then equal to:

$$CMOD_{i,t} = exp\left(X_{i,t} + res_{i,t}\right)$$

The two scenarios concentration and their intervals are then fed to the AHA function to determine the effects on human health:

$$AHA = Population \cdot Basline\ Hospital\ Admissions \cdot \left(1 - \frac{1}{e^{\beta \Delta C}}\right) with\ \Delta C = BAU - LOCK$$

where β is computed starting from the HRAPIE RR values for an increase of 10 $\mu g\ m^{-3}$ of concentration as natural logarithm of RR divided by 10. The data for the exposed population is obtained from the ISTAT database while the baseline for hospital admission is available in the "European Hospital Morbidity Database".

We analyzed the uncertainty for RR, BAU and LOCK individually by keeping the values of the other two sources of uncertainty fixed to the mean value, and then we estimated the combined effects of the two AQ scenario and the combined effects of the three uncertainty sources.

3 Results

In order to compare different station types, we performed a normalization of the AHA values, by dividing the 2.5–97.5% interval by the median value. In Fig. 2 and in Table 2 we present a graphical and numerical summaries of the Monte Carlo simulations results. The acronyms for the different uncertainty analysis are: RR for the ERF, BA for the BAU, LO for LOCK, AQ for the combined effects of BAU and LOCK, and CO for the combined effects of all three sources of uncertainty.

We found that for the Rural Background sites the main source of uncertainty is related to the AQ model, with the LOCK scenario being the main source of uncertainty. In Urban, both background and traffic, and Sub-Urban background AQMS the main source of uncertainty is in the HRAPIE exposure–response functions, followed

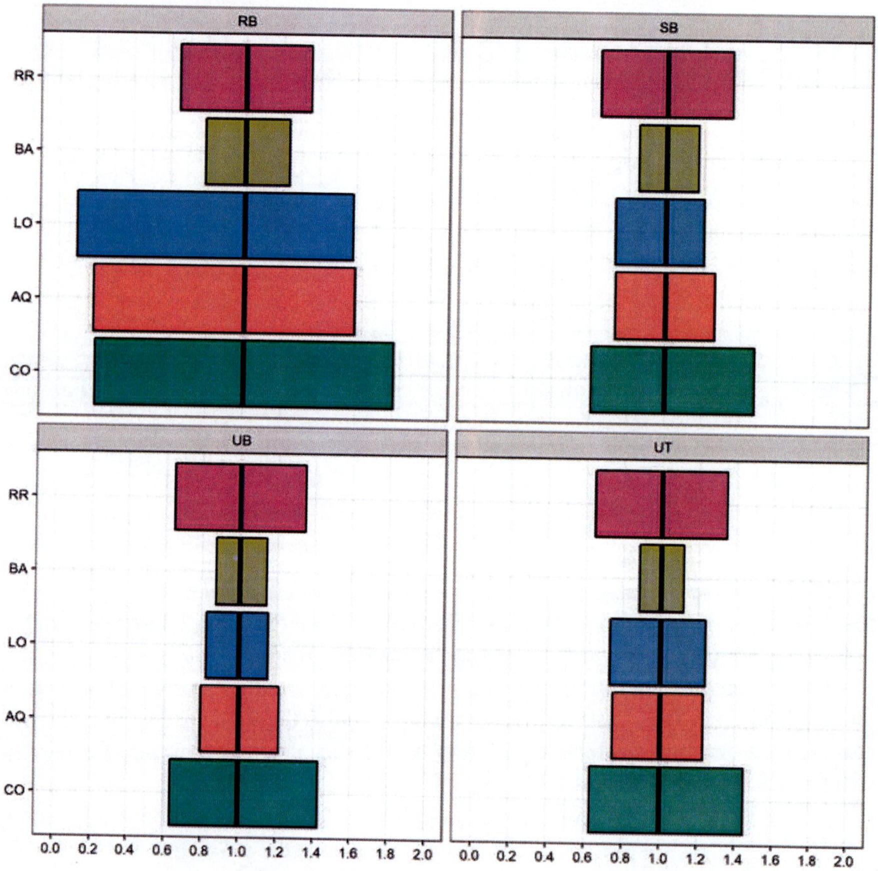

Fig. 2 Graphical summary of normalized AHA 95% interval values for AQMS type

Table 2 Summary of normalized AHA 95% interval values for AQMS type

	Normalized AHA lower limit (2.5%)				Normalized AHA upper limit (97.5%)			
	RB	SB	UB	UT	RB	SB	UB	UT
RR	0.646	0.647	0.648	0.648	1.352	1.351	1.349	1.349
BA	0.786	0.851	0.876	0.885	1.236	1.171	1.143	1.12
LO	0.098	0.731	0.824	0.732	1.582	1.204	1.153	1.238
AQ	0.19	0.734	0.795	0.747	1.595	1.263	1.215	1.229
CO	0.201	0.61	0.635	0.631	1.807	1.48	1.428	1.448

by LOCK and then the BAU simulation. Is worth noting that for the Urban and Sub-Urban sites the combined effects of the uncertainty in both AQ model runs, the size of the 95% interval of the normalized AHA, is still smaller the ERF related interval.

The greatest interval size for combined uncertainty CO is found for Rural Background AQMS (1.606) while the smallest is found for Urban Background AQMS (0.793).

Further analyses on the possible effects of the smaller sample size of RB AQMS on the AQ model uncertainty is needed, since the stark differences with the other AQMS types.

4 Conclusions

In this study we analyzed the effects of different kinds of uncertainty on the Avoided Hospital Admissions estimated for NO_2 during the 2020 spring lockdown in the Milan area by comparing a Business As Usual simulation and a lockdown scenario simulation. Exposure–response functions are the one proposed by the WHO's HRAPIE projects.

We analyzed results for different Air Quality Monitoring Sites types, using Monte Carlo simulations and we normalized the 95% confidence intervals of the health effects model of air pollution to increase comparability across AQMS types. We considered three types of uncertainty, one in the ERF and one for each air quality model simulation. We then considered also the combined effects of both AQ simulations and the combined effects of the three sources of uncertainty.

We found that, with the exception of Rural Background stations, the greatest source if uncertainty is linked with the exposure response functions, as found in a recent study (Khomenko et al., 2021). For RB AQMS the main source of uncertainty are the LOCK scenario concentrations. When comparing results across station type, the lowest AHA uncertainty is obtained for urban background stations while the highest for Rural stations.

References

ARPA Lombardia. *Dati sensori aria | Open Data Regione Lombardia*. https://www.dati.lombardia.it/Ambiente/Dati-sensori-aria/nicp-bhqi

ARPA Lombardia. *Inemar (Inemar.HomeLombardia)*. https://www.inemar.eu/xwiki/bin/view/Inemar/HomeLombardia

Collivignarelli, M. C., Abbà, A., Bertanza, G., Pedrazzani, R., Ricciardi, P., & Carnevale Miino, M. (2020). Lockdown for CoViD-2019 in Milan: What are the effects on air quality? *Science of the Total Environment, 732*, 1–9. https://doi.org/10.1016/j.scitotenv.2020.139280

EEA. (2018). *Air quality in Europe 2018*.

Environ, R. (2015). *Seasalt guide version 3.2*.

Guenther, A., Karl, T., Harley, P., Wiedinmyer, C., Palmer, P. I., & Geron, C. (2006). Estimates of global terrestrial isoprene emissions using MEGAN (Model of Emissions of Gases and Aerosols from Nature). *Atmospheric Chemistry and Physics, 6*, 3181–3210. https://doi.org/10.5194/acp-6-3181-2006

Guevara, M., Tena, C., Porquet, M., Jorba, O., & Pérez García-Pando, C. (2019). HERMESv3, a stand-alone multi-scale atmospheric emission modelling framework-Part 1: Global and regional module. *Geoscientific Model Development, 12*, 1885–1907. https://doi.org/10.5194/gmd-12-1885-2019

Khomenko, S., Cirach, M., Pereira-Barboza, E., Mueller, N., Barrera-Gómez, J., Rojas-Rueda, D., de Hoogh, K., Hoek, G., & Nieuwenhuijsen, M. (2021). Premature mortality due to air pollution in European cities: A health impact assessment. *The Lancet Planetary Health, 5*, e121–e134. https://doi.org/10.1016/S2542-5196(20)30272-2

Mareckova, K., Wankmueller, R., Anderl, M., Poupa, S., & Wieser, M. (2019). *Inventory review 2019—Emission data reported under the LRTAP convention and NEC directive: Stage 1 and 2 review*. Status of gridded data and LPS data. EEA and CEIP.

Marongiu, A., Angelino, E., Fossati, G., Moretti, M., Peroni, E., Pantaleo, A., Malvestiti, G., & Abbattista, M. (2020). *Stima Preliminare Delle Emissioni In Lombardia Durante L'emergenza Covid-19*.

Menut, L., Bessagnet, B., Siour, G., Mailler, S., Pennel, R., & Cholakian, A. (2020). Impact of lockdown measures to combat Covid-19 on air quality over western Europe. *Science of the Total Environment, 741*, 140426. https://doi.org/10.1016/j.scitotenv.2020.140426

Piccoli, A., Agresti, V., Balzarini, A., Bedogni, M., Bonanno, R., Collino, E., Colzi, F., Lacavalla, M., Lanzani, G., Pirovano, G., Riva, F., Riva, G. M., & Toppetti, A. M. (2020). Modeling the effect of COVID-19 lockdown on mobility and NO_2 concentration in the lombardy region. *Atmosphere (basel), 11*, 1–18. https://doi.org/10.3390/atmos11121319

Ramboll. (2020). *CAMx user's guide user's guide*. www.ramboll.com

Sicard, P., De Marco, A., Agathokleous, E., Feng, Z., Xu, X., Paoletti, E., Rodriguez, J. J. D., & Calatayud, V. (2020). Amplified ozone pollution in cities during the COVID-19 lockdown. *Science of the Total Environment, 735*. https://doi.org/10.1016/j.scitotenv.2020.139542

Skamarock, W. C., Klemp, J. B., & Dudhia, J. (2008). *A description of the advanced research WRF version 3*. Tech. Note NCAR/TN-475+STR. https://doi.org/10.5065/D68S4MVH

UNC. (2013). *SMOKE v3.5 user's manual*. http://www.smoke-model.org/index.cfm. http://www.smoke-model.org/i%0Andex.cfm

World Health Organization. (2013). *Regional office for Europe: Health risks of air pollution in Europe—HRAPIE project, Recommendations for concentration–response functions for cost–benefit analysis of particulate matter, ozone and nitrogen dioxide*.

Author Index

C. Mensink and O. Jorba (eds.), *Air Pollution Modeling and its Application XXVIII*, Springer Proceedings in Complexity, https://doi.org/10.1007/978-3-031-12786-1